THIRSTING FOR EFFICIENCY

THE ECONOMICS AND POLITICS OF
URBAN WATER SYSTEM REFORM

THIRSTING FOR EFFICIENCY

THE ECONOMICS AND POLITICS OF URBAN WATER SYSTEM REFORM

Edited by

Mary M. Shirley

The World Bank
Washington
USA

THE WORLD BANK

2002

PERGAMON

An imprint of Elsevier Science

Amsterdam – London – New York – Oxford – Paris – Shannon – Tokyo

ELSEVIER SCIENCE B.V.
The Boulevard, Langeford Lane
Kidlington, Oxford OX5 1GB, UK

First edition 2002

Library of Congress Cataloging in Publication Data
Thirsting for efficiency: the economics and politics of urban water system reform /
edited by Mary Shirley.
p. cm
Includes bibliographical references.
ISBN 0-08-044077-0
1. Municipal water supply – Economic aspects. 2. Water utilities. I. Shirley, Mary M., 1945-

HD4456 .T48 2002
363.6′1—dc21

2001058841

British Library Cataloguing in Publication Data
A catalogue record from the British Library has been applied for.

ISBN 0-08-044077-0

♾ The paper used in this publication meets the requirements of ANSI/NISO Z39.48-1992 (Permanence of Paper).

Printed in The Netherlands.

To Arturo Israel

TABLE OF CONTENTS

LIST OF FIGURES

LIST OF TABLES

TABLES IN THE TEXT

STATISTICAL APPENDIX TABLES

LIST OF CONTRIBUTORS

THE EDITOR

Mary M. Shirley — President, the Ronald Coase Institute and Consultant, The World Bank, Development Economics Research Department

THE AUTHORS

Manuel A. Abdala — Director, LECG, LLC and Professor of Regulatory Economics at Instituto Tecnológico de Buenos Aires, Argentina

Lorena Alcázar — Head of Research, Instituto Apoyo, Lima, Peru

Penelope Brook — Program Manager, Rapid Response, World Bank Private Sector Advisory Services Department

George R.G. Clarke — Economist, World Bank Development Economics Research Department

Keith Crocker — Professor of Business Economics and Public Policy, University of Michigan Business School

Luke Haggarty — Economist, World Bank Development Economics Research Department

Scott Masten — Professor of Business Economics and Public Policy, University of Michigan Business School

Claude Ménard — Professor of Economics, Université de Paris, Panthéon-Sorbonne

Roger Noll — Morris M. Doyle Professor of Public Policy, Department of Economics, Stanford University and Senior Fellow, the Stanford Institute for Economic Policy Research, Stanford University

Lixin Colin Xu — Economist, World Bank Development Economics Research Department

Ana Maria Zuluaga — Consultant, Japan Special Fund, Inter-American Development Bank-FSS

PREFACE

MARY M. SHIRLEY

WATER IS A WORLDWIDE ISSUE

One billion people in the world lack safe drinking water and almost 2 billion lack adequate sanitation services. As a result millions die every year from water related diseases. In many cases poor management and inefficient investment are responsible for this situation, and countless attempts at reform have accomplished little. Water may seem to be an archetypically local issue, since cities in the same country can suffer from shortages or pollution while a neighboring locale is awash in clean water. Yet because we draw water from a global system that serves a worldwide community, the waste, pollution and disease caused by water mismanagement concern us all. To improve the governance of our global water resources, we need to understand the management and mismanagement of water in the large cities of the world, such as those studied in this volume.

The challenge faced by large cities in developing countries can be severe, especially for those large urban conglomerations that have grown beyond hydrologically sustainable sizes for their locations, such as Lima (Chapter 4) or Mexico City (Chapter 5). Lima is in an arid region of Peru where the average rainfall is less than in the Sahara desert. Two thirds of its water come from a badly polluted river; the rest is from wells that become contaminated by salt water when increased extraction during the dry season causes the water table near the ocean to drop. Water is rationed during the dry season and interruptions are frequent. Mexico City relies for most of its raw water on deep wells, but its aquifer is nearing exhaustion. The extraction rate from the aquifer is currently 37 percent higher than the replenishing rate. The city is built on a dry lakebed and because of the nature of the soil the city is sinking with the depletion of the aquifer. The city has been relying increasingly on water pumped uphill from distant rivers. One site is 127 kilometers distant and 1,200 meters below the city. This book explores how such cities have coped with the challenge of managing consumption where the cost of raw water is high and alternatives are few.

A number of developing countries have tried to reverse years of mismanagement of their utilities by auctioning contracts to private operators, and water is no exception. [1] It is important to analyze these experiments in reforming urban water management, not only because water is vital for human health, but also because water reforms hold lessons for the administration of other infrastructure as well. Water systems are similar to electricity or telecommunications, in that they are networks with large sunk and durable assets, that in the past were typically owned or regulated by the state. Yet there have been relatively few attempts to study systematically the effects of private operation in urban water supply. This volume aims to fill this gap.

URBAN WATER SYSTEMS CAN BE HARD TO REFORM

Despite the simple and stable nature of the technology, it can be difficult to reform the management of water. For one thing, it is not enough to supply clean drinking water. Epidemiological studies of small children in urban areas of developing countries found that connecting households to piped water alone did not make children healthier if sanitation remained unimproved (Esrey 1996). [2] Moreover, the type of water and sewerage service mattered. Gains were realized when clean water was provided in the house or on the premises, i.e. not from standpipes or central wells, and combined with a move to better sewage disposal, i.e. from no facilities to pit latrines, or from pit latrines to flush toilets. This may be because water from central sources is often contaminated in handling or because easier access to water combined with better sanitation encourages better hygiene. The biggest gains came from a move to piped water combined with flush toilets or other water based sanitation, a costly option for many poor households.

This raises a second issue, affordability. Expanding and improving the system are also not enough; people must be willing and able to connect and consume the services. The poor's willingness to pay for water usage is often sufficient to cover costs; indeed many pay a great deal more than the cost of piped water to buy water from a vendor. In Lima, for example, water from a vendor was estimated to cost about six times more than piped water from the planned private water concession (see Chapter 4). Willingness-to-pay cannot be inferred by simply extrapolating from the high prices paid for vendor water, however. Studies show considerable variability in willingness-to-pay, depending on available alternatives, among other factors. Poor people may consume cheap but unsafe alternatives. In Conakry (Chapter 8), people responding to surveys

[1] A 1999 survey of the World Bank's Private Sector Development Department's database on private participation in infrastructure since 1989 found 53 contracts for private operators to manage urban water and/or sewerage systems in 27 developing countries. Private participation in water and sewerage was more common in the nineteenth century, and some developed countries (for example, France and the U.S.) have a long history of private participation, as do a few developing countries (for example, Cote d'Ivoire has had a lease contract for decades).

[2] The study measures effects of improved water and sewerage on diarrhea prevalence, height for age, weight for age and weight for height scores of 5,000 urban children and 12,000 rural children aged 3 to 36 months in 8 developing countries (Bolivia, Guatemala, Burundi, Ghana, Togo, Uganda, Sri Lanka and Morocco). The study controls for a number of factors such as age and education of mother, number of household members, income proxies such as whether the household has a bicycle or motorcycle, % of life child was breastfed, etc.

say they use contaminated well water knowing that it is risky because they feel they have no affordable alternative. Subsidies may be necessary where there are externalities from expanded access to water and sewerage, such as fewer communicable diseases or less pollution, but these are subject to well known distortions and administrative problems. Abidjan (Chapter 7) and Santiago (Chapter 6) have overcome the affordability problem, but it was a major stumbling block to the concession in Buenos Aires (Chapter 3).

Politics presents an even greater conundrum for reformers. Water is a necessary good that affects all voters and its management is highly politicized in most places. Water is often priced below its cost for political reasons, and new investments are directed to maximize political payoffs. The reforms we studied were all driven and shaped by political concerns, which we analyze in each of the country cases.

THIRSTING FOR EFFICIENCY

Why do countries that have tolerated mismanagement of their water and sewerage system for decades develop a thirst for efficiency? How should we evaluate their efforts to change? What factors determine success or failure? These questions are the subject of our analysis.

Urban water systems are politically sensitive, and many cities in developing countries have tolerated crisis conditions for years. Reforms are infrequent despite such poor conditions, because the beneficiaries from water reform are often poorer and less politically powerful than the opponents. The potential beneficiaries are people who cannot afford to connect or who have no alternative when municipal service is unreliable. Powerful interests stand to lose: state enterprise employees who might lose their jobs, contractors who might lose construction contracts, politicians who might lose votes and campaign funding if they no longer control investments, bureaucrats who might lose power, and wealthier consumers whose water bills might increase. Using standardized measures we show systematically which conditions are necessary for sustained reform, and analyze how they were met in some, but not all of our cases.

In a sensitive area such as water the best can be the enemy of the good. Critics of reforms sometimes fail to consider what the circumstances would have been without change; reform boosters sometimes fail to take account of the effects of reform on the poor or on pollution. We develop a comprehensive measure of the effects of reform on consumers, government, investors and workers and compare it with a counterfactual that assumes the status quo would continue. We also measure gains in productivity, access and investment. We assess the impact on the poor and the effects on waste and pollution.

This volume explains success and failure in urban water reform by examining the micro-institutions that governed the sector. These include the laws, regulations, norms and the implicit and explicit contracts between the operator and the state. We characterize the autonomy, power, capability and efficiency of the regulator and other key actors as well. The book also explores the role of macro-institutions in successful reform. Thus, we analyze the role of political structures and rules that check arbitrary action by one branch of government or curb capture of the regulator. We also examine legal institutions, such as the effective protection of property rights and bureaucratic and judicial norms supporting compliance with laws.

This book was designed as a coherent whole from its inception. The country cases are all based on the same questionnaire, employ a consistent methodology to construct similar variables measured the same way, and analyze the research findings within the same conceptual framework. The authors worked as a team, communicating often and meeting several times before beginning the research and during its implementation to assure consistency in approach and to understand differences in results.

This book assesses reforms in six urban water systems in Latin America and Africa and analyzes the political and institutional factors that lead to success or failure. It relies on New Institutional Economics, a field of economics that analyzes not only economic forces, but also the laws, rules and social norms that govern human interaction. Our goal was to fill a gap in the literature by providing an economic theory of water that covers issues of supply, demand, regulation, and competition, and by analyzing the experience with reform in developing countries, in contrast with the historical record for the United States.

Chapter 1 is a synthesis of the six country cases, focusing on their regulatory contracts, and drawing out the different reform outcomes and the policy implications. Chapter 2 provides a conceptual framework for the study of urban water systems. The next six chapters contain the country cases: Buenos Aires, Argentina; Lima, Peru; Mexico City, Mexico; Santiago, Chile; Abidjan, Cote d'Ivoire; and Conakry, Guinea. The sample is small and includes only capital cities, but covers the dominant contractual forms (management contracts, leases, concessions and sales) and contains a variety of reforms in water systems of widely varying size and coverage operating in very different economic settings.

The book concludes with a historical study of the United States experience in urban water systems, where reforms led to nationalization rather than private participation. The authors treat this as a cautionary tale for developing countries. Finally, a statistical appendix for the six developing country cases rounds out the volume.

ACKNOWLEDGEMENTS

The authors would like to thank their advisors, Mathew McCubbins, Roger Noll, Douglass North, and Dale Whittington for their invaluable guidance. A special word of thanks is also due to Raquel Alfaro for her comments and advice and François-Marie Patorni for his support and advice. The authors owe a large debt to the numerous individuals and organizations in the countries studied who generously gave their time and assistance; we cannot thank them enough. Special thanks are also due to Ana Maria Zuluaga for excellent research assistance on all the country cases, and to Zenaida Kranzer, Kari Labrie, Mary Caperton Means, and Agnes Yaptenco for their outstanding work on preparation of the manuscript. The authors would also like to thank Myrna Alexandra, Yao Badjo, Luis Chang, Harold Demsetz, Vincent Gouarne, Yoko Katakura, Philip Keefer, Paul Kriss, Brian Levy, Alain Locussol, Alberto Mejia, Ontoniel Ochoa, Sixto Requena, Liliane Saade, Sophie Sirtaine, Nicola Tynan and Pierre Werbrouck for very useful comments on earlier drafts of various chapters, and Mark Jones for his kind help with data on Buenos Aires. The book also benefited from comments of participants in meetings and seminars at the International Society for New Institutional Economics, Washington University — St. Louis, the Western Economic Association, the University of California at Los Angeles, and the World Bank. The authors gratefully acknowledge the generous support of the World Bank. All errors and omission remain our own.

Thirsting for Efficiency: The Economics and
Politics of Urban Water System Reform
Mary M. Shirley (Ed.)

CHAPTER 1

CITIES AWASH: A SYNTHESIS OF THE COUNTRY CASES

MARY M. SHIRLEY AND CLAUDE MÉNARD

INTRODUCTION

This chapter compares the experiences of the six country cases in the book, and draws lessons for other reforming countries. We analyze each of the six cases in terms of the 'regulatory contract', which is the explicit and implicit agreement that defines the relationship between a government and a regulated monopoly. We also investigate the macro- and macro-institutions that govern this relationship, including the regulatory, judicial and political rules. We have been as careful as possible to identify variables that allowed us to compare different contractual arrangements and different institutional environments without mixing the two.

To gauge how well the regulatory contracts performed we compare the improvements in efficiency, and in economic and social welfare. We used measures of labor productivity, non-wage expenses, profitability, investment, and a comprehensive measure of net benefits compared to a counterfactual to assess efficiency. Social gains were harder to measure since few of the cities collected health, income or similar information in sufficient detail to attribute changes to water system reform. The cases permit us instead to analyze what happened to access, prices, waste and pollution and infer the effects on social welfare and the poor.

The next section presents the analytical framework for this chapter; following that we describe initial conditions in the water sectors and in the economic and political environments. We then analyze how well the contractual frameworks dealt with issues of information, incentives and commitment, following which we assess the supportive institutions. Finally, we compare the performance of the six cases and consider some of the policy implications of our findings.

ANALYTICAL FRAMEWORK

Our premise is that a regulatory contract will be more likely to achieve its goals when three necessary conditions are met: (i) information asymmetries between regulator and manager are reduced; (ii) sufficiently high powered incentives are provided to motivate the manager to comply with the contract's goals; and (iii) both parties provide credible signals of their commitment to abide by the contract and credible enforcement provisions are established. The importance of information, incentives and commitment in the success of contracts in improving efficiency is explored at length in World Bank (1995), Alcazar and Brook Cowen (1996) and Shirley and Xu (1998), and we summarize the argument briefly here.

Information

Information asymmetries allow contracting agents to use their private information at the expense of their contractual partners. In regulatory contracts, asymmetric information arises because the firm can observe its costs and performance at much lower cost than the regulator (Laffont and Tirole, 1993). Information asymmetries can be reduced through competition or monitoring. Competition reveals information at lower cost than monitoring and allows the regulator to compare the performance of rival firms. Although product market competition is rare in water, as we discuss below, the regulator can use competitive bidding for the franchise to operate the system and yardstick competition to increase information.

Incentives

Incentives can be used to induce the firm to reveal information and not to act opportunistically. The main incentives in our contracts were penalties, assignment of risk and reward, and price regulation. Price regulation determines the operator's incentive to invest and improve efficiency, as well as the extent to which efficiency gains are passed on to consumers. Fees and penalties may also induce operators to comply with the contract's goals.

Problems arise in assigning incentives because the contract may not assign residual rights and returns to the operator and investor.[1] For example, if the contract is a lease it may be difficult to assign incentives to the contractor to keep the assets in good condition because of the separation of investment and operating decisions. A concession provides stronger incentives to maintain the assets because the investor is also the operator, although these incentives diminish as the contract nears its end.

Commitment

Commitment depends on the credibility of the contract. The operator may doubt that the rewards promised under the contract will be realized if the regulator is vulnerable

[1] These issues are dealt with at length in the literature on franchise bidding. See, for example, Laffont and Tirole, 1993 and Williamson, 1976, 1985.

to political pressures to set prices too low. Targets may not be credible if the regulator lacks the power to impose penalties sufficiently large to deter reneging or the regulatory body is not disposed to apply them because it has been captured by the operator.

Institutions

Four sets of institutions are particularly important for utilities. First, regulatory institutions, such as rules governing the status, power, autonomy and staffing of the regulator, determine the regulator's capacity to monitor and enforce the contract. Judicial institutions, including the laws supporting the contract and protecting property rights, and enforcement characteristics, such as the capacity and independence of the judiciary, play a key role in enforcing the contract and resolving disputes. Political institutions, such as checks and balances, reduce the scope for arbitrary actions by the regulator or the executive branch, while electoral and other political rules determine the relative strengths of parties favoring and opposing the contract. Finally, international constraints, such as openness to trade and capital flows or dependence on foreign aid, give governments a stake in maintaining their reputations with foreign investors or donors, thereby putting a check on arbitrary action.

A contract can meet the three conditions in a supportive institutional setting and still not improve efficiency or social welfare if these are not its objectives. Hence our concern is not only with how well the contract meets its stated objectives, but with how well these objectives coincide with the needs of the community, which we judge in terms of efficiency and social welfare.

Efficiency and Social Welfare

Efficient operation keeps costs down, thereby reducing dependence on government subsidies and freeing resources for investment in expansion and maintenance. It also entails prompt and reliable service and quick action on complaints.

Social benefits arise first because the supply of clean water for drinking and hygiene and the removal of waste are necessary for human health and the prevention of contagious diseases. Treating wastewater to remove pollutants can have important external benefits for downstream users, while reducing water consumption can prevent depletion of a water source where the replenishment rate is less than the extraction rate. Supplying water and sewerage has *equity* effects as well. Poor households may not be able to afford safe alternatives to piped water or sewerage, and pay a heavy cost for medical expenses and days lost due to illness from water-borne diseases. Time-savings from proximity of improved water can allow poor households to spend more time on child care and practice better hygiene, as well as allowing more time for earning income.[2] Social gains can also be realized by reducing the waste of water, where the

[2] Esrey, 1996 suggests that the time saved because water is nearby may allow more time for breast feeding, food preparation, income-generating activities that allow for the purchase of better health care and better food, more time for learning activities, such as visiting clinics to attend child care classes.

opportunity cost of usage is high. Usage costs are high where current and future rates of extraction exceed replenishment rates or usage has external effects such as causing settlement of land that damages buildings or standing water that breeds insects. Metering consumption and pricing water above its opportunity cost provides incentives to reduce waste, and taxing external effects motivates operators to reduce pollution or curb usage (see Chapter 2 for further discussion of these issues).

As we shall show in the next section, problems of efficiency, access, affordability, waste and pollution varied in our sample. Economic and political circumstances were as important as water sector problems in motivating reforms.

INITIAL CONDITIONS IN THE WATER SECTOR

The cities studied in this book differed in size and income levels, and had correspondingly different sized systems, yet they shared many of the same problems (Table 1.1). All of the cases were experiencing rapid growth in some parts of the city through migration of poorer residents from rural areas or from outside the country to the outer edges of the metropolitan area. This section compares the situation in the water sector in our cases as well as characteristics of the water resource that made it inherently more or less costly. In another paper (Ménard and Shirley, 2001) we deal at some length with these issues; we summarize that discussion here.

Assuring access to water and sewerage was a key concern in all the cities, even in those with high coverage rates, such as Santiago and Mexico City. Mexico City had to expand at a relatively rapid rate to keep coverage high.[3] Although Santiago was growing more slowly, several poor communities without service were just outside the concession area, and increasing numbers of people were moving from shared housing to separate dwellings. Rates of sewerage connection lagged behind water in all the cities and especially in the two African cases, while sewage treatment was minimal in all the cases.

Although magnitudes varied, service quality and waste of water were problems for the entire sample, with the possible exception of Abidjan. Lima and Conakry had the worst service before reform. In Lima 48% of the connected population received water service for less than 12 hours a day, 28% for less than 6. In Conakry, interruptions were also frequent and the quality of piped water was very poor. In both cities there were large losses of water because of leakage and low pressure. Performance was better in the other cities, but there too water was sometimes interrupted, the number of leaks was high and increasing, pressure was inadequate and/or response to complaints was slow. The only system without serious service problems at the start of the reforms under study was Abidjan, which was already functioning under a lease with a private operator.

[3] The numbers in Table 1.1 understate the amount of the population that was not connected because they exclude squatters and poor suburban areas that are outside the utility's water district. In addition, the meaning of a connection varies: in our Latin American cases it generally refers to a hook up to internal pipes within the house, while in the African cities a connection is usually a standpipe in the yard. The standard for such pipes was less than 200 m from the house, but that standard was not always met which can make control of access by the bill payer difficult. A connection in Africa typically serves a much larger number of people than in Latin America; for example, a connection in Abidjan serves an average of 15 people versus 5 in Santiago.

Table 1.1

Conditions and reforms in the sample water systems

	Buenos Aires	*Mexico City*	*Lima*	*Santiago*	*Abidjan*	*Conakry*
Year of start of reform	1993	1993	1992	1989	1988	1989
Type of reform:						
Planned	Concession	Management	Concession	Sale	Lease [a]	Lease
Implemented	Concession	Service contr.	SOE	SOE	Lease	Lease
Population in service area (millions) at time of reform	8.7	8.4	6.4	4.6	2.0	1.0
GDP per capita at time of reform (US $) [b]	8,861	7,647	3,462	7,101	1,582	1,398
Population connected (percentage) at time of reforms:						
Water [c] (%)	70	97 [d]	75	99	60	38 [e]
Sewerage (%)	58	86 [d]	70	88	35	10
Population growth rate (percentage), annual average 1980–1995	1.5 [f]	3.1	2.3 [g]	1.8	5.1	5.6
Water production (millions of m^3 per year) at time of reform	1,402	1,113	527	478	67	16 [d]

[a] Before reform the lease in Abidjan had some characteristics similar to a management contract.
[b] National.
[c] Includes private taps in yards of dwelling. These were predominant in Abidjan and Conakry, and important in Mexico (20% of connections) and probably in Lima; they were minimal in Santiago and Buenos Aires.
[d] 1990.
[e] Includes people with access to standpipes or neighbors' taps.
[f] 1980–1991.
[g] 1993–1996.
Source: Statistical Appendix.

Table 1.2

Principal externalities before reform

	Buenos Aires	*Mexico City*	*Lima*	*Santiago*	*Abidjan*	*Conakry*
Principal externalities	Pollution of aquifer by cesspools causing water-borne diseases among those using wells	Depletion of aquifer causing sinking of city; poor quality water causing water-borne diseases for those in periphery	Low coverage, rationing, and irrigation with polluted water causing water-borne diseases; depletion/contamination of groundwater	Irrigation with polluted water causing water-borne diseases, risk to crop exports	Lack of sewerage causing pollution of aquifer	Low coverage, lack of sewerage causing water-borne diseases and contamination of aquifer
Indicators of externalities		Extraction rate 42 m^3/s; replenish. rate 24 m^3/s; parts of downtown sunk by 2 m over last decade	Incidence of diarrhea in slums = 8 per person per year; cholera epidemic centered in Lima killed almost 3000 in 1991	Higher rates of typhoid than in rest of Chile (4100 cases in 1989) and other water-borne diseases		Diarrhea diseases caused 20% deaths children aged 0–4; cholera epidemic killed 330 in Conakry in 1994

Source: Country cases.

Table 1.3

Nature of the water resources in the sample cases

	Buenos Aires	*Lima*	*Mexico City*	*Santiago*	*Abidjan*	*Conakry*
Principal sources	River (92%)	Rivers (2/3) Wells (1/3)	Rivers (18–36%) Wells (82% in 1987; 64% in 1993)	Rivers (89%) Wells (11%)	Wells (100%)	Rivers (90%) Wells (10%)
Variability	Low	High	Low	Low	Low	Low
Sustainable [a]	Yes	No	No	Yes	Yes	Yes

[a] Current rates of extraction below replenishment.
Source: Country cases.

Low rates of water and sewerage connection, interruptions in service and poor water quality or contamination of food and water from untreated sewage caused health problems in all of the cities, as we can see in Table 1.2, and these reached crisis proportions in Lima and Conakry. Poor water quality, frequent interruptions, and water scarcity as well as irrigation of food crops with untreated wastewater and pollution of ocean fishing grounds with sewage have been blamed for a cholera epidemic that was centered in Lima and killed almost 3000 people in Peru in 1991. In Conakry, consumption of well water contaminated by latrines and storage of water in unsanitary conditions accounted for high rates of infant deaths from gastro-enteric diseases and periodic epidemics of cholera. Besides instances of mortality, morbidity was a frequent problem for unconnected consumers in all of our cases.

There is important variance in the cost and scarcity of raw water sources in the sample cities for reasons explained in the subsequent chapters. Most of the other cities we studied relied on relatively clean water from rivers or aquifers that they extract at a rate well below the rate of replenishment (Table 1.3). The exceptions were Lima and Mexico City.

Macro-economic Conditions Leading to Reform

A core impulse for major reforms is the combination of sector problems, such as those described above, with a macro-economic crisis. In particular, rising inflation rates combined with rising public deficits and debt motivate water system reform. Rising inflation reduces cost coverage in the sector, since increases in nominal prices for water services typically fall behind inflation rates. As a result, the sector requires higher subsidies at a time when government is under rising pressure to curb deficit spending.

In the years prior to reform, inflation was increasing in all of our sample, and especially in Argentina, Peru and Guinea. All of the sample were running budget deficits at the time of the reform decision, which usually predated the reform by several years. While we do not have numbers for the entire sample on government debt, it was high when the reform decision was taken in Peru, Mexico and Chile.

Political Conditions Leading to Reform

The political circumstances that matter most to the subsequent contract are the relative power of potential winners and losers (Ménard and Shirley, 2001). Typically water reform has high social benefits but low political benefits, especially when compared with reform of other utilities. One reason for this is that the transfer of management in the water sector usually produces fewer new resources with which politicians can reward supporters or compensate opponents than the sale of state owned assets. Another is that increases in water prices or cut offs of non-payers can be politically sensitive because water is an essential good. Layoffs of workers in the state-owned enterprise (SOE) can also have political costs. Hence reform is more likely where the political saliency of water is low and potential losers have low political influence or can be persuaded not to oppose reform.

Water system prices are likely to have lower political salience in cities where the opportunity cost of usage is low and the system is not near capacity (Buenos Aires, Santiago, Conakry and Abidjan), since these are cities where cost recovery would be less likely to require large price increases. As we will show, however, high prices are not the only factor determining salience. Water contracts will be controversial if consumers believe that the operator has been allowed price increases that are seen as unfair or in violation of the contract, or if they believe that their interests are not well protected in regulatory decisions.

Expansions in water and sewerage typically benefit those urban poor without connections who now have access to service, while improvements benefit those who are already connected and who now receive more reliable service, higher quality water, better pressure, or quicker attention to complaints. Contractors and investors can benefit from rents from building works and operating the system. Losers from new water contracts are usually any workers who are laid off and bureaucrats who lose power and rents. Consumers who are connected may also lose if the value of the service improvements is less than any price increases or the cost to them of better bill collection. As Ménard and Shirley (2001) suggest, reform occurred in our cases when a political change reduced the political influence of potential losers, in combination with sector problems and economic crisis.

Lobbying by those who stood to gain was not a factor since beneficiaries were disbursed and uninformed. Rather, reform happened when a regime change or other political event diminished the power of potential losers and/or their opposition was muted through compensation for their losses. For example, the original Buenos Aires concession did not require already connected consumers to pay all the cost of expanding the system; instead the design of the auction gave them a 26% decrease in prices. At the same time workers in the water system were offered generous and voluntary severance packages. Bureaucrats who might have opposed the concession had lost power with the change in regime, and were also subject to retrenching through early retirement. Later, when people without connections realized that they would have to pay much of the cost of expansion and when opposition politicians organized their discontent, the concession had to be renegotiated to spread this charge over all consumers.

INFORMATION IN THE CONTRACTS

This section considers ways the contracts reduced information asymmetry through three forms of competition — product market competition, competition for the market and yardstick competition — and through monitoring.

Product Market Competition

None of our cases tried to introduce direct product market competition beyond tolerating water vendors (Mexico City, Abidjan, and Conakry) and self supply (Abidjan and Conakry).[4] To some extent this was dictated by water's monopoly characteristics, but it was also due to the absence of models for competition in the sector (see Chapter 2). One reason for this is the high cost of pumping water long distances and laying pipes, as well as the increases in leakage with distance, all of which mean that water will remain a local or regional business (Cowan, 1993). Nevertheless, local systems with more than one reservoir or which rely extensively on wells, could potentially introduce competition to supply a central grid regulated as, for example, are power grids in the U.K.

What explains the lack of experimentation with product market competition in water compared to other network infrastructure? London Economics (1998) argues that because a greater percentage of the costs in water systems are in the noncompetitive network than in the potentially competitive areas, the relative efficiency gains from competition are small. Forty percent of the costs in electricity are in transmission and distribution, versus 60% in potentially competitive areas, while in water 66% of the costs are in the network and only about 34% in potentially competitive areas (London Economics, 1998). Nevertheless, there are gains to be had and it may be that experiments with competition in this sector will increase in the future.

Competition for the Market

By competition for the market we refer to the periodic auction of the right to operate the system under a management contract, lease or concession. Competition for the market has drawbacks that are not unique to water: it is intermittent and in the long intervals between bids the incumbent gains advantages that may make the threat of entry a feeble one. Competition for the market also faces a special hurdle in water systems because until recently few private companies have been willing to participate.

Five companies dominate the market for private contracts in water, although this may be currently changing.[5] Two of the dominant players are responsible for 70% of the projects managed by this group (Silva et al., 1998).[6] Since the five companies also

[4] Permits for wells are very difficult to get in Buenos Aires and Santiago. Mexico City and Abidjan charge for well usage but the charge is waved for many users in Mexico City and seldom enforced in Abidjan.

[5] The entry of U.K. operators and other utilities (especially electricity companies) into the market for water service contracts has increased competition in the sector.

[6] The top two are Suez Lyonnaise des Eaux and Vivendi (formerly Générale des Eaux); the rest are Aguas de Barcelona, Thames Water and SAUR International.

form consortia to bid on projects together, they are as accustomed to cooperating as competing.[7] Yet even with a limited number of players, a properly constructed auction with periodic rebidding could potentially reveal much information to the regulator and create a threat of entry that could discipline the incumbent operator as long as collusion can be prevented.

The cases we studied, however, made limited use of competition for the market and did not design the auction to enhance information. Santiago and Lima opted not to go ahead with plans to tender private participation. Chile's regulator also passed up opportunities to auction concessions for new areas or where the existing operator's license had been revoked or expired.

Mexico City contracted out services (meter installation, metering reading and billing) as a first step toward management contracts. Private consortiums were invited to submit bids, with unit prices based on a list of actions they must agree to undertake anywhere in the city for a specified number of people, with a premium or discount for fluctuations above 10%. These bids were then fed into a computer model that was to select the least cost number of providers (as long as it was more than one) and four companies were chosen. One drawback of this approach was that the bids were for the contracts in their entirety, including phases that have not been — and may never be — implemented. As a result, the winning bid for say, zone one, might include a charge for metering that would not be the least cost bid if metering had been auctioned by itself.

Côte d'Ivoire, which has had a lease with a private operator since 1957 did not rebid the contract when it expired in 1986.[8] Instead, the government used the threat of an auction to win a 20% real reduction in the average tariff. In exchange for this price reduction, the government gave the operator more control over investment, and this had the unintended effect of reducing competition by progressively eliminating bidding for investment.

Buenos Aires and Conakry were the only cases where competition for the market was effectively employed. Buenos Aires auctioned its concession and Conakry bid a lease for the lowest consumer tariff. Three bidders participated in Buenos Aires and four in Conakry, all private consortia affiliated with international firms. The Buenos Aires auction revealed more information than the bid in Conakry since the price reflected the bidders' assessment of investment as well as operating costs.

Yardstick Competition

Since water and sewerage are local in nature, there will be many independent water systems in most countries, and we might expect that there would be ample opportunity for comparative competition. One problem with yardstick competition in water, however, is that we cannot assume that the exogenous factors affecting companies' costs are correlated. Local differences in climate, topography and other endowments affect

[7] For example, in Lima, Vivendi teamed up with Thames Water and Lyonnaise with Aguas de Barcelona; in Buenos Aires, Lyonnaise, Vivendi, and Aguas de Barcelona are all part of the winning consortium.

[8] The company's principal strategic partner was a French private water company, SAUR.

underlying cost and demand conditions, and make comparisons difficult. Yardstick competition also requires that companies not collude, suggesting that operators within a country should have different owners. Achieving ownership diversity in water systems may also be a problem since the market for operation of water companies is so thin. In countries with low credibility it may not be feasible to attract different private operators to smaller urban areas. The explanation lease in Guinea covers all cities of a certain size precisely because the government expected that bundling in the capital was the only way to attract a private operator to the smaller cities.[9] The cases we studied made little use of yardstick competition. The most feasible environments for comparisons in our sample were arguably Mexico City and Santiago, since there was more than one operator in each city. The bids in Mexico City were not comparable. Companies were asked to bid on a basket of actions before they know which zone they would operate or which actions they might be specifically asked to do at different points in the contract, and the number of zones was then decided based on these bids. There were major differences between the four zones resulted from this process and the companies faced very different structures of pricing and incentives. As a result the price the city paid for metering and other services varied widely from zone to zone. In 1998 the new (and for the first time elected) municipal government demanded that all prices be equal to those of the least cost provider in the last stage of the contract, which might be considered a form of yardstick competition.

Although Santiago had more than one company they were not easily compared. The private company served only about 5% of the city's population, compared to the state owned enterprise that served 87% of the city. The regulator employed a form of yardstick competition since it modeled a hypothetical company with the assets, geographic conditions and markets of the real company, but with other parameters set according to best practice in Chile or worldwide. However, the model company was kept secret to reduce the risk that companies manipulate their reporting. Unfortunately, secrecy also reduced the incentive effect of comparison competition, since companies were not sure what they are competing against.

Monitoring

Another way to reduce information asymmetry is through monitoring. All of the contracts provided for the operators to report regularly on their performance and most had fines for failure to report. However, three of the cities — Abidjan, Conakry and Mexico City — did not have a separate body responsible for regulating the operator(s). In the three that do have a designated regulator, only Santiago's is adequately staffed and able to compel the operator to provide enough prompt and timely information to be able to reduce information asymmetries through monitoring. In some ways the Santiago regulator was not fully tested, since the operator was state-owned during the period

[9] Another contributing factor may have been the fact that Guinea, in contrast to the Latin American cases, had not decentralized water services before private participation. Côte d'Ivoire did have decentralized systems before 1974 that were centralized to assure that water prices across the cities were the same with Abidjan subsidizing the other urban areas.

under study and hence likely to have lower incentives to use its information advantage to maximize profit. The regulator's powers were strengthened prior to the sale of the water company in 1999.[10]

INCENTIVES IN THE CONTRACTS

The incentives in the contract affect operators' decisions to increase operating efficiency, expand coverage, prevent waste, discriminate among classes of users, maintain the system and share information with the regulator. The structure of incentives will also affect consumer motivation to connect to the system and to curb water usage. In this section we first describe the targets and the assignment of risk, then consider how tariff policy motivates operators and consumers, and conclude with a discussion of supportive institutions.

Targets

One way to motivate operators is to set explicit targets and penalize the operator upon failure to achieve them. As we discussed, these targets give us some sense of how well the contracts are aligned with the problems of the sector. If we compare the contract targets in Table 1.4 to the initial conditions described above, some striking gaps appear. The two cases with high opportunity cost of usage, Lima and Mexico City, effectively do not target waste: Lima because the concession contract was not implemented, Mexico City because the UFW targets are in a later stage of the service contract that was put on indefinite hold. The three cases with the lowest rates of connection, Abidjan, Conakry and Lima, did not have explicit access targets, despite serious health problems in the latter two. The two cases where contamination has had serious external effects, Lima and Conakry, have not addressed pollution controls.

The absence of an explicit target does not mean that the objective was not addressed, since the operator could be implicitly motivated to achieve a goal through other incentives, which we consider next.

Assignment of Risks and Rewards

The assignment of risk and rewards has important incentive properties. If companies bear the risk of investment, they will an incentive to conserve water when capacity is constrained by controlling leaks and maintaining the system. If they reap the financial returns from usage, they will have an incentive to invest in expansion.

The operator was fully responsible for investment only in Buenos Aires and Santiago. Although Santiago's operator was an SOE, government curtailed its borrowing and forced the company to finance most investment from retained earnings. It therefore faced an investment risk similar to the private operator under Buenos Aires' concession, and thus an incentive to use water efficiently. Lima's state-owned operator was

[10] Among other things, fines and penalties were raised, and the regulator was empowered to hire more people.

Table 1.4

Targets under the contracts

	Buenos Aires	*Lima*[a]	*Mexico City*	*Santiago*	*Abidjan*	*Conakry*
Coverage, universal service requirements	80% water, 64% sewerage in 5 years, yearly targets; regulator must approve 5-year investment plan	None	Not part of service contract	100% coverage required by law; regulator must approve 5-year investment plan	Not part of lease contract	Not part of lease contract
UFW, standing water requirements	UFW from 45 to 25%; 45% water network renovated by year 30; no sewerage overflows by year 5; non-residents must be metered[b]	None	UFW targets on indefinite hold; 100% metering target	Maintenance, waste targets; 100% metering required by law	Mainten. targets	Mainten. targets
Wastewater use, sewage treatment requirements	100% secondary treatment of effluents of those connected to sewerage system by year 13	None	Not part of service contract; farm use of untreated wastewater controlled	Farm use of untreated wastewater controlled; sewage treatment required by law, not enforced	Not part of lease contract	Not part of lease contract

[a] Draft concession contract required 95% water and 93% sewerage coverage, 25% UFW, and 95% metering in 10 years.
[b] This requirement was later reduced.
Source: Country cases and relevant contracts.

nominally responsible for investment, but in practice has relied extensively on grants channeled through government or government guaranteed loans from donors for most new investments.

The leases in Abidjan and Conakry provide for the operators to manage the systems and collect revenues, but not to bear any of the financial risk of investment.[11] In both, the operators are expected to turn over a portion of revenues to state-operated agencies for investment, and compete for contracts to implement the investments. In practice, however, both operators have often retained these funds to cover the cost of government's overdue receivables; they have also had de facto monopolies over most new construction. Thus, despite the contractual assignment of responsibilities, these companies have a stake in expanding access, since they have leeway in pricing their construction services, but this incentive is lower than their motivation to confiscate funds due for unpaid bills. The leases provide less incentive for maintenance than a concession would be expected to provide. The Abidjan lease has been under the same operator for more than 30 years, which gives it some of the stability of a concession, however.

The operators in Mexico City bear limited responsibility for investment. The Federal Government pays a significant share of the investment costs for large projects bringing water from outside the Mexico City basin. Responsibility for billing and collection and funding operational and local investment costs rests with the government of Mexico City. Thus the operators in Mexico City have little incentive to curb water usage, but some incentive to expand access, since the cost is subsidized by national taxpayers.

The Incentive Effects of Tariff Policy on Operators

Tariff policy strongly affects incentives to improve efficiency, expand access and reduce waste in ways described in Chapter 2. Table 1.5 compares three aspects of tariff policy for the cases we studied: the general policy and timing for setting tariffs, and how metered and unmetered tariffs are determined.

Two of the cities we studied had formulas for setting tariffs that are less vulnerable to information asymmetry than the others, Santiago and Buenos Aires. Santiago's tariff was inflation adjusted annually, and adjusted every five years to cover the marginal and average costs of an efficient benchmark firm. The benchmark was built using a mix of company information and national and international standards. The tariff in Buenos Aires was initially set by reducing the current tariff by a K factor based on the winning bid for the concession. Subsequently, K was adjusted whenever a weighted composite of many government indices and tariffs went up by more than 7%.[12] It was also subject to readjustment every five years excluding the first five years of the contract. As Chapter 3

[11] Sewerage expansion is excluded from the leases in both cities. Abidjan's operator won a separate bid for a management contract to operate the city's sewerage in 1987. When the government stopped paying the company's management fee in 1992, the company began to withhold revenues due the government, although it continued to operate the system. The contract expired in 1997 and had not been renewed by mid-1998.

[12] This was used instead of a simple price index, as in Santiago, because price indexation has been illegal in Argentina since a period of hyperinflation.

Table 1.5

Comparison of tariff regulation after reform

	Buenos Aires	*Lima*	*Mexico City*	*Santiago*	*Abidjan*	*Conakry*
Tariff adjustment	Adjusted to invest. plan every 5 years and when cost index increases by greater than 7%	Arbitrary cost plus; based on company requests	Cost plus based on commission proposal, municipal assembly must approve	Adjusted to cover marginal cost and average costs every 5 years; inflation adjusted yearly	Cost plus revision every 5 years	Cost plus revision every 4 years; interim revisions according to cost index
Metered tariff	Based on concession bid and 50% unmetered bill; first 30 m^3 free; most metered are nonresidential	Rates vary by type of customer (residential, industrial, commercial, etc.)	Rates vary by volume blocs	Rates vary between peak and off peak	Rates vary by volume blocs with minimum monthly consumption	Rates vary by volume blocs with minimum monthly consumption
Unmetered tariff (% of total connections that were unmetered in year)	Location, size and type of property, size of built area, age and type of house (92% of connections in 1995)	Estimated monthly consumption for location and type of consumer (92% of connections in 1996)	Location and type of property except where 70% are metered, then average metered charge (46% of connections in 1998)	No unmetered connections	Average monthly consumption estimate (2% of connections in 1996)	Average monthly consumption estimate (1% of connections in 1997)
Subsidy schemes in 1996/97	Nonresidential pays approx. double residential; residential properties of same size pay up to 7 : 1 depending on age, size and type of house	Industry and business pays up to 6 times residential; also subsidy to government and standpipes	Nonresidential pays up to 5 times residential (1998); cross subsidy large volume consumers to small volume	Direct, means tested subsidy up to 60% of first 20 m^3	Cross subsidy large volume consumers to small volume; rate for less than 19 m^3 approx. one third rate for greater than 300 m^3	Cross subsidy large volume consumers to small volume; rate for 1st 20 m^3 approx. 73% rate for greater than 60 m^3

Source: Country cases and relevant legislation.

describes, this cost pass through arrangement has been the subject of disputes, partly because it is complicated and results in lumpy price increases.

The other cities used cost/plus formulas that were more vulnerable to information problems and provided less incentive to increase efficiency. One notable exception was in 1987 in Abidjan when the regulator threatened to rebid the contract when it expired, which led the operator to offer a 20% reduction in tariffs for low volume users, 25% for high volume users to avoid a bid.

The second incentive effect of tariffs is on coverage. Tariffs that cover marginal cost and allow the operator a reasonable return on capital invested will provide the motivation and wherewithal to expand coverage in line with demand.[13] Buenos Aires, Santiago, Abidjan and Conakry had tariffs that covered marginal costs; they also rewarded expanded coverage. In Santiago and Buenos Aires the link was direct: tariffs were estimated to cover marginal cost and a return to capital and periodically reviewed in line with investment plans. In Abidjan and Conakry the link was indirect. The private operators had a stake in the expansion of the secondary network since their returns increased with new customers, and they could earn construction revenues. In contrast, in the other two cities tariffs were below marginal cost and hence provided no incentive to add customers.

Metering also has an effect on the incentive to expand coverage. Assuming prices cover marginal cost, where consumption is not metered operators have a somewhat stronger incentive to increase the number of users, since that increases revenues. With metering the operator can increase revenues by expanding service, but also by reducing interruptions in service and leakage and connecting large as opposed to small volume consumers. Cross subsidies have similar incentive properties to unmetered billing: some consumers do not pay the marginal cost of water and the utility has less incentive to serve those customers. Metered tariffs also give the operator an incentive to reduce physical losses, as long as the return is greater than the maintenance cost. The incentive is higher when the system is near capacity, as it is in Abidjan and Buenos Aires. Thus, we would expect that where the opportunity cost of usage is high, capacity to supply users is constrained, or usage has a high variable cost (because of disposal costs, for example) metered pricing would be preferable. As we see in Table 1.5, however, the opposite prevailed. Metering was low in the two cities with the highest cost of usage — Lima and Mexico City.

---

[13] There are no good recent estimates of marginal cost for the sample cities, but what estimates there are generally fit with our finding that the water system is more costly to operate in Lima and Mexico City. Analysis for a World Bank investment project estimated the marginal cost of water and sewerage in Lima at US $0.45. The average incremental cost (AIC) of groundwater and water from the Rimac River in Lima was estimated at US $0.25 per m^3 while water from the next available source (the Atlantic watershed) was estimated to cost US $0.53 per m^3 (Munasinghe, 1992). Groundwater in Mexico City was estimated to have an AIC of US $0.41 and water from the next available source, the Cutzamala River, at US $0.82 (Munasinghe, 1992). The rate setting agency in Mexico City (CAN) charged private, non-residential users $1.00 per m^3 to extract water from private bore holes in 1997 (field interviews). Nasser (1997) estimated the marginal cost of water in Buenos Aires to be as little as $0.15. World Bank estimates put the MC of water in Abidjan at $0.13 and Conakry at $1.00. The MC for Conakry seems high in light of what we know about costs there.

The Incentive Effects of Tariff Policy on Consumers

As discussed earlier, an optimal pricing and subsidy policy would give consumers an incentive to use enough water and sanitation for good hygiene, yet avoid waste or pollution. This suggests that most consumers should pay the marginal cost of water; only usage with external benefits or cost should be subsidized or taxed. Usage charges require metering, but since metering is costly, it may not make sense to meter where the cost of usage is low.

Another reason why consumers may not pay the marginal cost of water is the use of cross subsidies. Subsidized consumers have less incentive to conserve water, while consumers facing a higher rate, particularly large industrial users, have an incentive to drill their own bore holes. The cities in our sample used a number of pricing schemes with the intent of making water more affordable for the poor. Abidjan and Conakry used increasing bloc tariffs, where larger consumption blocs pay more, and the lowest bloc pays below cost. Buenos Aires, Lima and Mexico City charged customers classified as nonresidential or industrial more than residential users, and also had low tariffs for low usage. Chile used a means tested subsidy that covered up to 60% of the cost of the first 20 m^3 of consumption.[14] Only Chile's scheme seems to have made access more affordable for the poor, for reasons we document later.

COMMITMENT

Mechanisms to enforce the contract and resolve disputes play an important role in engendering commitment. The mechanisms used in the contracts we studied are summarized in Table 1.6. All the contracts except Mexico City and Lima specify penalties for reneging ranging from fines, forfeiture of bonds, and revocation of the contract.[15] Although the water operator in Santiago was state-owned, under the law its concession could be revoked for nonperformance and the company could be sued by consumers for failure to provide adequate service.

Four of our cases allowed operators to resolve disputes with the regulator by appealing to presumably neutral third parties: Buenos Aires and Santiago allowed appeals to the courts; Buenos Aires, Abidjan and Conakry allowed for international arbitrage. In addition, Santiago allowed disputes over price adjustments to be settled by an arbitration panel with one member each appointed by the operator and the regulator, and one member jointly agreed upon. International arbitration is generally seen as a last resort, and is not effective in resolving ordinary disputes. Appeal to the courts is also an extreme step for an operator and only useful if the courts are indeed independent of government

[14] The consumer must be current on the bill for the remaining water services. The municipalities determine who is eligible for this benefit (and other direct subsidies) on the basis of house visits and a set of nationally determined criteria. The federal government allots the subsidy to the localities on the basis of income data; the amount is set to try to assure that no more than 5% of household income goes for water services. The funds are paid to the water companies, which have been enthusiastic about connecting poorer customers and informing them of how to apply for the subsidy.

[15] The draft concession agreement for Lima would have required a $50 million security deposit.

Table 1.6

Contractual provisions for enforcement and dispute resolution

	Buenos Aires	*Lima*	*Mexico City*	*Santiago*	*Abidjan*	*Conakry*
Enforcement of water contract	Performance bond ($150 million), fines; can revoke concession	None	None	Fines; can revoke concession; consumers can sue operator	Fines; can revoke lease	Perform. bond $400,000, fines; can revoke lease
Dispute resolution	Courts; international arbitrage	None	None	Neutral arbitration panel; courts	International arbitrage	International arbitrage
Representation of consumer and other interests in regulatory procedures	Local government representatives on board; no consumer representation	None	Municipal government control; no consumer representation	None; tariff setting secret	None	None

Source: Author's calculations based on country cases and relevant contracts.

control, something we consider later. With the exception of Santiago, the only realistic course of appeal was to the federal (or in the case of Mexico City, municipal) ministry responsible for the water sector, seldom a neutral body. How the ministry reacted depended upon the institutional setting, which we consider in the next section.

None of the cases enlisted consumers in the regulatory process and few allowed other interested parties any official standing in decision making. Buenos Aires included representatives of the municipality and province on the board of the regulator, with perverse effects we consider later. Mexico City was the only case where the municipal water system was actually controlled by the municipal government, but as we shall see, this has not worked to improve the credibility of the contracts.

SUPPORTIVE INSTITUTIONS

In this section we describe the institutional factors that supported the contracts. In particular we analyze how regulatory, judicial and political institutions reduced information asymmetry, motivated the parties to comply and made it credible that commitment would be sustained and the contracts enforced.

Regulatory Institutions

A survey of the regulatory characteristics for the sample raises serious doubts about the ability of regulatory institutions to collect and monitor information and apply penalties and rewards; they were also inadequate support for the governments' commitment to enforce the agreements (Table 1.7). The exceptions are Santiago, and, possibly Buenos Aires.

Santiago's regulator was powerful, independent, politically insulated and guided by detailed laws that left little room for discretion. It paid above civil service norms and its staff was regarded by most observers as competent, honest and hard working. Chile's strong bureaucratic norms of an educated and honest civil service subject to rule of law supported the regulator's independence.

Buenos Aires set up an independent body that nominally had the power under the concession contract to compel information and fine noncompliance. However, the regulator's board was politicized and divided and that reduced its credibility, and led the executive branch to intervene in the regulatory process. Even though the regulator assessed fines and penalties, these were abrogated in a subsequent renegotiation of the contract with the executive branch. Aggravating the monitoring problems, the original concession agreement in Buenos Aires suffered from serious information problems, which led the drafters to require the operator to perform specified processes, as well as to meet specified goals. These contractual problems were exacerbated by the inexperience of the staff of the regulator who intervened in the engineering decisions of the firm and did not adequately monitor the achievement of broader goals. Nevertheless, the regulator was superior to the others in the sample.

Lima's regulator was nominally independent, but in practice dependent on the supervising ministry. Unlike other regulatory bodies in Peru, it did not have a board or governing council in which government was one of several stakeholders, as a way to

Table 1.7
Regulatory characteristics after reform

	Buenos Aires	*Lima*	*Mexico City*	*Santiago*	*Abidjan*	*Conakry*
Delineation of regulatory responsibility	Separate agency	Separate agency but under same ministry as company	Three municipal agencies regulate quality, contracts and bill collection	Separate agency	Four supervisory agencies plan and manage investment, debt, and supervision of contractor	State enterprise regulates and decides investment
Autonomy and insulation from political interference	Has budget autonomy, but under political board	Has budget autonomy but no political insulation	No autonomy or political insulation	Regulator's autonomy protected by legal status; high political insulation	No autonomy or political insulation	No autonomy or political insulation
Power, and if powerful, discretion	High nominal power, but by-passed in contract renegotiations; fines, pricing decisions overturned; high discretion	Low power	Low power	High power but subject to appeal; little discretion	Low power	Low power
Staffing skills	Inexperienced but improving	Weak	Weak	Strong	Split among many agencies	Very weak

Source: Authors' calculations.

insulate it from direct intervention by the ministry. Its staff was weak and focused more on control than regulation. Although the regulator could assess penalties of up to 30% of an enterprise's revenues for failure to comply with regulation, this power was not used because there was no provision on how to enforce the penalties.

Abidjan's regulatory institutions were ill designed to extract the information necessary to monitor compliance. The private operating company was monitored by several government bodies that were often in conflict with one another and lacked clear lines of responsibility. For many years the agency in charge of building the water infrastructure was also the de facto regulator, even though another ministry was formally the regulatory body. In 1996 power was shifted to the formal regulator, but the private operator was still able to by-pass the regulator, thanks to direct access to the president. The system worked well despite these weaknesses thanks to political and international institutions that we describe below.

In Conakry there was no neutral regulator independent of other actors in the sector and information on the private operator was very poor. Instead a state enterprise played multiple roles of investor, regulator and asset owner, and was in constant conflict with the private operator. The state enterprise's staffing was weak and poorly paid and information on the operator was very poor. Since the contract covered urban areas in the entire country it was difficult to sort out activities in Conakry from elsewhere, although the importance of this was somewhat reduced by the fact that Conakry represented over 70% of the operator's activities.

In Mexico City there was no single regulator or operator and no clear targets to monitor. The system was not operated as a single corporate entity, and different types of costs and revenues were the responsibility of different organizations. There was one municipal entity responsible for building new infrastructure, another responsible for administration, and sixteen political sub-units of the municipal government (*delegaciones*) responsible for the operation and maintenance of the secondary distribution network. The sixteen *delegaciones* and the other distribution and investment units operated as government departments with costs covered by government transfers. When the four private operators signed contracts to provide specific services none of this administrative apparatus was eliminated. Regulatory responsibilities were split among three municipal agencies and the data to determine even operating costs were not consolidated. The agency in charge of monitoring the contracts was the closest to a regulator, but it acted more as a procurement control agency than a regulator and had no responsibility for price setting.

Legal and Judicial Institutions

Legal and judicial institutions are important to the incentive properties of contracts in water and to credible commitment. Among the key institutions are the status of water property rights in law and judicial neutrality when government is a party to a dispute.

In our sample Chile represents one extreme in the protection of property rights. Water has long been treated as a private good in Chile, property owners could trade water rights, and the constitution explicitly protected private property. The water regulation long allowed Santiago's operator to cut off non-payers, which helps explain why collection rates were 96% in 1996. Finally, only in Chile was the court system regarded as independent, and capable of ruling against government in a dispute with a water operator.

At the other extreme are Mexico and Guinea. In Mexico a federal law banned completely cutting off service to residential users for nonpayment. In compliance with this law Mexico City's Financial Code stated that service can be reduced to minimum "vital levels" but cannot be completely severed. In practice, however, no consumers have ever had their service reduced for nonpayment in Mexico City. Cut-offs were allowed in Conakry, but Parliament failed to past a law allowing fines to be levied on persons who repeatedly steal water. Nor could the company rely on the courts to force government to pay its water bills promptly. Collection rates were 68% in Mexico City and 64% in Conakry in 1996.

Collection could have been a problem for Abidjan, since the courts were weak, although not as weak as in Guinea. However, private users in Abidjan had an incentive to pay their bills. Connection charges for small users (with access pipes less than 15 mm in diameter) were only subsidized once; a user cut off for nonpayment had to pay the full charge to reconnect. No such incentive operated for the public sector, however, and government delinquency was a recurrent problem in Abidjan.

In Argentina and Lima, laws permitted the operator to cut off non-payers but enforcement was weak; large arrears on connections charges were an important reason for the renegotiation of the Buenos Aires concession. The courts were not regarded as independent in either country.

Political Institutions

Political institutions can substitute for weak legal and regulatory support in some cases, although sustainability is less secure. For example, checks and balances can restrain regulatory discretion, while electoral laws and legislative rules can make it hard for opponents of reform to overturn the water contract or force the government to approve lower price increases than promised in the agreement.

Chile's water contract was credible because it was based on laws that were difficult to change. Chile's Constitution required a two-thirds majority to overturn legislation. The combination of electoral rules that tended to split votes among the two rival factions, and appointed Senators who tended to favor the status quo, made it hard to obtain enough votes.[16] In addition, features of the water contract served to reduce water's political saliency, despite relatively steep prices increases following reform. One was the technocratic, apolitical nature of the pricing regulation that reassured consumers that no special interests were being advantaged at their expense. Another was the means tested subsidy for water that benefited lower middle income consumers as well poorer users.

In contrast, Buenos Aires' water regulation was the subject of partisan dispute despite a large price drop at the time of reform and some of the lowest water prices in our sample. The initial credibility of Buenos Aires' concession with investors was built on

[16] See Baldez and Carey (2001).

the reputation effects of Argentina's sweeping and successful privatization and stabilization programs, the early public support for a reform that reduced tariffs and pledged to improve service, and the legislative consensus that accompanied Menem's ascension to power.[17] The likelihood that the regulation would be sustained was increased through constitutional amendments in 1994 that made drastic revision in utility regulation less likely by changing electoral rules in ways that made it harder for the majority to ignore the minority (Heller and McCubbins, 2000).[18] These safeguards functioned at the national level, however, and the Buenos Aires water regulation proved vulnerable to political interference from local government. This was partly because the regulator had a politically appointed board, with six directors, two appointed by the executive, two by the provincial governor, and two by the mayor of Buenos Aires. Partisan interests increasingly divided the board.[19] The company protected itself by appeal to the federal executive, which was influenced by international institutions, as we describe later.

The other cities had fewer institutional safeguards against expropriation, and instead designed reforms to reduce risk and relied on international institutions to give their commitment greater credibility. Côte d'Ivoire and Guinea opted for leases, which meant that the private operator did not put investment capital at risk. Ties between the government and local investors also reassured investors in Côte d'Ivoire.

Mexico City's reduced the financial exposure of private operators by using fee-for-service in the first two phases of three-phase contracts. Risk was also reduced by not contracting for service for the entire city. The gradual phasing-in of private operating risk was expected to contain political opposition and to build private "confidence in the government's commitment to commercially viable services" (Richard and Triche, 1994, p. 15). Ultimately, political circumstances became less, rather than more, favorable towards reform, after the introduction of direct election of the mayor and assembly. Mexico City's handling of the contractors, including numerous delays, reduced rather than enhanced government credibility as a private partner.

As in Argentina, Peru's program of economic reforms and privatization gave it initial credibility, but the durability of Peru's regulatory promises was less secure. The regulator was subservient to the central government and, unlike other regulator bodies in Peru, had no legal standing to protest interference or board or council representing nongovernment interests. There were checks on executive discretion, especially after President Fujimori suspended the legislature and courts in 1992.

---

[17] Not only did Menem's party win control of both houses of Congress, but the opposition allowed him to take power five months early rather than continue as a lame duck regime in a period of hyperinflation and declining growth (Hill and Abdala in Levy and Spiller, 1996).

[18] Prior to 1994 senators were elected by provincial legislatures from two-member districts by plurality rule. Since the 1994 electoral changes, senators are elected in three-member districts with the first two seats going to the party with the most votes and the third seat to the second-place party. This made it even harder for an extreme faction to capture both houses of Congress and created more pressure for compromise with the opposition.

[19] The governor was President's Menem's main opponent in his party and the Mayor's position was captured by the opposition party when the post became an elected one in 1994.

International Constraints

International constraints can support contracts and partially substitute for weak domestic institutions. Openness to international trade and capital flows increases a government's concern with its reputation with international investors, while loans from aid agencies may influence government's water policy through conditions on disbursement or by directly supporting expenses associated with the contract. Operators competing for contracts in global markets will be concerned about their reputation in other markets in the same country or globally, particularly if the contract is highly visible.

International constraints played an important role in some of our cases. Support from international and bilateral aid agencies were important to the credibility of the contracts in Guinea and Côte d'Ivoire. In Guinea, a price subsidy financed by the World Bank phased in the full increase in consumer tariffs over five years, which helped the private operator by keeping demand higher and receivables lower than they would been with immediate cost recovery. Support from the World Bank and the French government helped reassure the operators that they could enforce the contract despite weak local institutions. In Côte d'Ivoire the French government pressured the government to pay its water arrears.

Enforcement of the Buenos Aires contract was strongly affected by the federal government's concern with maintaining Argentina's reputation with foreign investors, especially to support Argentina's fixed parity with the dollar. The water concession was a highly visible contract, and this helps explain why the executive branch intervened in regulatory decisions and did not leave them to a regulator subject to pressure from local partisan interests. For example, the federal ministry by-passed the regulator in the 1997 renegotiations of the contract. Subsequently, in 1998 when the regulatory board awarded a price increase of 1.6% versus the company's claim of 11.7% under the cost pass through provision of the contract, the operator appealed to the supervising federal ministry and the increase was raised to 4.6%. These ad hoc interventions protected the company's interest but cost the concession credibility with consumers and made the contract more vulnerable to a change in executive policy.

As for the company's concern for reputation, interviews suggest that the Buenos Aires concession was seen as visible and important to the operator's reputation. Concern about their long term relationship and reputation may explain why the companies in some of the other cases stated that they never considered suing government for egregiously violating the terms of their contract by, for example, not paying its water bills.

COMPARISON OF CONTRACT CHARACTERISTICS

Based on the previous discussion we can assess the characteristics of the regulatory contracts in terms of information, incentives and commitment. Based on the information in Table 1.8, we would expect Santiago to have the best performance on all measures except pollution. Although competition was underutilized in Santiago, there was a strong monitor able to collect information on the firm. Incentives were efficiently assigned, the operator was rewarded when the system expanded and received a return on sewerage as well as water, and there were no disincentives to serve the poor. Poor consumers could afford to connect and use water and sewerage, and users were motivated to conserve water through metering and marginal cost-based pricing. The contract's legal basis, regulatory setup and enforcement mechanisms made it credible. The main weakness in the contract was the failure to make polluters bear the cost of pollution; something none of the other contracts we studied did either.[20] The other contracts did less well in meeting the three conditions and the institutional settings were weaker. Buenos Aires' contract was somewhat better than the rest, thanks to: (i) the contestability of the concession arrangement; (ii) the incentives to improve efficiency and expand provided by the assignment of responsibilities and rewards, average prices above marginal cost, price cap, bonding and targets; and (iii) the credibility afforded by the government's concern with its international reputation. Abidjan and Conakry had poorer information because no single agency had clear responsibility for regulation and monitoring and enforcement of reporting requirements was poor. Competitive bidding and metered prices above marginal cost provided incentives to expand and reduce waste, but these were diminished in Conakry by massive water theft. The African cities overcame their credibility problems by using lower risk leases, but paid a price in lower powered incentives for operators.

The contracts in Mexico City and Lima were ill designed to address the cities' problems. The incentives in Mexico City's contracts were mixed. On the one hand, users did not pay the full cost of service, but on the other, the operators and the city did not pay much of the cost of expansion either. Because the responsibilities of the private operators were curtailed and their competition was limited, their capability and motivation to improve efficiency were also limited. This situation gave the operator and residential consumers little incentive to curb waste, while high prices for large industrial consumers encouraged them to drill their own bore holes.

Lima's regulation provided few incentives to expand. Most users did not pay full cost for service, while large cross subsidies meant that poorer consumers would pay the least and most of those without service were poor. Ad hoc pricing rules and lack of competitive threat gave the operator little incentive to improve efficiency. Rationing of service was the main check on waste.

EFFECTS OF THE REGULATORY CONTRACTS ON ECONOMIC PERFORMANCE

Economic Welfare

To assess the economic effects of the contracts we compared the achievements of the reformed company with a realistic counterfactual based on conditions most likely to pertain in the same company and city absent the contract. To do this we measured the net economic benefits of the changes attributed to the contracts using a partial

[20] Instead farmers pay the cost of not being allowed to irrigate with untreated wastewater. To the extent farmers pass that on to consumers in Santiago as higher food prices, we could say that the polluters pay.

Table 1.8

Comparison of characteristics of the regulatory contracts

	Buenos Aires	*Lima*	*Mexico City*	*Santiago*	*Abidjan*	*Conakry*
Information:						
Competition for market?	Concession bid competitive	No	No	No	Contestable lease	Lease bid competitive
Yardstick competition?	No	No	No	Some	No	No
Separate monitor?	Yes	Yes	Only of service contracts	Yes	No	No
Incentives						
(a) Operator:						
Coverage targets?	Yes	No	No	Yes	No	No
Waste targets?	Yes	No	Stage III	Yes	Yes	Yes
Pollution targets?	Yes	No	No	Yes	No	No
Bears investment risk?	Yes	No	No	Yes	No	No
Responsible for sewerage operation?	Yes	Yes	Yes	Yes	No	No
Price cap or other efficiency tariffs?	Yes, but disputed index	No	No	Benchmark	Cost plus; periodic	Cost plus; periodic
Tariffs cover MC?	Yes	No	No	Yes	Yes	Yes
No or low cross subsidies?	No	No	No	Yes	No	No

Table 1.8 (continued)

	Buenos Aires	*Lima*	*Mexico City*	*Santiago*	*Abidjan*	*Conakry*
(b) Consumer:						
Affordable for poor to connect to water?	No [a]	No	Yes	Yes	Yes	Yes, usage not afford
To sewerage?	No [a]	No	Yes	Yes	No	No
Metered prices?	Few	Few	Increasing	100%	~100%	~100%
Pollution tax or other measures?	Treatment targets	No	Prohib. irrigation of food crops	Prohib. irrigation of food crops	No	No
Commitment:						
Performance bonds	Yes	No	No	No	No	No
Dispute resolution	Yes	No	No	Yes	No [b]	No [b]
Consumer representation	No	No	No	No	No	No
Institutions:						
Expropriation safeguards?	Reputation, checks and balances	Reputation, no checks and balances	No	Law, checks and balances	Donors	Donors
Cutoffs allowed?	Yes	Yes	No	Yes	Yes, except Government	Yes, but no penalty for theft
Independent, neutral regulator?	Yes, but political board	No	No	Yes	No	No
Neutral enforcement?	Courts, but not independent	No	No	Courts, arbitration	No	No

[a] Became affordable after the 1997 renegotiation.
[b] Does not include international arbitration.
Source: Author's calculations.

Table 1.9

Welfare gains from reform [a] (in 1996 US dollars)

	Buenos Aires	Mexico City	Lima without concess. [b]	Lima with concess. [b]	Santiago	Abidjan	Guinea [c]
Welfare gains NPV in 1996 US $:							
Total domestic (millions)	1419	n.a.	43	590	284	n.a.	23
Total per capita	150		6	85	64		12
Of which (millions):							
Government	−169		40	13	241		8
Consumers	1388		57	266	4		16
Workers	43		−55	32	39		0
Domestic investors	160		0	280	1 [d]		0
Consumer gains per capita	147		8	38	1		8

[a] Methodology is explained in the cases; sensitivity tests are also given.
[b] Lima without the concession is a comparison of the actual with a counterfactual of no reform; Lima with the concession is a comparison of the actual with a counterfactual of the concession implemented according to the draft contract.
[c] Numbers for Guinea are for all urban areas under the lease.
[d] Although Santiago's utility was state-owned at this time, a few shares were privately held.
Source: Country cases.

equilibrium, cost–benefit methodology developed by Jones et al. (1990) and previously applied to privatization (Galal et al., 1994). Using the same methodology (described in each of the cases), we projected the flows generated by the firm after reform to the different parties involved — consumers, workers, government, buyers — and compared them with the counterfactual flows. We projected the counterfactual by assuming that operations would continue along pre-reform trends unless there was a reason to assume otherwise.

Unfortunately we only had sufficient data for the five cities shown in Table 1.9. These magnitudes are vulnerable to the assumptions we used to describe the counterfactual and must be looked at with caution. However, they give a sense of the impact of reforms.

There are two calculations for Lima. The 'without concession' calculations compare the reforms actually introduced under public ownership with a counterfactual based on trends before these improvements were introduced. The 'with concession' calculations compare the actual situation in Lima after reforms with a counterfactual projected on the basis of the targets in the draft concession agreement. It is clearly unrealistic to assume that all of these targets would have been met. Nevertheless, since Lima's gains with a concession are five times larger than the reforms under public ownership, it seems safe to conclude that the city would have been better off even if the operator had renegotiated considerable reductions in targets or increases in prices.

As we would expect from the previous analysis, the largest gains were realized in Buenos Aires and Santiago. Since the systems vary in size we also calculated the gains per capita; these were largest in Buenos Aires, $150 per person, and in Santiago, $64 per capita. The Santiago gains are smaller but striking since the system was one of the best operated before reform so the room for improvement was limited. The distribution of the gains was radically different. Buenos Aires' gains went largely to consumers, thanks to expanded access combined with a major price reduction at the outset of the concession. Chile's government captured most of the gains by requiring the company to pay dividends.

The gains in consumer surplus in Table 1.9 are understated because we could not quantify improvements in the quality of service. We know from field interviews that the quality of service (pressure and reliability) remained very high in Abidjan, improved in Conakry, Buenos Aires, and Santiago, and continued poor in Lima and in parts of Mexico City. Water quality in Conakry and Lima went from questionable to safe; water was already safe in all the other cities, although there are problems with appearance in parts of Mexico City.

Lima's reforms produced much smaller gains, which reflect the failings in information, incentives and commitment described above. The proposed concession would have done much better, in part because we have assumed away many of these contractual flaws. Nevertheless, welfare would likely be higher with a concession even if we relax these assumptions, because the private operator would add a new stakeholder to the negotiations. In a setting such as Peru, where there are few checks on executive discretion, a state owned firm is subject to political manipulation. Private investors would demand to be safeguarded against such behavior. Since the regulation proposed under the concession would provide incentives to expand access, improve reliability and curb waste, we expect that these would have improved. It would have been harder, given the institutional weaknesses in Peru, to guard against excessive price increases since a private operator would take advantage of information asymmetries to earn monopoly rents. Also, the concession did not provide for subsidized finance to make connection charges affordable. Thus, many of the poorest consumers, who do not have access now, would probably not have had access under the concession either.

Guinea is doing better with the lease, despite its flaws and the weakest institutional environment in our sample. Consumer welfare went up because of the expansion of the system. Weak regulatory capacity meant that much of the gains were offset by higher prices.

Efficiency and Investment

Some of the welfare gains just described resulted from more efficient operation and expanded investment. Number of staff per connection provides an indication of the extent to which the operator is concerned about cost minimization.[21] As we can see

[21] This ratio overstates employment in the African cases since the company is responsible for all urban areas but connections outside the capital are not included in the denominator. Note that that would not explain why Conakry is so much higher than Abidjan. The numbers are not strictly comparable because, as we discussed, the definition of a connection varies. We also calculated staff per m^3 of water sold (production minus UFW). The only change in relative labor productivity was that Lima has a higher volume of sales per employee than Santiago.

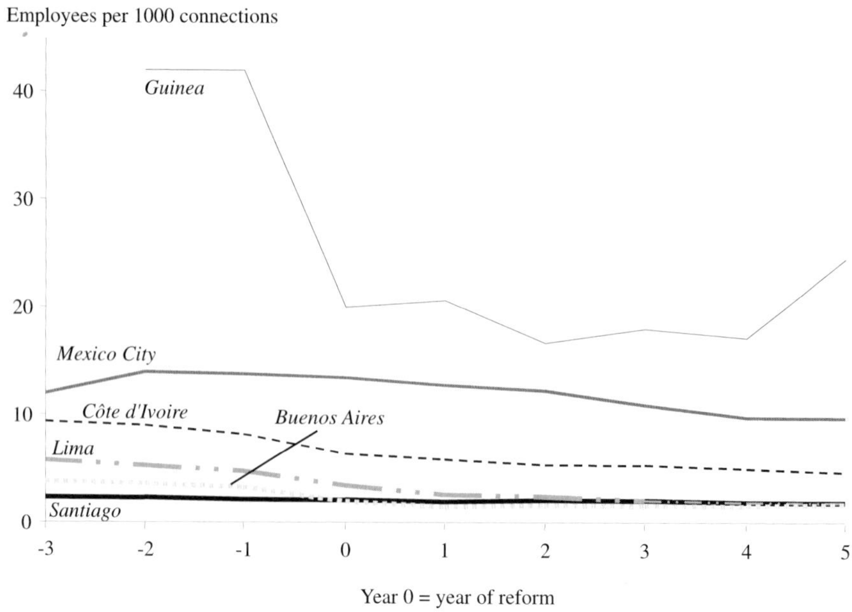

Figure 1.1. Labor productivity.

in Fig. 1.1 this measure was reduced in all of the cities: through outsourcing and staff reductions in Abidjan, Conakry, Lima and Santiago; through leaner private operation in Buenos Aires; and through registration of existing connections in Mexico City. Labor productivity remained low in Mexico City because private operation was overlaid over a group of existing state agencies that were given little incentive to reduce employment. (The staffing numbers include only the state operators.) It also remained low in Conakry, despite layoffs at the time of reform, followed by a faster rate of expansion in connections and production than in staff. Political pressures to maintain employment combined with cost plus pricing gave the operator little incentive to reduce staff.

Another indicator of efficiency is the extent to which revenues covered costs. As we can see in Fig. 1.2, costs fell below 100% revenues in all our cases except Mexico City. The improvement in cost coverage was especially steep in Lima and Buenos Aires. The Buenos Aires case is striking because prices were reduced at the start of the concession. Lima had some moderation of costs, but prices also increased. The opposite was the case in Abidjan and Santiago; these are low cost systems that continued to operate well throughout the period. There were some efficiency gains in Guinea but cost coverage improved mainly through large price increases.

A final measure is fixed investment, which we calculate per cubic meter of water produced to adjust for the differing sizes of the systems we studied. Investment increased sharply in Buenos Aires and Santiago after the contract was implemented, and expanded less rapidly in Lima and Côte d'Ivoire (Fig. 1.3). Mexico City's high investment levels reflect both the high costs of expanding the system because of its topography and the disincentive to minimize investment costs when these are largely paid by the federal

Figure 1.2. Operating costs and revenues.

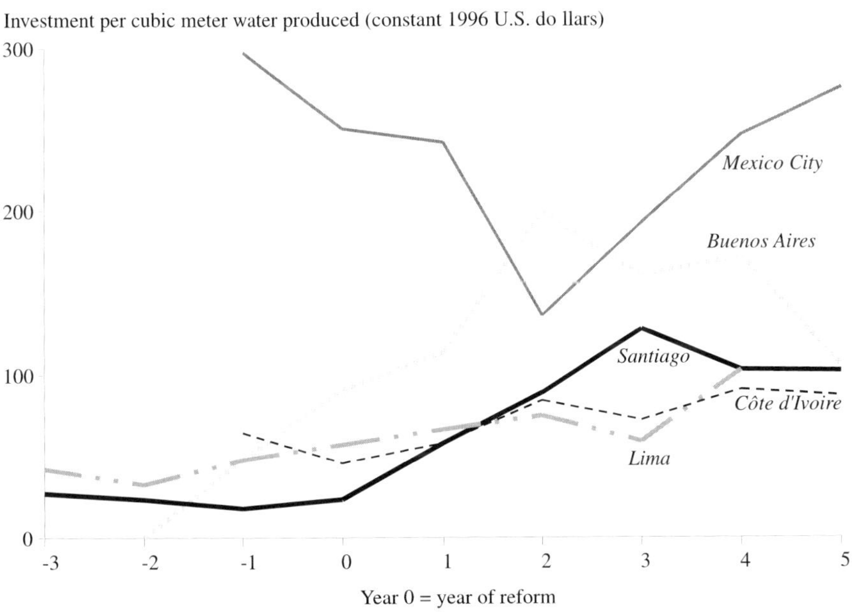

Figure 1.3. Fixed investment.

government. We exclude Guinea from the figure because its investment was much higher than the others largely because of the very small scale of its system and the foreign aid it received during the early years of the contract.

Social Welfare and the Poor

This section assesses the effects of the reforms on access to water and sewerage, waste and pollution. Unfortunately data on health and incomes are not disaggregated for our sample cities so we infer the effects by analyzing access and affordability.

Access to water and sewerage

The gains in consumer surplus described above gives some sense of the welfare effects of the contracts on consumers who were connected. This calculation, however, does not tell us how many people were not connected, nor does it address sewerage where that is provided separately. Since Esrey et al. (1991) and Esrey (1996) suggest that water alone will not improve health as much as water combined with sewerage, we expect better health outcomes from contracts that included sewerage.

The biggest improvements in coverage of water connections are in Abidjan, Conakry and Buenos Aires (Table 1.10). Santiago's expansion is less because it was at almost full coverage when the reform started. Lima and Mexico City provided low powered incentives for expansion and little change in coverage occurred. As we would expect from the contract characteristics, sewerage expansion was minimal in Lima and probably in Abidjan and Conakry, while coverage of sewerage expanded in the other cities.[22] Part of the growth in water and sewerage coverage in Mexico City shown in Table 1.10 was through the registration of existing but unregistered connections, which were about 30% of total connections before reform.

Affordability and willingness-to-pay are hard to judge because we have no survey data for our sample. Prices are also difficult to compare because the cities vary in how much of the bills is collected and in the use of fixed charges, metering, etc. Table 1.11 gives several approximations of consumer prices using total revenues billed and collected, as well as an average monthly bill for the first 30 cubic meters of consumption.

If we divide total water revenues billed per cubic meter produced, the resulting prices are much higher in the two African cities than in the rest of the sample (Fig. 1.4). This has created affordability problems in Conakry, but not in Abidjan. Despite major expansion in Conakry, an estimated 50% of the population are still not connected to piped water, almost none are connected to sewerage, and water prices are the highest in our sample by any measure in Table 1.11 even though incomes are the lowest.[23] Prices are high in Conakry because government and many prominent persons do not pay their water bills; for example in 1993 the government paid only 10% of its bill. As a result,

[22] The number of users served by a connection varies widely from 27 on average in Conakry to 5 in Santiago. Recall that the definition of a connection in some cases includes a standpipe in the yard. The cases also varied in how connections to multiple households were counted. For this reason we used volume of water produced rather than connections as the numerator in most calculations.

[23] We calculated effective prices in two ways in Table 1.11. First we report average revenues billed per cubic meter of water distributed (water produced minus UFW) and then we adjust average revenues for collection rates (using average collection rates since the lease began for Guinea). To take account of flat charges, we also calculated the monthly bill for the first 30 m^3 for a metered connection using the official tariffs.

Table 1.10

Effects of the reforms on access and waste (in percentages)

	Buenos Aires	*Mexico City*	*Lima without concess.*	*Lima with*[a] *concess.*[a]	*Santiago*	*Abidjan*	*Conakry*
Coverage							
Water:[b]							
Pre-reform[c]	70	95[d]	75	75	99	72[e]	38[f]
1996	81	97[g]	75	85	100	82[e]	47
Sewerage:							
Pre-reform[c]	58	86[d]	70	70	88	35[e]	*
1996	62	91[g]	70	83	97	*	9[g]
Growth in new connections (annual averages percent):							
Pre-reform[c]	2.1	n.a.	4.0		2.9	4.0	−0.1
Post-reform	2.8	5.1	4.0		3.8	6.7	8.5
Unaccounted-for water:[h]							
Pre-reform	44	37–47	42	42	34	13	35–60
1996	34	37	36	30	20	16	50

The asterisks indicate that the sewerage estimates in Abidjan and Conakry are not believed to have changed much.
[a] Estimated on the basis of draft concession agreement.
[b] Does not include public standpipes.
[c] Pre-reform dates are: Buenos Aires 1992; Lima 1991; Mexico City 1992; Santiago 1988; Abidjan 1987; Conakry 1988.
[d] 1990.
[e] For all urban areas under private operation.
[f] 1989.
[g] 1995.
[h] Unaccounted-for water (UFW) is the difference between volume of water distributed to the delivery system and water sold, and includes physical losses from pipe breaks and overflows as well as commercial losses from illegal use, under registration of users, etc.).
Source: Country cases and Statistical Appendix.

those who do pay must cover the investment costs for a system sized for a much larger number of consumers, and cross subsidize the government as well as many wealthy, non-paying users.

Another reason why poor people cannot afford water in Conakry is that many people share a connection, and the price increases with the volume consumed.[24] There are estimates that in Conakry as many as 27 persons use water from a connection, which is usually a pipe in the yard (see Chapter 8). Household surveys suggest that there are strong social mores against charging neighbors and relatives for water in Conakry, in contrast to Abidjan where most connections are also yard pipes but charging is well accepted. Even though per capita consumption in Conakry is the lowest in our sample,

[24] For more discussion of the problems with increasing bloc tariffs see Boland and Whittington (2000).

Table 1.11

Average revenues per m³ billed and distributed[a] and average bill for 30 m³ (in US dollars)

	Buenos Aires	*Lima*[b]	*Mexico City*[b]	*Santiago*	*Abidjan*	*Conakry*
Average price per m³ water using revenues						
Billed:						
Pre-reform[c]	0.21	0.21	n.a.	0.09	0.85	0.30[d]
Post-reform[c]	0.18	0.20	0.37	0.11	0.75	1.00[e]
1996	0.24[f]	0.35	0.32	0.30	0.55	1.19
Collected:						
Pre-reform[c]	0.18	0.15	0.22	0.08	0.81	0.13[d]
Post-reform[c]	0.16	0.18	0.27	0.11	0.71	0.62[e]
1996	0.23[f]	0.32	0.22	0.29	0.51	0.74
Average monthly bill for 30 m³ for metered household, 1996	n.a.	7.87	5.79	7.69	13.08	23.66

[a] Revenues net of indirect taxes and adjusted for collection rates per m³ produced adjusted for UFW.

[b] Includes sewerage.

[c] Dates are: Buenos Aires pre = 1992; post = 1993; Lima pre = 1991; post = 1992; Mexico City pre = 1992; post = 1996; Santiago pre = 1989; post = 1990; Abidjan pre = 1987; post = 1990; Conakry pre = 1987; post = 1995.

[d] 1984.

[e] Excludes World Bank subsidy. With subsidy prices were $0.34 billed and $0.19 collected.

[f] 1995.

if one consumer must pay for 27 persons, the bill quickly becomes unaffordable for much of the population. The failure to connect and the lack of sewerage have had serious health consequences since most unconnected consumers in Conakry rely on contaminated water from shallow, hand-dug wells.

Had the concession gone forward in Lima, the likely rise in prices combined with connection fees could have made service unaffordable for many low-income people. The water and sewerage connection costs plus higher usage tariffs under the concession, would have resulted in a water and sewerage bill as high as 22% of household income, which seems unaffordable for many poor families (43% of Lima's households earn US $200–$240 a month). As it is most of those poorer households are not connected because of the failings of the current regulation to address the problems of the system. Cross subsidies provide no relief to users who cannot afford connection. Unconnected consumers in Lima rely on public standpipes or vendors. Typically water in poorer communities is stored in uncovered or partially covered metal drums, cement tanks or plastic tubs which easily permit contamination; water can also be contaminated by dust during the process of collection at public standpipes; and the quality of water from vendors is not high (see Chapter 4).

In Buenos Aires the concession contract charged new customers the cost of the connection itself and part of the cost of expanding the secondary network, which resulted in a very high access charge, between $1100 and $1500. The contract required

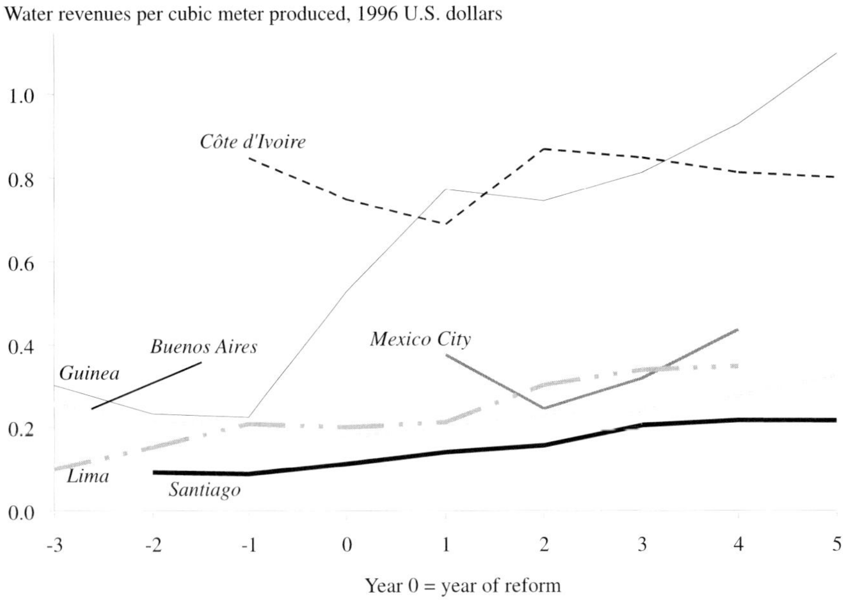

Figure 1.4. Average price of water.

the firm to provide financing but allowed it to select the term. The operator opted for two years, the minimum the contract allowed. Many unconnected customers in poor sections of Buenos Aires would have had to pay 18% of their income just for the connection. After many refused to pay, the charge for the secondary network expansion was eliminated in the 1997 renegotiation of the contract and the cost was spread to all consumers through a surcharge.

Curbing waste

Another measure of social welfare is the extent to which the system provides incentives to conserve water where the opportunity cost of usage is high. One mechanism to curb UFW is enforceable targets for waste or maintenance. As we have seen, targets are enforceable only in Santiago and to a lesser extent in Buenos Aires. A second tool is for the operator to bear the cost of expansion; this was used in Santiago, Buenos Aires and Lima. Prices that reflect the opportunity cost of usage combined with metering, give consumers an incentive to curb usage. The contract led to prices above opportunity cost in Buenos Aires, Santiago, Abidjan and Conakry, and metering was widespread in the latter three cities.

As we can see from Fig. 1.5, UFW declined significantly in Santiago, and remained very low in Abidjan. Ironically these are cities with much lower opportunity and external cost of usage than Lima or Mexico City, where rates did not decline as much.

The pre-reform numbers for Conakry are very poor, but it is clear that it continued to have very high UFW; five years after the lease began almost half of the water is

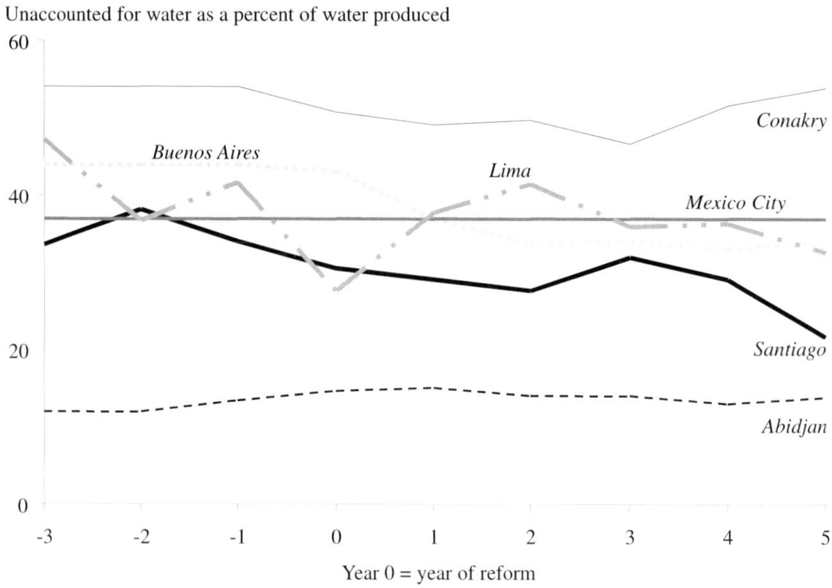

Figure 1.5. Unaccounted-for water.

unaccounted for. This is not surprising given the lack of operator incentives to invest in maintenance and the lack of institutions to combat theft and nonpayment. The operator also has no incentive to curb waste in order to bring down operating costs since this might lead to lower prices. Although lower prices might expand demand, this has less incentive effect in a city with Conakry's collection problems.

Pollution

Another cost of usage is pollution. Sewage treatment was minimal in all of the cities, and there were no pollution taxes or penalties that were effectively enforced. Because of the location of many of our sample cities near oceans or rivers without downstream development, downstream users have not been much affected. Instead, the main externalities caused by pollution have been illness from contamination of well water because of under-investment in sewerage in Buenos Aires and Conakry, irrigation of food crops with untreated wastewater in Lima, Mexico City and Santiago (until 1991), and contaminated fishing grounds in Lima and Abidjan. The failure to treat wastewater had another cost for Mexico City and Lima. Even though their groundwater was being diminished, they could not recharge it because untreated sewage was mixed with rainwater. In Mexico City rainwater and sewage was pumped away; in Lima it was dumped into the ocean.

Sewage treatment can result in large increases in tariffs. In our sample cities such rate increases would have been politically difficult to enact, especially where sewerage was absent and the cost of water alone was high, as in Abidjan and Conakry. Santiago and Mexico City were able to reduce the risk to health and food exports by prohibiting irrigation of certain crops with untreated wastewater. This indirectly applied the principal of polluter pays, since the cities' residents paid for their pollution through higher food prices, although not fully since much food production was exported. These measures are cheaper than sewage treatment, but do not abate the esthetic costs, the harm to tourism and fishing (in Chile), the risk to food exports and possible long run problems from accumulating contamination.

LESSONS AND CONCLUSIONS

Our cases illustrate how contractual design combined with supportive institutions can motivate major improvements in operating efficiency, social welfare and equity. They also suggest that countries with weak institutions can design contracts that compensate for some of these failings, much as Levy and Spiller (1996) found in telecommunications. The contrast of Chile and Guinea is an example. In Chile, bureaucratic norms, rule of law, checks and balances, and protections of property rights were all strong. These institutions enabled the Santiago water system to perform well under public ownership before the new regulations were implemented. The introduction of a regulatory contract that met the three conditions of reducing information asymmetry, heightening incentives and signaling credible commitment generated significant additional gains.

In Conakry the water system was delivering unsafe water to a small proportion of the population with high rates of leakage. Although the contract in Conakry could not overcome all of Guinea's institutional weaknesses, it was still an important improvement over any realistic counterfactual. Under private operation capacity more than doubled, water quality and service improved dramatically, the population served almost doubled and coverage expanded from 38 to 45%. About half of the city's population is unable to access piped water (legally) under private operation, but we cannot assume that they would have had access to piped water under any reasonable counterfactual. On the other hand, it is reasonable to assume that Lima could have had better performance under the concession than was possible under public operation, even if the gains were a fraction of those the draft concession targets would predict. The experience of Abidjan further suggests that institutionally weak countries can improve water operations through contracts, although Mexico City's experience shows that when only small responsibilities are contracted out, only small gains can be expected.

Our sample is small so results must be treated with caution, but the contrasting experiences of our cases imply that contracting out water services to private operation may be most useful to countries with weak institutions. The concern of global operators with their worldwide reputation and local relationship and their responsiveness to the incentives under the contract can propel improvements that would not have occurred under a public operator not subject to these motivations. Governments' concerns about its international reputation function more strongly when there is an international operator present, as illustrated by the Argentine support of the operator in Buenos Aires or Conakry's changed relationship with the water operator.

The risk of private operation in the presence of weak institutions is that consumers will be poorly protected and an operator motivated by profit will make service unaffordable. In Guinea, however, the high prices are not primarily due to the operator's need to earn a return, but to the government's refusal to pay its water bill or to allow the company to collect from well connected private consumers. This situation existed before the lease was signed, but with private operation the source of problems is more transparent which may begin to create pressures for its solution.

This is not to suggest that the contracts could not have done better in Guinea as in the rest of our sample. We conclude with some suggestions of ways to heighten the gains from reform by: (i) increasing social and external benefits; (ii) taking full advantage of competitive opportunities; (iii) designing regulation to reduce information asymmetries; and (iv) assuring political support by spreading benefits more widely.

How could Equity and External Benefits be Increased?

Judging from our sample, more attention needs to be given to the external and equity effects of water services, including the social benefits from improving sanitation. Our cases suggest that it is particularly important to lower the cost of connection through financing and reduce free riding by government, rather than subsidizing the usage price. Cross subsidies did not benefit the poor who were unconnected or consumed too much to receive a subsidy, and gave subsidized, metered consumers less incentive to preserve water and reduced operator's incentives to connect lower income households. In addition high tariffs for industrial users in Lima and Mexico City encouraged larger industries to drill their own wells, further threatening aquifers.

The Santiago case provides evidence on how affordability can be addressed with fewer distortions, although only in countries with strong institutions and metered systems. While this might not be relevant for our entire sample, it would be applicable to, for example, Buenos Aires. The Chilean government gave the utility a direct subsidy to cover on average 60% of the first 20 cubic meters of consumption for those consumers certified as poor by their municipality; the consumer paid the balance and could be cut off for nonpayment.[25] This motivated the utility to serve the poor and the poor, to avoid waste.

Although in theory metering should only be introduced where the cost of installing meters is less than the opportunity cost of usage, in practice there are other, non-quantifiable considerations. Meters provide a mechanism for consumers to reduce their usage charges, which can be important for poorer households. Santiago's utility launched a campaign to help consumers reduce their water bill by detecting and fixing leaks. Since a toilet left running in a house for two days wastes 28 m^3 of water, equivalent to the average monthly consumption of a family of five in Chile (Alfaro, 1997), the savings from simple repairs can be considerable. In addition, metering gives consumers more information that can serve as a check on operator behavior and increases the political acceptance of water reform.

[25] The proportion of the bill covered by the subsidy varies. The amounts are allocated to municipalities with the intent that no household will pay more than 5% of its income for water services. Thus, the amount covered depends on income and prices in a given water district.

Reducing government nonpayment is difficult but not impossible, as experience in other infrastructure suggests. It seems prudent to make cut offs automatic after a certain period, penalize nonpayment with fines and secure promises of prompt government payment before the contract is signed. How well this will work where enforcement institutions are weak is an open question. Projects in private power generation sometimes secure payment with international guarantees or by making a source of funds (such as oil revenues) hostage to the agreement. In a sense funds were held hostage in Guinea and Côte d'Ivoire, since the operators withheld to cover government arrears that proportion of the revenues they collected to finance government investment in water. The problem with this remedy is that it has high economic and social costs, since it reduces the money available to maintain the system and connect new customers, and creates uncertainty and distortions in management. Ideally a hostage can be found where the costs of nonpayment are higher for those who would otherwise be tempted to free ride.

Enhancing Competition

There were frequent, missed opportunities to expand competition in the water systems we studied. Failure to bid or rebid contracts, to procure investment competitively where government bears the risk, or to design a proper auction were costly mistakes in our cases. On the other hand, yardstick competition proved difficult, perhaps because of the information problems we described, which may explain its virtual absence in our sample. Splitting up large city systems might improve the chances for yardstick competition, if the division could produce reasonably comparable circumstances, as opposed to the very different zones in Mexico City. Third, it seems advisable not to give the successful bidder an exclusive franchise. Solo (1998) suggests that competition among water vendors can result in reasonable prices and reduce the cost of access for the poor. Regulation and better consumer information could reduce the tendency of vendors to collude or use Mafias to eliminate competition; it could also curb the quality problems observed in Lima.

Improving Regulation and Reducing Information Asymmetries

Theory and practice in infrastructure privatization in developing countries have focused on reducing regulatory risk to operators, not surprising given past histories of underpricing and current vulnerability to politically mandated regulation. In the cases we studied, however, most governments did not try to confiscate returns to private capital by pricing below cost recovery, and monopoly rents were as serious a risk as expropriation.

To protect against monopoly rents, the regulators in our cases needed much better information on costs then they usually had. While it might be justifiable for a new regulator and operator to be ignorant about costs, it is harder to understand the lack of data on operating costs, investment and capital stock in a case such as Abidjan, which had had a private operator for over 30 years. Information requirements and penalties for noncompliance should be spelled out before a contract is signed. When a private operator contracts to provide service in more than one urban area, the government should insist that city information be kept separate, to be able to assess performance against other municipal water companies and to hold the company accountable to the citizens in each service area. Pricing formulas such as price caps that are inflation adjusted annually but only raised every five years, might also help reduce the information required during the period between adjustments. Regulators could contract out data gathering and processing, which might also enhance the information available and make it more credible to the public.

Greater involvement of consumers and other interests in regulatory decisions might be another way to reduce the burden on the regulator. This was not tried in the cases we studied, although the practice is common in developed countries. None of the cities gave consumers or other affected interests, such as downstream users or farmers buying wastewater, opportunities to comment on regulatory decisions, and information sharing with the public was minimal. Even in Chile, where rules were clear and transparent in other ways, tariff setting was secret and consumers were ill informed about their rights and had no standing in the regulatory process. In Buenos Aires, public outcry was partly responsible for the regulator's decision to drastically reduce the proposed increase of water prices in 1998, showing that consumer activism can be a powerful check on company power. But it also illustrates the risk that consumer opposition will prevent price increases needed to assure investors a reasonable return. This risk is not a necessary result of consumer activism per se, but of a weak regulatory framework that gives consumers little information about the process, few opportunity to provide comments before decisions are taken, and low confidence that regulatory procedures will be applied fairly and systematically. Comparative data about costs and customer services such as time to repair a burst pipe or respond to a billing mistake could be standardized and published.

Metering is another device to increase regulator and consumer information and give consumers more control over their bills. As we have discussed, metering is usually not considered worth the expense in cities with ample water and no capacity or disposal problems. This ignores the loss of consumer control over the total bill when billing is based on consumption estimates, as in Lima, or a complex series of property characteristics, as in Buenos Aires. Complex unmetered billing also creates more opportunities for the utility or the consumer to manipulate information and gives more leeway for arbitrary action by the regulator.

In several of the cases we studied the regulators were weak because the underlying bureaucratic and legal institutions were vulnerable to political interference and had few safeguards against corruption or rewards for competence. Experience with public sector management reforms suggests that weak rule of law or weak bureaucratic norms of honesty and competency can usually only be overcome in the long run. What can be done immediately is to try to simplify the regulatory burden. For example, transparent rules and rule making and standing for those affected by regulatory decisions in the decision process might help, as would more competition. Simplification of mechanisms governing the inevitable revisions and renegotiations of contracts and more transparent and published rules for revision should also help reduce the risk that the operator or the government tries to capture rents to the disadvantage of the consumer.

REFERENCES

Alcázar, Lorena and Penelope Brook Cowen, 1996. "Research Proposal. Institutions Politics and Contracts: Private Sector Participation in Urban Water Supply". Unpublished Research Proposal.

Alfaro, Raquel, 1997. Competition and Social Extension in a Water Supply and Sanitation Company. Unpublished Report for the World Bank.

Baldez, Lisa and John Carey, 2001. "Budget procedure and fiscal restraint in post-transition Chile". In: Stephan Haggard and Mathew D. McCubbins, eds., *Presidents, Politics and Policy.* Cambridge: Cambridge University Press.

Boland, John and Dale Whittington, 2000. "The political economy of increasing block water tariffs in developing countries". In: Ariel Dinar, ed., *The Political Economy of Water Pricing Reforms.* Oxford University Press, Oxford.

Cowan, Simon, 1993. "Regulations of Several Market Failures: The Water Industry in England and Wales." *Oxford Review of Economic Policy* 9 (winter): 14–22.

Esrey, Steven A., 1996. "Water, waste, and well-being: a multicountry study". *American Journal of Epidemiology* 143: 608–623.

Esrey, S.A., J.B. Potash, L. Roberts et al., 1991. "Effects of improved water supply and sanitation on ascariasis, diarrhoea, dracunculiasis, hookworm infection, schistosomiasis, and trachoma". *Bulletin of the World Health Organization* 69: 609–621.

Galal, Ahmed, Leroy Jones, Pankaj Tandon and Ingo Vogelsang, 1994. *Welfare Consequences of Selling Public Enterprises: An Empirical Analysis.* New York: Oxford University Press.

Heller, William B. and Mathew D. McCubbins, 2000. "Political institutions and economic development: the case of electric utility regulation in Argentina and Chile". In: Stephan Haggard and Mathew D. McCubbins, eds., *Presidents, Politics and Policy.* Cambridge: Cambridge University Press.

Laffont, Jean-Jacques and Jean Tirole, 1993. *A Theory of Incentives in Procurement and Regulation.* Cambridge, MA: MIT Press.

Levy, Brian and Pablo Spiller, eds., 1996. *Regulations, Institutions and Commitment: Comparative Studies of Telecommunications.* New York: Cambridge University Press.

London Economics, 1998. "Competition in Water". Paper prepared for the Department of International Development, London.

Ménard, Claude and Mary M. Shirley, 2001. "Reforming Contractual Arrangements: Lessons from Urban Water Systems in Six Developing Countries". Unpublished Article.

Munasinghe, Mohan, 1992. *Water Supply and Environmental Management. Developing World Applications.* Boulder, CO: Westview Press.

Nasser, Thomas-Olivier, 1997. "Water Concession in Buenos Aires: Issues and Propositions". Unpublished Consultant Report.

Richard, Barbara and Thelma Triche, 1994. *Reducing Regulatory Barriers to Private-Sector Participation in Latin America's Water and Sanitation Services.* World Bank Policy Research Working Paper 1322, Washington, DC: World Bank.

Shirley, Mary and Lixin Colin Xu, 1998. "Information, incentives and commitment: an empirical analysis of contracts between government and state enterprises". *Journal of Law, Economic and Organization* 14(2).

Silva, Gisele, Nicola Tynan and Yesim Yilmaz, 1998. *Private Participation in the Water and Sewerage Sector-Recent Trends.* Finance, Private Sector and Infrastructure Network. Viewpoint Note No. 147, Washington, DC: World Bank.

Solo, Tova Maria, 1998. "*Competition in Water and Sanitation. The Role of Small-Scale Entrepreneurs.* Finance, Private Sector and Infrastructure Network. Viewpoint Note No. 165, Washington, DC: World Bank.

Williamson, Oliver, 1976. "Franchise bidding for natural monopolies — in general and with respect to CATV". *Bell Journal of Economics* 7(1): 73–104.

Williamson, Oliver, 1985. *The Economic Institutions of Capitalism.* New York: Free Press.

World Bank, 1995. *Bureaucrats in Business: The Economics and Politics of Government Ownership.* Washington, DC: Cambridge University Press.

Thirsting for Efficiency: The Economics and
Politics of Urban Water System Reform
Mary M. Shirley (Ed.)

CHAPTER 2

THE ECONOMICS OF URBAN WATER SYSTEMS

ROGER G. NOLL

INTRODUCTION

This chapter examines the relationships among the underlying economics of urban water systems in low-income countries, the design of optimal water policies, and the political economy of urban water supply. The role of this chapter is to provide a common analytic framework for the case studies in this book and for evaluating other urban water systems in the developing world. Of course, the underlying economic conditions of urban water systems exhibit substantial regional variation, due to differences in climate, topography, and other endowments that affect the structure of a regional economy. Because of these differences, the optimal design of an urban water system varies enormously from locality to locality. Nevertheless, the economic principles for evaluating water systems and policy alternatives are general.

The economics of urban water systems usefully can be separated into the issues that pertain to the four main components of a water utility: the private supply cost of water delivery to various types of customers; the demand for water by major customer groups; the externalities associated with urban water delivery and use; and the market and political institutions that allocate water among competing uses. These four components interact to create the incentives that are faced by water users and water supply organizations, and thereby to determine the efficiency of urban water systems.

The structure of these four aspects of the economics of urban water systems lead to five important and not widely recognized conclusions about the choice of allocation institutions. First, the marginal private cost of water frequently is so low that metering and billing costs can be a substantial fraction of marginal cost. Hence, the second-best optimal price structure for recovering private costs plausibly is to set the usage charge to zero. Second, a very large proportion of the cost of urban water supply is accounted for by fixed and sunk capital investments that have very long useful lives. Consequently, to induce adequate private investment in water systems at prices that are realistic for low-income consumers government must create an institutional environment in which investors can expect to recover annually a small fraction of their fixed costs for several decades. Third, the externalities associated with urban water supply frequently are more important than private cost and demand (willingness to pay) conditions. In particular, a major issue in designing urban water systems is how to dispose of water after it is delivered to users. In some cases, the externalities from low-valued uses are sufficiently large that internalization of these externalities provides a justification for introducing usage charges, despite high metering and billing costs. Fourth, some externalities associated with urban water use are very difficult to internalize in water prices, and also are very difficult to regulate effectively. Consequently, the magnitude of the optimal usage charge can be highly uncertain. Fifth, despite the institutional design problems that are created by externalities and durable capital assets, and notwithstanding the universal reluctance of policy makers to permit competition, some elements of water systems can be organized efficiently as a competitive private market.

The remainder of this chapter considers the four elements of urban water system economics and provides the basis for these five conclusions.

WATER SUPPLY COSTS

The costs of a water system are usefully separated into its private cost and the costs associated with its externalities. Following convention, the term direct social costs is used here to refer to the sum of private costs and externality costs, and the term total social costs refers to direct social costs plus the opportunity cost of water when the resource is not sufficient to satisfy all demand at prices that reflect the direct social costs of production.

Private water supply costs refer to the resources that are used by a water supply system to deliver water from natural sources to urban customers. The externalities of water systems fall into two categories. One type of externality is the effects of expropriating water from its natural source on the surrounding environment, including the ecosystem associated with the natural water system and industries that depend on that ecosystem, such as agriculture and fisheries. The other form of externality is the effects of disposing of used or wasted water, including adverse effects on human health as well as on the environment. The first externality is considered in this section. The second is discussed separately following the discussion of water demand because of the relationship of both to public health concerns.

A water delivery system has five components that together determine its direct social costs: capturing and storing the natural resource (diversions, reservoirs and wells); transporting the water to areas where it will have an economic use (aqueducts, penstocks and mains); treating the water to raise its quality; delivering water to users (ditches, pipes and taps or tank trucks and bottles); and disposing of both dirty water after its use and clean water that flows through the system but is not used. Except for delivery systems using tank trucks and bottles, all components of water systems require very long-term investments in capital assets that are sunk costs in that they cannot be redeployed to supply water somewhere else and have little or no scrap value. Investments in capture, storage, transportation and piped distribution have especially long useful lives, with some components operating for hundreds of years.

The most important variable costs of water systems consist of the energy that is used in pumping and purification systems, repair and maintenance services, and services associated with marketing, monitoring and billing. For treatment facilities and for wells and transportation investments that involve extensive pumping, operating costs can be substantial. Moreover, the annualized cost of the long-term capital investments include a very tiny depreciation component because of their extraordinarily long useful lives. Nevertheless, the fixed capital costs of a water system usually are much higher than variable costs. For example, over 80% of private water supply costs in the United Kingdom are accounted for by fixed capital investments (Armstrong et al., 1994).

Water typically is regarded as a *renewable resource* in that natural rainfall and river flows replenish expropriated water, but this is not always the case. A water resource has bounded renewability in that it is refreshed by finite inflows, and in many populous areas usage is near or exceeds the refreshment rate. For some underground aquifers in arid climates, water inflow is minimal, so that the underlying resource is essentially a *nonrenewable resource*, much like deposits of minerals and hydrocarbon fuels. The boundary between renewable and nonrenewable water resources defines the extent to which expropriation of a resource is *sustainable*. Utilization of a resource is sustainable if the sum of all current uses (including protecting ecosystems that depend on the resource) does not exceed the rate at which the water system is refreshed.

The distinction between renewable and nonrenewable resources and the concept of sustainability have important implications for ascertaining the total social cost of water supply and hence for optimally allocating the resource (Gisser, 1983). If the net usage of water from a resource (including the maintenance of the quality of the water that remains in the natural system and the associated ecosystem) is now and in the future less than the inflow, the opportunity cost of incremental water use is zero. In this case, the only relevant costs of the resource are the private cost of water distribution plus the external costs of exploitation and use. If water use at prices reflecting these direct social costs exceeds inflow, either temporarily or permanently, then some uses that would be attractive at these prices must be curtailed, and usage in one period has an opportunity cost equal to its net value of foregone use in the future. Thus, a price system that allocates water efficiently under these circumstances must set prices above the direct social costs of supply (private costs plus externality costs).

Private Costs of Renewable Resource Systems

The private cost of capturing water depends on the short-term variability in the flow of water into the resource. If water usage for all purposes always is less than inflow, water system design is greatly simplified. Pumps and diversion canals provide a steady flow of water to satisfy all demands, including protection of the environment. The following discussion focuses on the easy, baseline case in which externalities are unimportant, usage is sustainable, and so all water supply costs are private. The discussion then proceeds to add the complexities associated with externalities and unsustainable use.

Water systems exhibit engineering scale economies in some of their components, much like other utilities such as railroads, electricity grids, and telecommunications networks. For water systems, scale economies arise from two sources. First, the capacity of a canal or pipe is governed by its cross-sectional area, which increases in proportion to the square of the external dimensions of the conduit. While larger pipes must withstand more weight, within the relevant scale range strength requirements do not offset the volumetric gain from expanded circumference. Second, larger canals and pipes have less turbulence, which expands effective capacity more rapidly than in proportion to the cross-sectional area of the transport medium.

Of course, engineering scale economies do not necessarily translate into organizational scale economies that cause the most efficient market structure to be a monopoly. Three factors can offset engineering scale economies: diseconomies of scale in organizational management; quality differentiated demand that cannot be satisfied most efficiently in a single network; and production inefficiencies that sometimes arise when firms are insulated from competition. The conventional view about water systems is that the simplicity and low variable cost of the technology cause water systems to be the most likely utility to be a natural monopoly; however, this conclusion has not been the subject of extensive empirical testing, and, for reasons discussed below, probably is not accurate for all aspects of urban water systems in all locations.

Seasonal and annual variability in water flow introduces another source of costs. Intermittent flows due to either seasonal or annual variability in rainfall, can cause temporary circumstances in which usage exceeds inflow, even though over a longer period usage is less than total inflow. For subsurface aquifers, pumping that exceeds inflow causes the water table to drop, thereby increasing pumping costs. For all types of water sources, storage systems can be used to transfer water from periods of excessive supply to periods of excess demand.

Storage systems typically are extremely durable, long-term capital investments with negligible variable costs. The classic example is a large reservoir. Another common storage method is to flood the surface above a large underground aquifer to increase the water table for later well extraction. For example, to insure against drought, the City of Los Angeles stores about a three-year supply of water in the aquifers beneath the city. In this case the storage system itself is costless; however, aquifer storage requires a system for capture, transportation and percolation, which consists of extremely durable assets that exhibit scale economies.

Limited economies of scale in storage arise at the level of a single reservoir. Typically a reservoir exhibits substantial economies of scale to an upper bound that is determined by the geology of the reservoir site and intertemporal variation in water flow. For example, a dam to impound a river with a highly variable seasonal flow at the bottom of a deep canyon typically will have huge economies of scale up to a storage capacity that is many times annual flow.

Whether economies of scale in a single reservoir translate into system-wide economies of scale depends on the number of storage elements in the system. In multi-reservoir water systems, which are common for large urban areas, a single reservoir is analogous to a single electric generation facility in a large electrical grid (which frequently is the other use of a single large reservoir). In this case, economies of scale for a single reservoir do not imply a natural monopoly in either the entire system or even in the part of the system that captures and stores water. As with generation for electric utilities, in a multi-reservoir system one can imagine a decentralized, competitive wholesale water market in which competing reservoirs deliver water to either long-distance transportation facilities or distribution utilities (London Economics, 1998). Indeed, as documented in Roth (1987), a few urban systems do rely on multiple independent sources.

Likewise, a water system that relies extensively on pumping is not a natural monopoly at the level of capture and storage. Pumping from many wells, whether from a fully renewable underground aquifer or an aquifer that serves as a reservoir through injections in periods of excess flow, can be organized into multiple, independent sources; however, because pumping from one well can cause an externality by causing a drop in the water table so that pumping at other wells is more expensive, decentralization of well sources may require centralized management of pumping activities. Whether managing the externality is more efficiently accomplished through owning all the wells is an empirical question, but in the petroleum and natural gas industries, in which the pumping externality is important, decentralized ownership of wells combined with centralized field management is widespread, indicating that a similar structure might be efficient in water systems.

Piped distribution systems are the best candidate for natural monopoly, but even here a plausible alternative exists. In some communities, nearly all water for residential use is distributed by vendors who sell water from tank trucks or large bottles. Whereas this form of water delivery typically is far more expensive than piped distribution, in some cases it is a viable alternative, especially in circumstances where the main sources of natural water are impure and a piped distribution system for low-quality water is available for uses that do not demand pure water (Deb, 1978; Zaroff and Okun, 1984; Roth, 1987).

Policy Implications of the Private Cost Structure

A water system's high ratio of fixed to variable costs has important implications for the pricing system. Extensive investments in transportation, storage and distribution facilities do not affect the first-best price of water usage. If water usage is sustainable, the optimal pricing structure is a two-part tariff. The first part is a capacity charge that determines a user's maximum usage during periods of excess demand plus any other costs that are not recovered from the second part. The second part, assuming no significant externalities, is a usage price equal to the private marginal cost of supply, which is likely to be very low compared to average private costs.

Two alternatives to the optimal two-part tariff are theoretically less efficient but may be practically attractive because they are easier to implement. One is a two-part tariff in which the first part is a fixed hook-up charge that covers the annualized infra-marginal capital costs. The second is a variant of peak-load pricing, in which usage prices include variable costs and, during periods when usage equals system capacity, the marginal capital costs of transportation and distribution and, during the low-flow period, the marginal cost of storage capacity. These price structures are less efficient than the capacity-charge approach because they do not permit consumers to determine their maximal claim on capacity in low-flow periods, but they are easier to implement because they do not require the technical capacity to cap individual usage during periods when demand equals supply capacity.

Because the private marginal cost of water usage frequently is very low, the implied usage charge in a two-part tariff also is very low. Indeed, in many systems, especially ones that tap nearby abundant resources, the costs of metering and usage-based billing are high compared to the private marginal cost of water, in which case the question arises whether metering is actually worth the cost. The advantage of a fixed charge for service is that it avoids metering. For metering to be justified, the efficiency gains from giving customers an incentive to curtail usage must offset the transactions costs of metering. The private marginal cost of usage is unlikely to be high enough to justify metering unless marginal costs are atypically high. Examples in which the marginal source of supply is very high in periods when demand equals capacity are circumstances in which the marginal water source is very distant, or requires extensive pumping from very deep wells or uphill long-distance transportation.

Usage-based prices also can be justified as an efficient means to internalize relatively high costs of externalities from water use. This issue is discussed at length below. The important point is that if externalities are a substantial part of the cost of urban water systems, the benefits of a usage tax can tip the scales in favor of a metering system in cases in which the direct private cost of supply (delivery and storage) is too low to justify metering.

Costs, Rents, and the Political Economy of Water

Regardless of the details of the price structure, in a self-financing water utility a high ratio of fixed to variable costs implies that a large proportion of revenue is quasi-rents. In the short run, these quasi-rents do not need to be collected to induce continued supply. Because capital investments in water systems are extremely durable, a water utility may be able to operate for many years, even decades, without recovering any fixed costs, which even on an annualized basis comprise nearly all of its private costs. But if quasi-rents are not paid, investors will not recover their investments. In this case private funds for new investments will not be forthcoming, and even government may be unwilling to allocate general tax revenues to these investments. And, if prices do not include quasi-rents, water customers will have an incentive to engage in inefficiently great use of water for low-valued purposes that they would curtail if prices accurately reflected long-run costs of supply.

If a society decides to rely on either a private water utility or a self-financing state-owned enterprise, government must be able to make a credible promise that it will not prevent the utility from earning quasi-rents. This commitment does not mean that governments commit never to intervene in water markets, but it does mean that government must commit not to place so low a cap on water prices that investments cannot be recovered and not to expropriate water assets without compensation. We will return to the problem of commitments after discussing other issues that can give rise to regulatory supervision of water utilities.

Quasi-rents are not the only rents arising in water systems. Two others are monopoly rents and Ricardian rents. Whereas prices that include recovery of quasi-rents have important implications for efficiency, these other forms of rents do not contribute to efficiency, and in most societies carry little, no or even negative normative weight as distributional entitlements.

Monopoly rents arise when a water supply utility has the ability to set prices in excess of the average cost of production, and derive from the absence of effective competition. Monopoly rents are not limited to private entities, where they are normally associated with excess returns to investors. Public monopolies can and do collect monopoly rents to finance other government activities, and both public and private monopolies may dissipate monopoly rents by paying above-market wages to employees or bribing corrupt public officials to obtain and protect the monopoly.

Monopoly is not necessarily inefficient, but usually it is. Whereas monopolists theoretically can extract excess profits through an efficient system of price discrimination or by setting excessively high inframarginal prices in multi-part tariff, the more common circumstance is for at least some marginal prices to exceed corresponding marginal costs, and hence inefficiently to exclude some users whose valuation of output exceeds the cost of supplying it. For both private and public monopolies, the absence of competition may lead to lax operations, in which case monopoly rents are dissipated by inefficiency. Hence, policy intervention to limit monopoly profits can improve efficiency as well as cater to a societal norm against monopoly exploitation.

If a water system has multiple water sources or storage sites, individual components of the water system can generate substantial Ricardian rents under an optimal pricing system. For example, consider a water system in a region with a large river and an underground aquifer. This system may be optimally served by one large reservoir and several wells. In this case, the long-run average cost of water from the reservoir is likely to be substantially below the average and marginal costs of well water. In this case optimal system design is to rely primarily on the reservoir and to use wells only to the extent needed to balance supply and demand. Hence, ignoring possible pumping externalities, the optimal first-best usage price will equal the marginal costs of well water (which typically is also the average cost). This optimal usage price causes revenues for reservoir water to be above costs, and so creates Ricardian rents for the reservoir.

Unlike monopoly rents but like quasi-rents, Ricardian rents do not arise from market power, but from the superior productivity of better resources. Unlike quasi-rents but like monopoly rents, investors normally do not need to receive Ricardian rents to induce future investments. The exception occurs when the productivity of a resource is uncertain until an attempt is made to exploit it, in which case the recovery of Ricardian rents *ex post* may be necessary to induce optimal exploration for new resources *ex ante*. This case is illustrated by oil wildcatting, in which investors in exploratory wells expect that only a small fraction of wells will strike oil. Thus, for wildcatting to be profitable, the wells that do strike oil must generate sufficient Ricardian rents to offset the costs of the unsuccessful wells. In the case of water supplies, unlike hydrocarbon fuels, exploration incentives are unlikely to be very important, so that the recovery of Ricardian rents is unlikely to have significant efficiency consequences.

Ricardian rents, because they are unnecessary to induce efficient supply, are vulnerable to a political demand to prevent suppliers from extracting them. The main problem arising from an attempt to extract Ricardian rents is that the method employed frequently is to set prices equal to the average cost of water. Such prices are an average of low-cost reservoirs and high-cost wells, and so are below the marginal cost of water, which is based on well costs. If price is below marginal cost, utilities are forced to satisfy demand by exploiting expensive resources for which the cost of water is below its value to consumers. The result is excessive consumption of water for low-valued uses that do not justify their high costs of supply.

The Regulatory Design Problem: Commitment without Capture

The presence of potential monopoly rents, high fixed costs, and different costs from different resources creates a potential political problem of commitment to allow efficient prices and full cost recovery. This problem has been examined in detail in studies of telecommunications (Levy and Spiller, 1996; Noll, 2000); however, the problem appears to be even more important (and more complex) in water systems.

The presence of substantial rents of any form can give rise to two types of political demands to extract rents from suppliers.

First, user groups may organize to force down prices to average total cost (to extract Ricardian rents) or even average variable cost (to extract quasi-rents). In either case, price is below long-run marginal cost, and some investments are not recovered. The effect of this policy is to encourage consumption while discouraging investment, thereby creating excess demand. In the case of water systems, this incentive is especially strong because assets have such long lives. A water utility with revenues below private long-run costs can continue to serve existing customers adequately for decades before significant replacement investments are required. Hence, current users face only a small discounted present value of future problems that may arise from inefficiently low prices compared to a huge price reduction that can be obtained by taking out all forms of rent.

Second, water utilities may be able to siphon substantial cash from the system before any visible degradation of service occurs. A water utility that sets prices to recover quasi-rents will earn a huge net cash flow in the short run. If this cash flow is siphoned by operators (whether private or public) rather than used to pay off investors or to make necessary repairs and replacement investments, the water system will not be able to expand to supply new users and eventually service will deteriorate for old users. Again the result is substantial long-run excess demand, but in this case at high, rather than low, prices.

The possibility of monopoly rents is especially important, not just because monopoly pricing (and, in some cases, lax monopoly management) is inefficient. Politically, monopolistic practices give rise to pressures for regulation or other government interventions to reduce prices and improve service in a manner that is efficiency enhancing.

Thus, the presence of a natural monopoly in some elements of water delivery increases the credibility problem. Given that political pressures to control monopoly are widely regarded as normatively valid, and so are probably impossible and certainly difficult to resist, how can government commit to limit monopoly rents, while not expropriating quasi-rents and inefficiently extracting Ricardian rents?

Governments have no perfect way to make these commitments, but they can increase the likelihood that a commitment will be honored by increasing the difficulty of abrogating it (Lupia and McCubbins, 1998). The likely durability of a commitment can be increased in four ways.

First, government can insulate governance institutions from short-term political forces that influence elected officials and the civil servants that serve at their pleasure. Examples are regulatory authorities and judicial overseers with long-term appointments and decision-making autonomy. Second, governments can create 'high law' (like the U.S. Constitution) that is very difficult to amend and that protects the rights of investors. Examples are rules stipulating that government cannot unilaterally abrogate contracts with private entities and cannot expropriate assets without paying to owners the replacement cost of the assets. Third, governments can adopt a system of representation that reduces the likelihood that advocates of expropriation will gain power. Examples are proportional representation and governments in which policy change requires unanimous consent among several political bodies, each having a different base in the electorate. Fourth, governments can commit financial resources to large projects, such as by making loans, guaranteeing debt, or issuing contingent liabilities on the government that mature only if assets are expropriated. Government also can maintain a large but not controlling equity stake so that it bears part of the financial cost of unremunerative rates.

The problematic aspect of commitment mechanisms is that they also can give rise to regulatory capture. If the governance of a utility is too distant from broadly based political pressures or too vigilantly protects government's financial stake in the utility, it more easily can become a mechanism for serving the interests of the regulated entity. Indeed, excessively protective regulation has been at least as big a problem as expropriation, as reflected by the fact that most of the research literature on the politics of regulation focuses on capture, not expropriation (for a survey of this literature, see Noll, 1989).

The institutional design problem is a two-edged sword. Regulators must have enough sensitivity to the interests of the utility to guarantee that a well-managed entity will recover its costs, but not be overly protective of its interests. The primary mechanism for tackling the capture problem is to design regulatory agencies to make them responsive to the regulated utility, its customers, and its suppliers by institutionalizing open information, procedural transparency, standing for anyone affected by a regulatory decision and competition policy advocates, and a broad scope for judicial review that goes beyond the legal formalisms concerning whether the agency has a valid mandate to act and followed procedural requirements to include review of the adequacy of the evidentiary basis for a decision in protecting both consumers and suppliers from exploitation (McCubbins et al., 1987, 1989; Noll, 1993).

A subtle advantage of vertical separation of a water system that can be competitive at some levels is that it reduces the risk of both expropriation and regulatory capture. If a local retail utility buys water at market prices from multiple sources, the distribution system is likely to be the only significant source of monopoly rents, and so can be directly regulated or publicly owned and operated to deal with this problem. The remainder of the system can be freed by statute from the authority of the regulator, and instead governed solely by competition policy. The distribution utility will have high fixed, sunk costs, and so will need to collect quasi-rents to maintain efficient investment. But most of the quasi-rents and most likely all of the Ricardian rents will be spread over many independent actors in the capture and transport system, rather than concentrated in a single, vertically integrated entity. This structure of the industry can allow policy to focus on the monopolized sector, in which case the competitive sector may be insulated from policies to expropriate its quasi-rents and Ricardian rents, as has usually (but not always) been the case for hydrocarbon fuels that are sold to regulated electric utilities.

Extensions to Nonrenewable Resources with Downstream Externalities

The preceding analysis deals with a water system in which exploitation does not cause a secular decline in the quality of the resource for either urban customers or others who are not part of the water system. A water resource can be exploited beyond the point of sustainable use. For example, a large lake or underground aquifer may be a nonrenewable resource. Using part of that resource today has the opportunity cost of preventing the use of water in the future. In addition, extracting more water now can raise the cost of extracting a given quantity tomorrow as a lake becomes smaller or an underground water table sinks. Likewise, greater exploitation today can lower water quality for other uses either now or in the future. First-best usage prices must take into account all of these effects so that each user faces the marginal direct cost plus the opportunity cost of water usage.

Decentralized water markets can work well in allocating water efficiently when the resource is not fully renewable, even when the effect is a quality externality. In principle, users who suffer from deteriorating quality, higher costs, or lower availability can outbid other users to retain the size and quality of the resource. Unsustainable usage and intertemporal variation in the market value of usage typically can be dealt with efficiently if permanent, transferrable property rights in the resource are well defined, for then the holder of the resource can balance the financial benefits of short-term exploitation against the return on the implicit investment of postponing exploitation to a period when the market price is higher. Likewise, if property rights are well defined and transferrable, users who suffer deterioration in supply when other users increase their consumption can simply buy water rights from the latter. In either case, water will be allocated to the highest valued uses both contemporaneously and intertemporally. As long as water prices fully reflect all of the social costs of water exploitation and usage, markets for a finite, diminishing stock of water are likely to function as well as markets for other nonrenewable resources.

In practice, Coasian transactions costs may be sufficiently different among categories of users so that the market on its own does not adequately deal with the problem of intertemporal allocation. If some users are not organized to participate effectively in a market for water supplies, their valuations of water will not affect water allocation. For example, people who are engaged in water transportation on a river or lake could be adversely affected by a drop in the water's edge, but these users are not likely to be part of the market for the affected water source. Moreover, incorporating these demands in a water market faces serious problems because attributes such as the location of the shore and the depth of the water are pure public goods. If a full-blown water market that includes such users fails to emerge, either contemporaneously or intertemporally, then market allocations will not be efficient.

Consider two examples of downstream externalities in which some users of a river flow are adversely affected by upstream uses. One is a fishery in a downstream lake, and the other is a downstream urban system that relies upon the river for its water. These examples, of course, are simply illustrations. Diminished downstream flow is only one form of externality that can be caused by a water system. Others are the effects of dams and reservoirs on the variability and temperature of river flows and the ecosystems of the area surrounding the impoundment. The examples of downstream urban systems and fisheries are developed at length to demonstrate principles that apply more generally.

The case of two water utilities on the same river is amenable to a market solution. The water utility that serves a downstream urban area could, in principle, become a customer of the upstream system. The downstream utility can guarantee an adequate downstream flow by simply outbidding the other users of water from the upstream system. The downstream urban system also could engage in more expensive treatment, or could turn to another water resource, so that its private incentive is to strike an efficient balance between bidding water away from customers of the other system and expanding the combined capacities of the two systems. Likewise, the upstream system has appropriate incentives to expand its capacity to accommodate demands from the downstream system in ways other than curtailing its own uses. In this case, the water market allocates water efficiently between the two sets of customers of the water utilities.

The problem becomes more complex when many systems attempt to exploit the same resource. In this case, the water available to one system is affected by the consumption decisions of all other systems that are upstream from it, so an effective market mechanism requires an agreement among all users about the rights that each has in the water resource. The essence of the agreement is an allocation of the rights to either the sustainable usage or the consumption in the future of a nonrenewable (unsustainable) usage among all users, which serves to provide a clear definition of the water rights that can be transferred through the market.

Of course, the efficiency of the market to balance conflicting demands among organized water users does not imply that political pressures will not arise to interfere with this allocation. In the case of the two conflicting urban water systems, the decision by the upstream community to expand its usage imposes a cost on the downstream community. In anticipation of this possibility, the downstream community (perhaps in an earlier era before the upstream community becomes a significant population center) has an incentive to ask government to prohibit any diminishment of its own water supply. For example, it can ask the government to adopt an appropriative water rights system in which no later user can undertake any action that would diminish supply to anyone who began to use the water resource at an earlier time. This policy change switches the effective property rights to water from rights that are based on location (upstream versus downstream) to rights based on sequence of use.

Theoretically, 'grandfathering' water rights according to historical use does not cause distortions in water allocation. If a growing community has a higher marginal evaluation of water than the community that first expropriated the water resource and the resource is at or above sustainable use, the latter can profit from selling some water to the former. But in reality a problem emerges because the community that was the first to expropriate the water supply is likely to experience a windfall of rents from the adoption of a water market when exploitation of the resource reaches sustainable use.

Note that at the time of first water expropriation, the use of the water resource is likely to be small compared to its sustainable use, so that the first user will face a very low cost (and a zero opportunity cost) of using water. If so, the first user will engage in some very low-valued uses for which, in an appropriable water rights system, it has a permanent property right. Thus, a water market inevitably will allow Ricardian rents and conceivably could lead to monopoly rents for the initial water user. In this case, later users, when they become a political majority, may vote to substitute administered allocations for market exchanges as a more attractive means to control the first-in users because of its redistributive effects. This argument applies to either of the above systems for allocating rights — whether the first user is first in location or first in time.

The attraction of prohibitions on sales to some users will be further enhanced if a water resource has more than two users. For example, a downstream water district might sell water to an upstream district, and the sale could impose an uncompensated external cost on a community that was located upstream from the seller but downstream from the buyer. Whereas the community in the middle could outbid the upstream user for some of the downstream user's water supply, in so doing it will begin to pay for water that it had been receiving for free before the upstream user came into existence. The attraction of a sales prohibition to the intermediate community is that it prevents this windfall loss from transactions in which it is not involved. As the number of separate participants in the market for a water resource increases, the significance of windfall losses of this form increases, as most of the redistributive effect of each transaction occurs to water users who are not parties to the transaction.

Legal prohibitions against water markets are far from a theoretical concern. During the 19th century, the appropriative rights system was introduced in the arid American west, and it was usually accompanied by a prohibition against transactions combined with a 'use it or lose it' requirement for grandfathered users. This system encouraged wasteful use of water by early users (typically, agriculture), and later caused newly emerging urban centers to make use of very costly sources of supply rather than purchase agricultural water that had a very low opportunity cost. Even though some limited scope for market transactions in water is now permitted, they usually are subject to complex rules and regulatory oversight, and frequently proposed transactions still are ruled to be unallowable.

The case of the fishery illustrates another problem that can prevent efficient water markets. A regional fishing industry may consist of many independent fishermen who are not organized into a cooperative for bidding for water to maintain a natural water system that can sustain the fishery. Urban and agricultural water users usually are effectively organized through water utilities that were created to distribute water to them, perhaps before water demand at average cost of supply reached sustainable use. In this case, urban and agricultural users will succeed in acquiring an inefficiently large allocation of water relative to the fishery if the allocation is determined through bidding in a market.

In principle, government can compensate for this form of market failure by either taxing usage by well-organized groups or directly allocating usage independently of the price system, such as by setting minimum flow requirements for the downstream course of a river. In practice, efficiently intervening in either way is itself subject to distorting influences.

The political influence of various user groups (here urban dwellers, fishermen and farmers) will not necessarily accurately reflect the opportunity costs of water to each group. Indeed, organizational and institutional factors (such as market concentration) that can distort the market also are likely to distort political influence. Moreover, for groups for which the market allocates water reasonably well, creating the opportunity for political intervention to reallocate water may create new distortions. For example, different types of urban users may have equal and undistorted access to the market, but unequal political influence. If so, a regulatory intervention to cope with the fisheries problem may give one urban user group (say, a large firm in a water-intensive industry) the means to induce the same regulators to distort urban water allocation in its favor. For these reasons, the best policy response to water market distortions may be no policy at all, or it may be to facilitate the creation of an organization to aggregate the demands of the excluded users (e.g., fishermen).

WATER DEMAND

The significance of water derives in part from its importance to human life. Human survival depends on access to water, and human health requires water that is free of unhealthful pollutants. Hence, at some level, the demand for clean water must be highly inelastic; however, this observation is not practically significant. Water is the classic example of the 'paradox of value' — the observation that a commodity can generate enormous total welfare but have a very low price.

As a practical matter, water demand is reasonably elastic, with estimates commonly in the range of 0.3 to 0.7. Moreover, these estimates probably understate the sensitivity of water demand to price because in many developing nations water utilities do not attempt to collect unpaid bills or to prevent unauthorized taps into the system.

The reasons that water demand is likely to be reasonably price elastic are simple. Even in the poorest urban areas, many uses of water are not purely for subsistence. Moreover, humans pick areas for urban settlement on the basis of the economic value of the location, including whether water supply is sufficient to make human life sustainable.

Thus, even where water is relative expensive, the price is still low enough that marginal uses among low-income households are far more prosaic than sustaining life, such as washing, cleaning, and gardening.

In many urban areas water quality is sufficiently low that small amounts for direct human consumption are purchased from other sources (tank trucks or bottles) at very high prices. Piped water is used only for these other purposes.

Sources of Price Elasticity

For essentially all uses, water consumption can be varied by small changes in usage methods. Water is usually one input in a household production process, with opportunities for shifts in input proportions in response to relative price changes. Moreover, part of demand is euphemistically called waste, i.e., a failure to fix leaking pipes or to turn off taps after use. The price of water strongly influences so-called water losses. An aspect of 'waste' that is not widely recognized is that if water is very inexpensive, 'wasting' water is perfectly rational. If the price of inexpensive water simply reflects the fact that it is plentiful and its delivery costs are very low, then 'wasting' water is inefficient only if it creates a significant externality, an issue that is explored in the next section.

This conclusion applies to both consumers and water utilities. Frequently 'unaccounted-for water' (UFW) is used as an indicator of poor management. One source of UFW is a leaky distribution system. Fixing these leaks is a substitute for expanding diversion, storage, and transportation capacity. Repairing leaks is worthwhile only if the value of the water saved justifies the investment in new pipes, which will not be the case if cheap water is plentiful.

The other important implication of elastic demand is that a small mistake in pricing can have large consequences for water use. Prices that are too low create a demand to expand water delivery substantially beyond the efficient point. If the costs of the water system are not primarily financed by water revenues, but by general taxes or revenues from another utility (such as electricity, which is a common joint product of reservoirs and gravitational penstock and aqueduct systems), under-pricing water can be very expensive to society by leading to massive, uneconomic expansion of the water delivery system. Alternatively, if water prices must cover all costs and if costs are high due to inefficient or corrupt management, much economically warranted water usage can be cut off, and a small improvement in the efficiency of the operation of the water utility can yield large economic benefits.

Water Quality

The presence of dual water systems (small quantities of expensive but clean water, large quantities of cheap but low-quality water) provides some insights into the demand for quality. Although urban water systems can provide water of different quality for different purposes, as is the case in some communities where recycled waste water is supplied separately for some uses, this type of system is impractically expensive in many areas, and virtually unknown in developing nations. Economies of scale in distribution must be sacrificed if demand is split between two systems, so that the best choice usually is either a low-quality municipal system that is supplemented independently by high-quality, commercial sources, or a high-quality system that involves some combination of treatment and transportation from areas where water quality is higher than near large urban centers.

Which of the three types of systems is optimal depends on a comparison of the costs with the marginal valuation of quality by consumers. A significant issue is how to estimate the value of quality, and whether revealed preference is reliable as an indicator of its value. For business uses and aspects of quality that are related to either aesthetics (e.g., turbidity, taste, and aroma) or household production (e.g., dissolved minerals that affect cleaning and washing), the valuations of quality by users based on usage experience plausibly are reliable indicators. For health effects, market behavior may not be reliable because of problems of information imperfections. Some users may lack understanding of the relationship between water quality and health. Others may understand these relationships in principle, but they cannot costlessly observe aspects of quality such as the presence of microorganisms or trace chemicals that are health hazards.

The two arguments about the nature of demand for quality have conflicting policy implications. Because much water use does not demand high quality, dual use systems can be efficient. The incremental cost of quality improvements for all uses may exceed the costs of acquiring from private vendors a small quantity of high-quality water at a high price. But if the effective demand for quality is too low because of information imperfections, consumers may not optimize their consumption of high-quality and low-quality water, instead consuming too much low-quality water from the urban system. One can interpret the public health justification for high-quality municipal water systems as a judgment that the latter effect is more significant than the former, so that the theoretical advantage of a dual system is overcome by the practical reality that consumers will not behave optimally in such a system.

USAGE EXTERNALITIES

The preceding discussion deals with one form of externality: costs of extracting water resources that are imposed on other users. This section deals with two additional externalities associated with usage: pollution and spillage.

Pollution and Sewage Treatment

Virtually all uses of water do not consume water, but instead diminish its quality for subsequent use. Water usually is not actually transformed into some other substance (as steel is transformed into an automobile or sand is transformed into glass). As a result, waste water is the inevitable result of water usage.

The purpose of urban water systems is to increase the availability of water at relatively low cost, thereby expanding the uses of water — and the amount of polluted water that is produced by a community. Different uses produce different types of water pollution, which in turn impose different costs on different groups. Whereas these distinctions are important for policy purposes, the more general point is that expanded urban water delivery causes increased pollution.

In low-income countries a common policy for dealing with waste water is basically to do nothing. Water that is polluted through use is simply returned into the natural environment by dumping it into a nearby stream, lake, ocean, dry canyon, cesspool, or vacant plot of land. In some cases, the user does the dumping, but usually user disposal is associated with situations in which the user also supplies the water (no water utility is present) or the community is too poor to afford a sewer system. In other cases, users must dispose of used water in a sewer (sometimes supplied by the water utility and sometimes by a separate, usually government-owned sewer utility), and the sewer than dumps the untreated waste or even sells it to other users who do not demand high-quality water.

The external costs imposed by water pollution depend on the circumstances surrounding its disposal. Whereas all pollution will have some effect on the natural environment, the magnitude of the cost depends on the value of the environment for other uses (including preserving the original state of nature). The most important source of variability in pollution costs arises from the exposure of humans to unhealthful pollutants, either directly by polluting the environment in which they live or indirectly by polluting their food and water sources.

Pollution creates two significant policy issues, the first pertaining to the efficiency of water systems and the second pertaining to the equity effects of uncompensated and unmitigated pollution. If users and water utilities are allowed to dump polluted water, they are not held accountable for the costs that they impose on others. Hence, they will overproduce pollution, and use an inefficiently large amount of water for polluting purposes. The economist's stock solution for this problem is to impose a tax on pollution that fully reflects the marginal cost of pollution on others. Users and water utilities will then have an incentive to cut back on polluting uses of water, and to treat waste water to remove pollutants, until the marginal cost of abatement equals the marginal cost of pollution.

The stock solution does not deal with the equity issue. The optimal policy usually leaves some pollution, and hence some uncompensated costs for those who experience it. Moreover, the decision to decide how much of a tax to impose (or, more generally, how much abatement to require) is subject to other distortions in the political process. Part of the stakes in pollution policy is how to allocate benefits and costs among groups, which usually will continue to motivate further political action even if policy initially is efficient.

Nevertheless, the policy principles for efficiency and equity converge: polluters should face a marginal cost of pollution that reflects its external cost, and the revenues should be used to compensate losers for efficiently unabated pollution. A water system that satisfies all demand at a price equal to the marginal cost of water supply and that faces neither a pollution tax nor a compensation requirement has too low a price and causes too much pollution.

In the presence of distorted incentives regarding water pollution, a water utility has no financial reason to invest in either sewers or water treatment facilities. Moreover, if required by government officials to make such investments, the utility has a financial incentive to delay taking actions (to postpone costs) and to undertake the minimum actions that are consistent with the law. Of course, distortions in the political process may cause the government to demand too much rather than too little abatement, but an equally important feature of these distortions is that they are as likely to misallocate the burdens of pollution and pollution abatement as to pick the wrong level of abatement.

As with the case of commitments to protect the cash flow of investors, the pollution issue raises a problem of institutional design: how can government commit to a long-term pollution policy that works reasonably well at reducing the harms from pollution while avoiding overzealous pursuit of environmental policy that indirectly expropriates the investments of the water utility? The optimal design of environmental policy institutions and instruments is beyond the scope of this paper; however, an essential part of the assessment of the performance of an urban water system is whether this problem has been addressed in a reasonable way.

Water Spillage

Another externality associated with water use is not associated with pollution of the water itself, but with uncontrolled release into the environment. In the simplest case, suppose that water is available in unlimited quantities at zero price. If so, there is no economic incentive to repair a leaky tap, or to turn off a garden sprinkler when water floods the street.

In a society with an extensive and mandatory sewage system, or with good natural drainage due to topography or soil conditions, uncontrolled water is not likely to create an important externality. But in some communities water spillage neither runs off, percolates into the soil, nor evaporates quickly, but instead remains in standing puddles or very slow-moving open waterways. In this circumstance, water spillage creates a serious external cost by serving as a breeding ground for disease-carrying insects and disease-causing micro-organisms (Esrey, 1996).

The implication of this externality is that in some conditions water should be priced even if delivery costs are low compared to metering costs and consumers have very limited incomes. Failing to price water gives consumers no incentive to prevent spills that cause community health problems. Fortunately, because water demand is elastic while the cost of epidemics of infectious diseases is high, usually very large benefits can be obtained with relatively small price increases.

Regulatory Instruments for Coping with Externalities

Several implications for regulatory policy were derived from the discussion of costs, demand, and the externalities associated with extraction of the resource. These discussions focused on optimal pricing theory, in which the prices are based on total social costs, and on avoiding expropriation and capture. Specifically, relative prices should reflect the social costs of production, and departures of cost from the associated marginal social cost should be the consequence of budget balancing, transactions costs, and other social constraints on revenues, not distortions created by governance institutions.

This discussion implies cost-based price regulation of water utilities, which is at variance with another strand of scholarly literature that advocates decoupling price regulation from the attempt to measure cost (Baron, 1989). The essence of this argument is that overall price-index caps on a firm should be established (perhaps based on historical costs), but that future adjustments should not be based on any measurable item, such as costs, that the firm can influence. If firms possess more information about their operations than regulators and if regulators base prices on indicators that a firm controls, firms can increase their profits by distorting the performance indicator. The firm thereby achieves higher prices (which harm consumers) and does not produce at maximum production efficiency (the right cost–quality combination).

Price regulation that adjusts price index ceilings without considering costs solve several problems. First, it assures future price performance at least as good as expected prices under cost-based regulation. Second, it leads to a second-best optimal price structure, given the firm's overall price ceiling, because a firm facing a price-index constraint maximizes profits by setting second-best optimal prices in relation to its privately known costs. Third, it eliminates the incentive of the firm to distort its production behavior and generates an unconstrained incentive to minimize costs, given its quality decision. Fourth, by eliminating the requirement to measure costs, this form of regulation avoids an expensive regulatory system that employs large numbers of professional accountants, engineers, and economists, an especially important attribute in small, poor countries where individuals with these skills are few in number.

So how does one reconcile the discussion of cost-based pricing with the consensus that cost-based price regulation is not the preferred system? The question is relatively easy to answer with respect to the relationship between private costs and prices. For both components of the two-part tariff, price ceilings with automatic adjustments that are infrequently updated by cost audits can be adopted. But, for reasons given above, these frequently are dominated by the costs of externalities, and here the standard economic prescription is to use compensating taxes to internalize external costs (Cropper and Oates, 1992).

The magnitude of externality costs can be influenced by the firm through decisions about the form of investments in capacity, the usage resulting from its pricing and investment decisions, and investments in water treatment facilities for either wholesale water supplies or used water. Initially, water prices will reflect historical investments in pollution-reducing activities plus a tax based on the remaining external cost of water usage. The firm's general incentive to minimize cost will lead it to reduce costly efforts to reduce external costs after the price index formula is adopted. If the tax is not adjusted to reflect actual changes in the external costs, the end result will be inefficiently large pollution.

The solution to the problem begins, but does not end, with simply adjusting the tax much more frequently than the price index formula, based on the external costs of the water system. But the incentive effects of this simple tax are not as sharp if the firm is subject to price index caps. An unregulated firm would see the tax as an increase in its marginal cost, which would cause the firm to raise the after-tax price by less than the tax. For a regulated firm, if the index cap is a binding constraint, the incidence of the tax will fall more on consumers and less on the water utility than would be the case if the utility were an unconstrained profit-maximizing firm. In this case, the tax effectively works as an increase in the overall index cap, and regardless of the elasticity of demand, simply will be added to the after-tax retail price of water. This will reduce usage, and so give the firm some incentive to make investments to reduce pollution in order to restore sales by reducing the tax, but the incentive is weaker than if the cap did not exist.

Regulators can compensate for this effect by taxing the firm by a greater amount than the external costs of pollution, but doing so has serious drawbacks. First, consumers are penalized because of the pass-through of the tax, and the overall efficiency of the water system is forced to depart even more from efficient usage. Second, to calculate the proper add-on to the externality tax requires massive amounts of information about demand elasticity, the true marginal cost of water, the cost of pollution abatement by the water utility, and the costs imposed by externalities.

This problem can be avoided by forcing changes in the externality tax to fall within the price index constraint. This policy causes the full incidence of the tax to fall on the water utility, which is also inefficient because it does not discourage use as a means to curtail pollution.

The theoretical solution to this problem is intermediate: to have part but not all of the tax enter the price constraint, based on the elasticity of demand and the marginal cost of both water supply and externalities. Practically speaking, even if this formula is based on historical costs and takes the form of an automatic adjustment formula, it has the liability of imposing a substantially greater information burden on the regulators each time the price formula is adjusted. Under cost-based regulation, regulators typically do not need to know the details of the components of marginal and total costs nor demand elasticity to know how to set an overall revenue requirement. In practice, regulators rarely try to measure elasticities, and cost-based regulation gives firms considerable influence in setting prices once the overall revenue requirement is determined.

The upshot of this discussion is that the likely outcome of well-intentioned attempts to control externalities of water systems is likely to include performance standards as well as economic incentives through the price structure. Efficient regulation of prices makes efficient use of tax incentives in environmental regulation more difficult, and so increases the attraction of performance standards over pollution taxes. In addition, the difficulty of efficiently dealing with externalities through the regulatory system constitutes an argument in favor of integrating sewers into the water supply system.

CONCLUSION

The optimal design of urban water delivery systems (including their governance institutions) is deceptively complicated. Whereas the standard technology for constructing water systems is simple, and private costs relatively easy to measure, policy also should take into account two very important features of water systems: their unusually high ratio of fixed, sunk costs to variable costs, and their unusually high ratio of external effects to private costs. The first leads to political pressures to reallocate various forms of rents that are potentially a significant part of the revenues of a water utility, which in turn can cause serious distortions. The second frequently leads to circumstances in which externalities are the dominant factor in determining optimal prices and the optimal investment portfolio of a water utility. Moreover, because both features (and especially the second) vary enormously across geographic areas, the optimal policy for a water utility is location-dependent, and so not amenable to a solution that is based on general principles that can be reduced to formula.

These features of water utilities inevitably lead to the conclusion that extensive government intervention in the industry is inevitable, and likely to take forms that are at variance with the conventional prescriptions from the research literature. Moreover, government interventions are especially likely to be distorted by distributive politics because of both measurement problems (and hence the difficulty of resolving controversies through evidentiary inquiries) and the competition for the enormous rents that are available from the system in an environment in which many of the social costs of inefficient policies that redistribute rents are likely to be delayed and difficult to observe.

One implication of these observations is that the introduction of competition may be far more attractive than thus far has been recognized by either policy makers or researchers. Competition is feasible in capture, storage and transportation for an urban area that relies on multiple sources of water. Likewise, competition from tank trucks and bottled water suppliers is feasible for high-quality uses, and in some cases is an alternative to piped distribution. Whereas decentralized distribution usually has substantially higher private costs than piped distribution, conceivably its relative costs would not look so bad when one takes into account usage externalities and regulatory distortions.

REFERENCES

Armstrong, Mark, Simon Cowan and John Vickers, 1994.*Regulatory Reform: Economic Analysis and British Experience*. Cambridge, MA: MIT Press.

Baron, David P., 1989. "Design of regulatory mechanisms". In: Richard Schmalensee and Robert Willig, eds., *Handbook of Industrial Organization*, Volume 2. New York, NY: North-Holland.

Cropper, Maureen L. and Wallace E. Oates, 1992. "Environmental economics: a survey". *Journal of Economic Literature* 30(3): 675–740.

Deb, A.K., 1978. *Multiple Water Supply Approach for Urban Management*. Publication ENV-76-18499, Washington, DC: National Science Foundation.

Esrey, Steven A., 1996. "Water, waste and well-being". *American Journal of Epidemiology* 143(6): 608–623.

London Economics, 1998. *Competition in Water*. Report to the Department of International Development, London.

Levy, Brian and Pablo T. Spiller, 1996. "A framework for resolving the regulatory problem". In: Brian Levy and Pablo T. Spiller, eds., *Regulations, Institutions and Commitment*. New York, NY: Cambridge University Press.

Lupia, Arthur and Mathew McCubbins, 1998. *Political Credibility and Economic Reform*.

McCubbins, Mathew D., Roger G. Noll and Barry R. Weingast, 1987. "Administrative procedures as instruments of political control". *Journal of Law, Economics and Organization* 3(2): 243–277.

McCubbins, Mathew D., Roger G. Noll and Barry R. Weingast, 1989. "Structure and process, politics and policy: administrative arrangements and the political control of agencies". *Virginia Law Review* 75(2): 431–482.

Noll, Roger G., 1983. "The political foundations of regulatory policy". *Zeitschrift für die gesamte Staatswissenschaft* 139(3): 377–404.

Noll, Roger G., 1989. "Economic perspectives on the politics of regulation". In: Richard Schmalensee and Robert Willig, eds., *Handbook of Industrial Organization*, Volume 2. New York, NY: North-Holland.

Noll, Roger G., 2000. "Telecommunications reform in developing countries". In: Anne O. Krueger, ed., *Economic Policy Reform: The Second Stage*. Chicago, IL: University of Chicago.

Roth, Gabriel, 1987. *The Private Provision of Public Services in Developing Countries*. New York, NY: Oxford University Press.

Zaroff, Barbara and Daniel A. Okun, 1984. "Water vending in developing countries". *Aqua* 5: 289–295.

Thirsting for Efficiency: The Economics and
Politics of Urban Water System Reform
Mary M. Shirley (Ed.)

CHAPTER 3

THE BUENOS AIRES WATER CONCESSION

LORENA ALCÁZAR, MANUEL A. ABDALA AND MARY M. SHIRLEY

INTRODUCTION

The signing of a concession contract for the Buenos Aires water and sanitation system in December 1992 has attracted worldwide attention and has been the subject of considerable controversy in Argentina (see for example, Idelovitch, 1994, Artana et al., 1997, *The Economist,* various issues). In addition to being one of the world's largest concessions, the case is interesting for a number of other reasons. First, the rapid implementation of the Buenos Aires concession is in contrast to the slow moves towards private participation in other water systems, such as Santiago or Lima. In this chapter we explore the circumstances in the water sector and the political and institutional factors that led Argentina to implement the concession. Second, the reform has been widely acclaimed for generating major improvements in the sector, including wider coverage, better service, more efficient operation of the company, and a reduction in waste. What makes these improvements especially striking is that the winning bid brought an immediate 26.9% reduction in water system tariffs.[1] The chapter analyzes the features of the contract that explain performance improvements, and considers how institutions affected outcomes. Finally, the Buenos Aires case is of interest because the contract was renegotiated in 1997, provoking criticisms of the original bid and the regulation. We describe these revisions and measure the extent to which Buenos Aires was better off because of the concession, with and without the renegotiation. The chapter concludes with lessons for reform design.

As we will show, water and sewerage services improved dramatically thanks to the concession. Consumers benefited from the large drop in real prices, which was only partly reversed by subsequent changes in tariffs and access charges, and from a large expansion of the system. Quality and reliability of service also improved, and wastage

[1] This decrease was subsequently eroded, as we explain later.

of water was reduced. We will argue that these improvements would not have occurred under any reasonable assumptions about continued public operation of the water system.

These gains notwithstanding, we will demonstrate how information asymmetries, perverse incentives and weak regulatory institutions could threaten the long-run sustainability of the concession. In particular, politicization of the regulator, a poor information base, serious flaws in the concession contract, an obscure tariff system and a general lack of transparency in the regulatory process, have created opportunities for the company to act opportunistically, and the regulator, arbitrarily. These circumstances have led the federal executive to by-pass the regulator, and public confidence in the process has eroded. The Buenos Aires concession shows the importance of transparent, rule-based decision making in maintaining public trust in regulated infrastructure.

CIRCUMSTANCES LEADING TO REFORM

Circumstances in the Water Sector

Reform is more likely to occur when the water sector is in crisis, but only if the causes of the crisis are tractable (Alcázar and Brook-Cowen, 1996). Problems that result from a failure to expand supply as fast as demand or from mismanagement are easier to address than those that stem from scarce and costly raw water. Buenos Aires did not have a public health crisis, but it did have serious problems. We argue in this section that it was ripe for reform because its sector problems were highly tractable, given the right political circumstances.

Cost and scarcity of raw water

Buenos Aires sits on the west bank of the Rio de la Plata, which has an abundant flow of 29,000 m^3 per second (World Bank, 1994a) and supplies 92% of the system's water needs.[2] As a result the supply of raw water is ample throughout the year and transport costs are low. Nasser (1997) estimated that the marginal cost of water in Buenos Aires could be as low as US $0.15 per cubic meter, even taking into account the investments being implemented under the concession. This is cheap compared to cities with scarce water supplies, such as Lima, where the marginal cost of water and sewerage was estimated at US $0.45 per m^3 in 1994 (World Bank, 1994b).

Supply and demand

Notwithstanding the ample availability of cheap raw water, before the concession only 70% of the population in the metropolitan area were connected to the water system, and only 58% to the sewer system. The shortfall was in the poorer, suburban areas, which had been growing by 5 to 6% a year prior to the concession and totaled 5.6 million

[2] The rest is from about 300 wells that are becoming increasingly contaminated by untreated sewage.

Table 3.1

Water and sanitation coverage in Buenos Aires before privatization

	Water	*Sewerage*
Capital district (%)	99	99
Suburban areas (%)	55	36
Total metropolitan area (%)	70	58
Number of connections (millions)	1.2	0.7

Source: Concession Contract.

people at the time it was signed in December 1992.[3] Only 55% of persons in these areas had access to water and only 36% to sewerage. Almost all of the 3 million inhabitants in the older center of the city, the Capital District, were connected to the municipal system (Table 3.1).

The state-owned water utility, OSN (Obras Sanitarias de la Nación), had not been investing enough to keep up with the pace of growth in these outlying areas of the city, and the bulk of new connections had been financed by neighborhood associations. The growth of connections was very slow, and below Buenos Aires' low population growth rate in 1991 (Fig. 3.1). Indeed, OSN's level of investment was not even enough to maintain existing assets. For example, average annual investment from 1987 to 1991 was only US $11 million, which did not cover depreciation. Much of the water and sanitation system was more than 60 years old and badly in need of repair and replacement at the time of the concession in 1993 (FIEL, 1996). The deterioration of the system caused water shortages in parts of the city during the summer, as well as frequent breaks and interruptions, low pressure, poor water quality, and sewer flooding during rainstorms.[4] Unaccounted-for water (UFW), the difference between water production and the volume billed to customers, calculated as a percent of water produced, reached 45% before the concession, compared to 10 to 20% in OECD countries (Idelovitch, 1994).[5]

One reason for OSN's low investment rates was its lack of funds because of its inefficient operation, combined with declining real low water tariffs. Water tariffs had declined in real terms during much of the preceding thirty years, falling by almost 70% from 1960 to 1976. The Ministry of Economy and Public Works was nominally responsible for setting prices and investment policy, but in practice pricing decisions were often influenced by political considerations.

Consumers had little incentive to conserve. Not only were water prices low, but also nearly 99% of consumption was unmetered, billing was based on factors with little relation to consumption, and only 80% of billed values were collected.[6]

---

[3] Approximately, 2 million people are in the province of Buenos Aires but are not in the concession area. They are served by municipalities.

[4] Only 17% of the population in the suburbs received adequate pressure before the concession, defined as greater than 8 wcm or water column meters (Abdala, 1996).

[5] UFW represents losses due to leaks, theft, and failure to register users.

[6] Although 20% of customers had meters, half of these were inoperative.

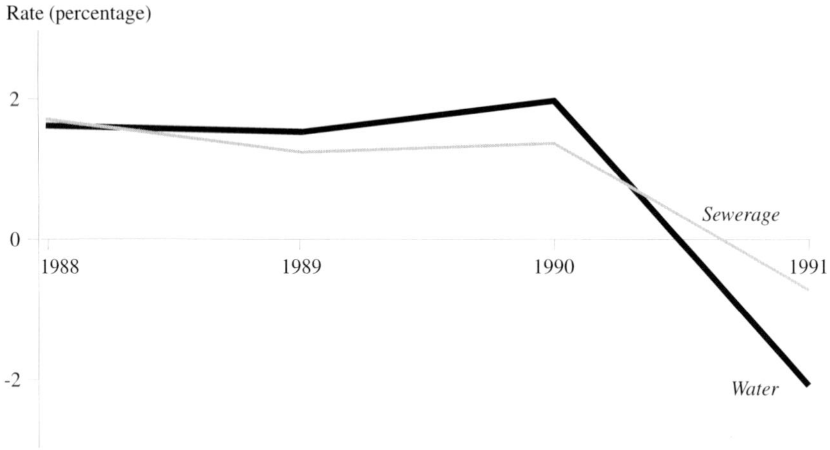

Source: OSN.

Figure 3.1. Rate of growth in population served by OSN.

Besides poor billing and collection because of OSN's failures, the law did not permit the company to cut service to households who did not pay. Water consumption in 1993 was estimated at 352 liters per capita per day, compared to, for example, 173 liters in Santiago, another city with ample raw water.[7]

Management

Besides defects in collection and investment, OSN was generally mismanaged. The company's auditor cited OSN's poor internal organization and administration and unreliable accounting information in its annual reports. The company was overstaffed with 8 workers per 1000 connections compared to 2 in Santiago.[8] Salaries were low, turnover of skilled personnel, high, and the entire workforce suffered from low productivity and lack of discipline (Idelovitch, 1994; FIEL, 1996).

Top management was politically chosen and priorities were set according to non-commercial criteria. As a result OSN's scarce funds were focused on more visible new construction projects rather than maintenance, even though many of these new works were never completed. In addition, OSN was widely regarded as unresponsive to customer complaints; at one point the backlog of water breaks awaiting repair was close to 7000.

OSN lost money in three of the five years leading up to the implementation of the concession in 1993. It made money in the other years only thanks to inflation accounting and nonoperating income. Indeed, OSN was not designed to be financially autonomous; its act of incorporation provided for the company to make use of Treasury subsidies

[7] Based on water distributed divided by connected consumers.

[8] One reason for this was that OSN had once been responsible for water throughout Argentina and when the sector was decentralized in 1980, OSN failed to reduce its staffing.

whenever its own revenues did not cover its costs (Republic of Argentina, 1949). Nevertheless, many observers believed that an efficiently operated OSN could have made money and expanded coverage, given the low cost of water in Buenos Aires (field interviews). The government had increased prices in the years before the concession, and even though the increases were less than inflation, it became more likely that an efficiently operated company could earn a reasonable return on capital.[9]

Externalities and health problems

The opportunity cost of water usage in Buenos Aires was and is low. Usage costs are low because the city's extraction rate, while relatively high, is far below the river's replenishment rate and drinking water is not in competition with other uses. Externalities due to pollution of the river have also been low. Even though only 5% of sewage is treated before being dumped in the river, this contamination has had low external costs thanks to the self-purification capacity of the river and the fact that downstream users are not affected.[10]

The principal externality from the system resulted from contamination of groundwater and rivers by septic tanks, cesspools and direct discharge of untreated sewage and industrial effluent. At the time of the concession 88% of those not connected to the sewer system used cesspools and septic tanks and the remainder used direct disposal such as discharges into creeks. Since 95% of the 30% of the population without connection to piped water relied on shallow wells, waterborne diseases from bacterial contamination were believed to be important among this population.

Thus, in many ways Buenos Aires' sector circumstances seemed exceptionally well suited for successful reform. Despite an abundant and forgiving source, water had been made artificially scarce by mismanagement and poor policy, resulting in preventable externalities. Of course the fact that problems have a solution does not mean they will be solved; that occurs when political and institutional circumstances are favorable, as we consider next.

Political Circumstances

Our premise is that reform becomes politically desirable when the political benefits outweigh the political costs, and feasible when reformers have the power to overcome opposition and implement changes (World Bank, 1995). It is credible when the reformers' promises are believable, and sustainable when those promises are hard to overturn. This section explains how water system reform became politically desirable, politically feasible and credible in Buenos Aires.

---

[9] Water tariffs were increased by 25% in February 1991 and by 29% in April of 1991. Although the CPI increased by 84% in 1991, by the time of the concession the rate was sufficient to cover OSN's operating and maintenance costs.

[10] The construction of a 2.5 km offshore pipe at one of the effluent plants (Berazategui) eliminated problems for downstream users served by Obras Sanitarias de Buenos Aires near the city of La Plata.

Economic crises made reform feasible

Water reform became politically feasible in Argentina because economic crisis changed the rules of political competition, encouraging cooperation after years of bitter conflict. To understand the unique circumstances that made reform feasible we must understand Argentina's history. Three political forces vying for power have dominated Argentina's political history since the early 1940s — the military, members of the Radical Party (mainly the middle class) and the primarily working class members of the Peronist Party. These groups failed to develop stable, crosscutting coalitions that could bring together different interests to support policies that would encourage economic growth (McCubbins, 1993). Rather, when one group was in control it used its power to redistribute wealth and income to the benefit of its supporters, and to reverse the actions of its predecessors (ibid., Hill and Abdala, 1994). As a result successive regimes tended to expand the role and employment of the state, control prices and salaries, and restrict trade, contributing to deficit public spending, rising inflationary pressures and falling production.

The frequent intervention of the military and the rise of Peronism contributed to the instability of Argentine politics. From 1930 to 1983 there were five military coups that resulted in a total of 23 years of military rule. The rise of Juan Domingo Perón in the 1940s with his base in urban, unionized labor, led to further instability by polarizing politics between the Peronist Party and the opposition Radicals. During the period when Perón and, following his death in 1974, his wife, were in power (1946–1955 and 1973–1976), the role of the state in the economy was increased through nationalization of public services, or direct interventions, such as fixing prices and wages. Both of these Peronist regimes were characterized by increasing political polarization and both were ended by military coups.

The instability and economic deterioration continued under the military governments that followed. The military regime from 1976 to 1983 focused mainly on defeating the urban guerrilla movement through increasingly repressive tactics (during their rule thousands of citizens disappeared). Although they followed more liberal economic policies with some initial successes, the conflict with the British over the Malvinas Islands in 1982 greatly increased the external debt and deficit spending, contributing to rising inflation that reached 15% a month by the end of 1983 (Snow, 1996). The economy did not improve with the restoration of democracy and the election of the candidate of the Radical Party, Raúl Alfonsín. Continued political polarization made meaningful structural reforms impossible. The executive and one house of congress were held by the Radical Party while the other house was controlled by the Peronists.

By the time of the election of the Peronist candidate, Carlos Menem, in 1989, Argentina was in the midst of a severe economic crisis. In the previous five and a half years real salaries had decreased by 50%, unemployment and underemployment had doubled, annual inflation had increased to more than 3000% and the economy was in recession (GDP fell by 1.3% in real terms). Menem won 47% of the votes and control of the Senate. Although Menem himself was a moderate, he ran on a traditional Peronist platform that stressed populism and won the support of most of the poor. His Radical

Table 3.2

Popular approval of certain groups in Buenos Aires (percentages)

Rate of approval of: [a]	*February 1989*	*June 1990*	*November 1990*
Military	26	24	22
Entrepreneurs	38	32	31
Political parties	31	23	14
Unions	21	15	8

[a] Average for greater Buenos Aires area.
Source: Edgard Catterberg and Associates.

opponent's platform, in contrast, called for stabilization and liberal market reforms, policies that Menem adopted after taking power.

The severity of the economic crisis prompted the moderate wings of the two major political parties, who represented middle to upper income voters, to build an informal coalition in order to bring the Menem government into office quickly and give it the power it needed to address the accelerating hyperinflation and economic deterioration.[11] The moderates supported compromise because of concerns that further deterioration of the economy might tempt the military to intervene, as had so often happened before. The low public support for the traditional parties in opinion polls (Table 3.2) created further pressure for consensus. Accordingly, the moderate wings of the opposing political parties agreed that Menem could take power five months ahead of schedule to avoid a long hiatus between the election and his inauguration.[12] The opposition Radical Party also agreed to support a series of laws giving the president sweeping powers that, in effect, allowed him to legitimize frequent use of emergency decrees during the crisis and early reform period (McCubbins, 1993).[13] Menem made wide use of this power: while his predecessor, President Alfonsín, had issued 10 decrees, Menem issued 308 decrees from 1989 to 1993 (De Michele and Manzetti, 1996).

The Menem administration strengthened the chances for reform

Reform was also politically feasible because the Menem administration was able to neutralize two formerly powerful factions that might ordinarily have opposed his free market reforms, the military and unionized labor. The military had been weakened after

[11] Menem's party, officially called the Justicialist Party, included not only the urban working class and the middle class of poorer provinces, but also newer industrialists. Some of his appointees were members of other parties.

[12] Menem won the election on May 14, 1989 but would normally not take power until December 10; instead he was inaugurated in July. The CPI rose by 78% in May, 114% in June, 197% in July, and 38% in August. The exchange rate devalued by about 3700% in the five months following the election.

[13] Law 23696 gave the executive power to reorganize, privatize or close any public agency or corporation by decree, and Law 23697 granted the executive power to modify, eliminate, or create regulations that affect almost any aspect of business or labor relations. The executive also had strong powers when congress was not in session or during exceptional periods according to the Constitution then in force (De Michele and Manzetti, 1996).

its defeat in the Falklands war with the U.K. and because of its use of repressive tactics to control guerilla violence when it was in power from 1976 to 1983. This weakness reduced the likelihood of military intervention, but did not eliminate it. By granting amnesty to low level participants in insurrections against Alfonsín and pardons to military personnel involved in some past violations of human rights, President Menem removed a potential source of tension.

Unionized labor, important to the extreme wing of Menem's own political party was also much weakened. Unions' credibility with the public was very low, as evidenced by polls after Menem's election. The approval rating of unions in Greater Buenos Aires (the Capital District and surrounding suburbs) dropped from 21% in February 1989 to 15% in June 1990, to 8% in November 1990. Although other groups also received low approval ratings, none were as low as the unions (Table 3.2). Electoral defeats in later congressional elections brought the number of union representatives down from 35 deputies in 1983 to 8 in 1993. Furthermore, since labor had backed Menem's election, they had few attractive alternatives to which they could switch support when he enacted policies they opposed.

The Menem administration was thus well situated for enacting radical change. Although Menem's platform had not stressed free market reforms, hyperinflation and a rapidly deteriorating economy demanded a new approach. Furthermore, most people favored or were at least neutral towards privatization in the first years of his government. Polls suggest that opposition to privatization in Buenos Aires was only 16% in 1989, and subsequently averaged about 35% between 1990 and the signing of the water concession in 1993 (Table 3.3).

Nevertheless, the administration had doubts about whether the political coalition that made privatization possible was robust or durable. On the one hand, most politicians and ordinary citizens, fearing that a failure to go along with change might bring back the status quo of hyperinflation and decline, were ready to give Menem a long honeymoon period. On the other hand, the government was enacting a very ambitious program that had not been part of its electoral platform. The new administration began in 1990 to liberalize trade, lay off government employees, curb state spending and privatize state-owned enterprises; then in 1991 it instituted a currency board pegging the peso to

Table 3.3

Public opinion of privatization in Buenos Aires[a] (percentages)

Opinion	*1988*	*1989*	*1990*	*1991*	*1992*	*1993*	*1994*	*1995*	*1996*	*1997*
Positive (1)	25.5	59.4	38.3	39.0	26.4	36.4	29.9	24.3	16.7	18.1
Neutral	20.6	15.9	24.4	26.5	25.0	25.6	29.2	31.6	27.7	26.9
Negative (2)	38.6	16.4	34.4	29.3	43.6	33.3	36.2	42.7	52.2	51.9
No opinion	15.4	8.4	3.0	5.3	5.0	4.7	4.7	1.4	3.4	3.1
(1)/(2)	0.7	3.6	1.1	1.3	0.6	1.1	0.8	0.6	0.3	0.3

[a] Greater Buenos Aires area.

Source: Centro de Estudios Unión para la Nueva Mayoría.

the dollar.[14] Consequently, the government launched privatization with a strong sense of urgency (field interviews). Speedy privatization was also seen as key to the success of the government's economic program since economic reforms depended on privatization bringing new private capital, reducing fiscal hemorrhaging, shrinking the debt overhang through debt: equity swaps, and endowing the economic program with greater credibility with investors and debt holders. This focus on speed was to have important implications for the water concession, as we discuss later.

The political desirability of water system reform

From the beginning private participation in the water system was part of the Menem administration's privatization program. Argentina's quick action on the Buenos Aires water system concession is in contrast to two of its neighbors, Chile and Peru. Chile had planned to sell the assets of Santiago's water system and Peru had planned a concession in Lima, also as part of large privatization programs. But both governments treated water privatization as low priority, delayed the transaction, and, after initially contacting bidders, ultimately failed to privatize. These delays are not surprising. Water does not usually have high saliency in privatization programs, especially a concession, since it brings no new capital to reward supporters or compensate losers and affects a single city rather than the nation. This low saliency helps explain why Chile and Peru delayed and ultimately dropped their privatization plans; the puzzle is why Argentina went forward. We might expect that the different decisions were prompted by a need to improve the poor performance of the Buenos Aires water company. But bad as OSN was, things in Lima were much worse (see Chapter 4).

The Buenos Aires concession went forward as part of the block of transactions because the net political benefits to Menem appeared to be larger than the net benefits from similar action in Peru to Fujimori or in Chile to Pinochet. In the 1989 election Menem received solid support from those most adversely affected by the hyperinflation (lower and middle class, salaried workers, self-employed and unemployed). Thanks to this support, he carried the poorer suburban areas of Buenos Aires, but did poorly in the upper middle and upper income districts. Coincidentally, the latter were the same districts that had the highest rates of connection to the water system (Table 3.4).

The design of the concession rewarded both supporters and opponents. Middle and upper income households who were already connected to the water system would benefit from the concession since the tender was awarded on the basis of the lowest water tariff and designed to bring substantial improvements in service. The suburban poor were also expected to benefit from increased access, as Table 3.5 shows. Although the poor were required to pay much of the cost of the expansion of the secondary network through a so-called infrastructure charge, the government staff who helped draft the contract thought that the poor could afford this infrastructure charge since the concession required the company to provide financing assistance for two years at

[14] Government employment was reduced by 60% during the first five years of the administration; 15% through layoffs and 85% through transfers to jobs in health and education in the provinces.

Table 3.4

Votes for Menem in the 1989 presidential elections compared to water and sewerage connections and income group

	Percent of votes for Menem	*Percent connected*		*Income group* [a]
		Water	*Sewerage*	
Capital District	36.6	99	99	
North Zone:		69	44	
Vicente Lopez	29.8	98	95	1, 3
San Isidro	34.1	81	35	1, 3
Tigre	53.7	24	8	4, 6
San Fernando	50.3	64	24	4, 6
West Zone:		54	45	
Moron	48.1	30	25	4, 6
La Matanza	60.2	54	55	4, 6
Tres de Febrero	50.8	69	57	1, 3
San Martin	50.1	77	40	1, 3
South Zone:		49	21	
Avellaneda	46.7	83	37	1, 3
Lanus	52.8	76	17	1, 3
Lomas de Zamora	52.8	53	30	1, 3
Almirante Brown	60.0	13	9	4, 6
Est. Echeverria	58.6	10	8	4, 6

[a] 1–3 = upper and upper middle income; 4–6 = lower middle and low income.
Sources: Election results: *Diario El Clarin*, May 16, 1989. Income groups: INDEC 1985/1986. Water and sewerage coverage: Republic of Argentina, 1993. Concession Contract.

Table 3.5

Projected socioeconomic status of households with new water connections (first five years of concession)

	Number of new connections	*Percentages*
High/middle high	90,200	15.44
Middle low	282,250	48.31
Low	211,800	36.25
Total	584,250	100.00

Source: Undersecretariat of Water Resources, 1996. Based on actual and projected works.

12% interest (field interviews).[15] As we shall see, this proved to be a serious source of problems later. Water worker opposition to the reform was reduced by the pledge of shares in the privatized firm. The close involvement of the union officials in the process

[15] The nominal interest rate on debt at the time was over 20%.

also helped win their acquiescence. (The head of the union was part of the privatization committee.)

Thus, in contrast to Santiago and Lima, the political economy of the Buenos Aires concession appeared to be a win–win decision politically with benefits for core constituents in the suburbs and potential swing voters in the center. Both Santiago and Lima would have had to raise prices substantially to cover their costs, while in Buenos Aires prices could be reduced through more efficient operation. Coverage could be expanded in Buenos Aires without putting much of a burden on existing customers, or so it seemed at the time of the concession, something that was clearly impossible in Lima. And the Buenos Aires concession might help the Menem administration's standing with important swing constituents, whereas Pinochet had little expectation of improving his support in Santiago when he faced the plebiscite in 1988, regardless of water reforms (Chapter 6).

Credibility of water system reform

Even though the Buenos Aires concession was politically feasible and desirable, the government's promise to honor its contractual obligations also had to be credible. Otherwise, potential investors would not participate or would demand a large risk premium (Levy and Spiller, 1994; World Bank, 1995). Investors also had to believe that commitments would be honored by future governments or they would try to front load their returns and invest less in expanding and maintaining assets. Three factors made the Buenos Aires concession agreement appear credible and a constitutional reform made it more likely that it would be sustained. Credibility was strengthened first by the fact that by the time the water system concession was signed in 1993, the government had already implemented a market-oriented reform program and privatized majority ownership of large state enterprises in telecommunications, electricity, gas and airlines. A total of US $18 billion in federal assets had been sold by 1993 (Shaikh, 1996) and by 1994, 90% of all federal enterprises had been sold.[16]

A second feature which enhanced the concession's credibility was the process of privatization, which was designed to build a broad base of support by involving both houses of Congress, all affected federal ministries, provincial and local governments, and the staff of the affected SOE. It also built workers' support by pledging to transfer 10% of shares in the new water company to them whenever the dividends paid to the government for these shares covered their book value (US $12 million). A bicameral congressional committee supervised the privatizations including the concession. An 11-member government committee was in charge of coordinating the privatization of OSN with representatives of most interested parties: the Ministry of Economy and Public Works, the Privatization Secretary, the municipality and province of Buenos Aires, OSN managers and labor unions (Idelovitch and Ringskog, 1995).[17] Transparency was en-

[16] In addition, the privatization of provincial water enterprises had already started (in Entre Rios).

[17] OSN workers were members of the National Federation for Sanitary Workers and the Union for Sanitary Workers in Greater Buenos Aires.

hanced through the extensive use of technical advisors, the involvement of international agencies (IDB, UNDP, and World Bank), and by monitoring by the government office responsible for SOE audits (Sindicatura General de Empresas Publicas).

Third, the favorable political situation described above also strengthened the credibility of the concession, as did popular attitudes. Shortly before the concession was signed 43% of the population of the Capital District and Greater Buenos Aires favored privatization of water and sewerage (Estudio Mora y Araujo, Noguera y Asociados in field interviews, 1997). This is similar to support for other privatizations such as electricity and gas (47%), railways (47%), roads (43%), airlines (41%) and oil (37%). Since observers consulted for this study agreed that water was not a highly salient issue for most of the city's population, investors could reasonably expect little opposition to a reform that was billed as one that would reduce prices, improve service and increase access.

Thus, there was little reason to doubt the short run credibility of the reform, but the prospects that it would be sustained over the longer term were less secure. The popularity of privatization and the extraordinary political circumstances that made it possible could easily change over time, as indeed they have. We consider this issue in the next section.

CHARACTERISTICS OF THE BUENOS AIRES CONCESSION

The contracting literature suggests that a contract such as the Buenos Aires concession can be expected to improve firm performance only if it (i) reduces information asymmetries, (ii) gives the firm incentives to comply with the contract, and (iii) provides credible signals that the contract will be enforced and that both parties will adhere to their commitments (see for example, Williamson, 1985; Laffont and Tirole, 1993; World Bank, 1995; Shirley and Xu, 1998, 2001). In the case of water, we are also interested in how well the contract addresses social issues, in particular access of the poor to water and sewerage. This section considers how the tender offer and the concession design determined the incentives, information and commitment of the contract, and affordability of access.

Information

Competition for the market and tender offer

Competition is an important mechanism to reveal information, and in that regard the Buenos Aires concession scores relatively well. The privatization committee opted for a 30-year concession for all of Buenos Aires, whereby a private or mixed company would assume responsibility for operation, maintenance and investment.[18] The number

[18] The government chose a concession over a management contract or lease because they wanted the private investor to take responsibility for the massive amounts of new investments needed to meet their goals. Sale of assets was rejected because the government did not want to delay the tender in order to collect the information necessary to value the assets, furthermore OSN's assets were not believed to be worth much given their poor condition (field interviews).

of likely bidders was limited by the size of the concession, which was one of the largest ever bid in a developing country. As a result, the pre-qualification requirements limited tender offers to consortia with experience in operating very large-scale systems.[19] Some consideration had been given to dividing the concession area in two, which might have increased the number of potential bidders as well as allowing yardstick competition, but this was rejected because it would require time and investment to separate the system. Since the least costly division would be between the central and suburban parts of the system, there would be few advantages to a split. The parts would not be particularly comparable because the suburban system would be so much less profitable and it would be harder to attract bidders for the less profitable area.

In the event, the bid was competitive. Five consortia prequalified.[20] This was reduced to three in the final phase of bidding because two French consortia decided to make a joint bid and the Spanish consortium's technical proposal failed to qualify.

As discussed, the concession was awarded to the consortium that offered the largest tariff reduction. The winner was Aguas Argentinas, headed by Lyonnaise des Eaux-Dumez, which offered a 26.9% reduction.[21] This bid was followed closely by an offer of a 26.1% reduction from the group headed by Thames Water. The third bid from North Water offered 10.1%.

The introduction of competition notwithstanding, a defining feature of the tender process was poor information. The emphasis on speed, discussed above, meant that weak and sometimes-wrong information could not be improved before the auction. Although the government commissioned several studies, the concession contract explicitly stated that none of the government signatories assumed any responsibility for the quality or accuracy of the information provided. Indeed, the winning bidder used the first four pages of the bid document to describe the serious lack of information and the sad state of OSN. The fact that bids were based on such poor information suggests that the information content of the bid was not reliable. Bidders might have been misinformed or assumed that they could renegotiate as new information came to light; a subject we return to later.

Tariff regime

One striking flaw in the concession was the decision to leave in place an inefficient and opaque tariff regime. Once again, the government actors did not think they had the time to develop a more transparent regime. They also feared that if they introduced a

[19] Specifically, the minimum requirements included more than US $750 million in equity, ability to borrow at least US $200 million, annual revenues above US $600 million and experience in providing water and sewerage services for a population of at least 10 million.

[20] The five prequalifying companies were: the two largest French water companies, Lyonnaise des Eaux-Dumez and Compagnie Générale des Eaux; the British companies, Thames Water and Northwest Water; and the Spanish state-owned enterprise Canal Isabel Segunda.

[21] The ownership of Aguas Argentinas was 53% foreign (besides the two French companies the other owners were the Spanish Aguas de Barcelona and UK's Anglian Water), 37% Argentine (including Sociedad Comercial del Plata, Meller, and Banco de Galicia) and 10% workers. In 1994 the IFC acquired a 5% stake.

new regime it would be hard to evaluate the tender offers and to ascertain whether the operator was indeed implementing the agreed reduction in tariffs.

The tender regime left in place by this decision increased government and consumer information asymmetries vis-à-vis the company. Most consumers in Buenos Aires were and are charged a flat rate; at the time of the concession only one percent were metered. The unmetered tariff is set by taking the existing flat rate for the type of consumer (residential, nonresidential, or real estate) and service (water only or water and sewerage) and multiplying it by a *K* factor. Thus, right after the concession was won with a bid to reduce tariffs by 26.9%, the *K* factor was 0.731. The resulting product is then adjusted up or down depending on five variables: where the property is located in Buenos Aires, the total area of the property, the square meters of the area that have construction on them, and the type of construction (six categories from low budget to luxury), and its age. The coefficient on location has varied between 0.8 and 3.5 and the coefficient on type of construction, between 0.6 and 2.6.[22] The metered tariff is no less complicated. It consists of a flat rate equivalent to 50% of the tariff that an unmetered customer with the same characteristics would pay, plus a variable charge calculated by multiplying the old OSN tariffs per cubic meter by *K* (there is no variable charge for the first 30 m^3).

The information asymmetries inherent in the tariff regime open the way for opportunistic behavior by the company. The company can change its tariff by negotiating not only an adjustment in *K*, but also by reclassifying consumers from residential to nonresidential, since nonresidential pay twice as much as residential. It can also propose adjustments in users' lot size or building type or size, or changes in criteria for age or location (Artana et al., 1997). In every case the company will necessarily have better information on the revenue and consumer surplus implications of such changes than the regulator. Another drawback to this complex formula is that it is almost impossible for consumers to monitor their billing and protest arbitrary increases in factors other than *K*. On the other hand, information asymmetry might also prompt the regulator to exercise discretion. This issue has already come up; ETOSS disallowed a reclassification of 80,000 users from residential to nonresidential, a decision that the company contended cost it over US $44 million over the first three years of the concession (Nasser, 1997).

Regulatory weaknesses

Information asymmetries can be reduced through strong monitoring, but two factors worked against this in Buenos Aires: the inexperience of the regulator and the poor information base on which the contract was designed.

The regulator's inexperience was natural since the agency was created quickly as part of the general rush to privatize and set up without a clear regulatory framework or established procedures. Prior to the concession there was no independent regulatory body for water; OSN as we have seen, was largely self-regulated. There was the Sub-secretariat of Water Resources but its responsibilities were to develop and enforce a natural water

[22] There is also a minimum bimonthly charge of US $4 times *K* for residential water and sewerage and US $8 times *K* for nonresidential water and sewerage.

resources policy and coordinate investments; there was also the Secretariat for Natural Resources and Human Environment, but it was in charge of environmental regulation.[23]

In May 1993 ETOSS (Ente Tripartito de Obras de Servicios de Saneamiento, Tripartite Entity for Sanitary Services) was created to regulate Aguas Argentinas and monitor its compliance with the terms of the concession. Besides determining the tariff, ETOSS approves and monitors the company's five-year investment plans, hears and investigates complaints by consumers or others, and establishes penalties and corrective measures should it find that the company is out of compliance with the regulation. Not only did the new regulatory agency have to learn on the job; it was staffed largely by former officials of OSN who were not experienced with arms length regulation of a commercial venture. Furthermore, according to water experts in Argentina and the World Bank, ETOSS staff is not highly qualified compared to other regulatory bodies in Argentina. The staff skills may not be the best for a regulator whose most important responsibility is tariff setting; ETOSS had only four economists and four accountants versus twenty engineers in 1995, for example (ETOSS, 1994, 1995).

The second problem was information. As we have seen, the tariff structure was nontransparent and there were mistakes and gaps in information in the concession about OSN's assets, customers, billing, etc. This opened the door to opportunism by the firm or the regulator. Moreover, because the drafters did not always have enough information to specify targets or end goals in the concession, they spelled out specific actions or projects that Aguas Argentinas would need to implement in order to achieve the goal the government had in mind. For example, instead of just specifying what targets the concessionaire must meet to provide optimal service, the contract provides that "the regulated firm must maintain and renovate all pumping materials, valves, hydrants, connections and other elements that are needed for the optimal provision of the service, regardless of their life span." This sort of drafting invited ETOSS to intervene unduly in the operational decisions of a commercial company (field interviews), and, according to some in Aguas Argentinas, empowers it to make arbitrary requests (field interviews).

Incentives

Incentives and the operator's obligations under the concession

The operator's risks and obligations under the contract are important in determining its incentives. The contract set a number of targets for the company, some of which are shown in Table 3.6. Besides the targets in the table, the contract had goals for service quality (water pressure and quality) and metering. These targets implied that approximately one million people would be connected every 5 years for the first 15 years for a total investment cost of about US $4 billion. Investment during the first 5 years would amount to some US $240 million annually compared to OSN's average annual investment of only US $10 million during the preceding decade.

[23] Until February 1998 these two entities were part of two different ministries and coordination between them was a problem. Now both depend on the Natural Resources and Sustainable Development Secretariat of the Office of the Presidency.

Table 3.6

Performance targets in the Buenos Aires concession contract (percentages)

Year of concession	*Population coverage*		*Sewerage treatment*		*Network renovation (cumulative)*		*UFW*[a]
	Water	*Sewerage*	*Primary*	*Secondary*	*Water*	*Sewerage*	
0	70	58	4	4	0	0	45
5	81	64	64	7	9	2	37
10	90	73	73	14	12	3	34
20	97	82	88	88	28	4	28
30	100	90	93	93	45	5	25

[a] UFW = unaccounted-for water and is defined as the difference between volume of water distributed to the delivery network and volume sold as a percent of volume produced. It includes physical losses from pipe breaks and overflows and commercial losses from illegal uses, underregistration of users and the like.
Source: Republic of Argentina, 1993. Concession Contract.

While the contract set significant goals, it also lifted some obligations from the company. First, none of OSN liabilities were transferred to Aguas Argentinas except for an undisbursed Inter-American Development Bank loan of almost US $100 million.[24] Second, before the tender the government reduced OSN's work force by 1,618 through a voluntary retirement program that cost about US $32.7 million or about US $ 20,000 per worker. (During the first year of private operation Aguas Argentinas offered retirement to another 2,000 employees at a cost to the company of US$ 50 million.)

To guarantee its performance the contractor had to post an inflation-indexed performance bond of US $150 million that was forfeit to the government if the concession were terminated for lack of performance.[25] ETOSS can also penalize the enterprise if it fails to fulfill its commitments. The fines are substantial and escalate with increases in water system tariffs.[26] The incentive of the company to achieve these targets depends not only on the penalties it may face if it reneges, but also on the credibility of the commitment it faces. Some experts, who predicted the need to renegotiate the contract later, considered the targets optimistic at the outset. If the company anticipated that it could negotiate lower targets, it would not be motivated by the threat of sanctions. We return to this issue later.

---

[24] The book value of the liabilities was US $743 million, of which US $502 million were contingent liabilities. So the value of realized liabilities was US $241 million, almost half of which were owed to government.

[25] If the government terminates the contract Aguas Argentinas is entitled to its bond, compensation for new assets that have not been fully depreciated and loss of revenue for the duration of the contract (based on the minimum profits in the five years preceding termination).

[26] Mismanagement and failure to provide requested information and reports are subject to a fine of US $100,000. Failure to supply service or meet quality standards and violations of tariff rules are penalized with fines amounting to US $500,000. Additionally, delays in the fulfillment of agreed five year investment plans warrant fines from US $100,000 to US $1 million. Any revenues from fines do not become part of ETOSS' budget, to avoid giving the agency an incentive to penalize the firm in order to collect the fine.

Guarantees and penalties can be important, but ultimately a successful contract is one that aligns the incentives of the operator with the goals of the agreement. Aguas Argentinas has an incentive to expand service since, as we will show, it earns a return on capital from its customers.[27] Also, under the original concession agreement, the infrastructure fee covered part of the cost of expansion, and could only be collected by building the network. The company also has an incentive to reduce UFW and wasteful consumption since this reduces the need for costly new investments in supply. By eliminating past liabilities and overmanning, the concession also freed Aguas Argentinas from some of the costs of past mismanagement. The tariff regime distorted the operator's incentives somewhat, however, which we consider next.

Incentives and the tariff regime

Buenos Aires' tariff regime had perverse effects on incentives as well as on information. First, the tariff embodied large cross subsidies between customers. For example, a customer with a new luxurious home would pay a tariff seven times larger than a customer with an older, low budget home on a property of similar size. Large cross subsidies create an incentive for the firm to expand services first to those consumers who will pay more. In addition, the tariff provided no incentive to curb usage for the vast majority of customers since they pay a flat rate regardless of volume. As the system expands, Buenos Aires' high rates of consumption will begin to push against the capacity of the system to supply the service. This could motivate the company to reduce the pace of expansion if the cost of expanding supply to meet additional demand is larger than the returns to new investment plus any anticipated penalties for failure to comply with the contract.

The concession contract created perverse incentives for metering. Metering is expensive, especially in Buenos Aires where the cost of a meter was estimated to be US $200, so the cost to meter all customers would be US $300 million (Abdala, 1996). It may not be economical to meter where water supply is ample, as it is in Buenos Aires. However, metering does make sense if metering would reduce the requirements for new capacity, and that new capacity costs more than the price to meter. Moreover when consumers are metered, they can adjust the size of their bills by curbing consumption. The original contract made it mandatory to meter nonresidential consumers, but left it optional for residential connections. 98% of all connections were classified as residential when the concession was signed in 1992; only 1% of connections was metered. Either the consumer or the company can request a meter, in which case a meter must be installed and whichever party made the request pays the cost. Unlike Aguas Argentinas, most consumers are unlikely to have the information to know when metering would be advantageous for them. Metering becomes profitable for Aguas Argentinas when the variable charge for metered water is more than half of the fixed charge (Abdala, 1996). Since those with low fixed charges are likely to be poor, in effect the tariff regime

---

[27] Changes in the law also enable it to cut off service for nonpayment.

provides Aguas Argentinas with an incentive to meter those households that are the least able to afford the higher water bill.[28]

The procedure to adjust prices is a further instance of perverse effects on incentives. Buenos Aires uses a mix of price cap and cost plus pricing regulation. Similar to a price cap, the *K* factor is subject to adjustment every five years, based on the investment plan for the following five years. *K* can go up or down, except in the second five-year period when prices can only be reduced. Between these periodic adjustments prices are not inflation indexed, as is usually the case with a price cap. Instead, whenever a composite cost index increases by a cumulative real rate of more than 7% in one year, the company can file for an equivalent increase in *K*.[29] This index is constructed by developing price indices for ten categories of costs, such as fuel, chemicals, electricity, labor, debt service, etc. Each cost category's index is based on one or more published price indices, such as the wage index for unskilled and skilled labor in industry and consumption, which are 20 and 80% respectively of the cost index for labor. The cost categories are then assigned weights for each five-year period of the concession, depending on their incidence in Aguas Argentinas' operating costs as estimated in the firm's tender and five year plans. For example, personnel costs are 39% of the index and electricity is 11% during the first five years. The increase is not automatic; the request is studied by the regulator and has to be approved.

Theory tells us that price caps have an advantage over other forms of regulation: they give the company an incentive to reduce costs in the period between price adjustments since it can keep any additional profits. However, it is not clear that Buenos Aires' hybrid price system captures this advantage. The regulation reduces Aguas Argentinas' incentive to control costs whenever the increase in the cost index nears 7%. The complexity of the cost index also allows both the company and the regulator to behave opportunistically, as was evidenced in a dispute over the size of this cost pass-through in 1998, which we discuss later.

Credibility

Although political circumstances made the initial concession contract credible to the bidders, whether credibility will be sustained as political circumstances change depends on the mechanisms available to enforce government's adherence to the contract. As Levy and Spiller (1994) point out, if investors fear administrative expropriation they will limit their exposure. For example, if Aguas Argentinas fears that its prices will

[28] Abdala (1996) simulated what the unmetered and metered tariffs would be for two properties of the same size where one is new, well located, and luxurious and the other is old, poorly located and low budget, and assumes that the household in the more luxurious property is high income and the other is low income. In this example, metering reduces the bill for high income households who consume less than about 125 m^3 and raises it for low income households who consume anything more than 50 m^3.

[29] The company or the regulator can also request price adjustments should costs go up because the government requested changes in product quality or works, or because of changes in the tax, convertibility, environmental or other laws that affect the company.

not be adjusted according to the contract or that ETOSS will act arbitrarily, it will want to delay or reduce its investment commitments. In this section, we discuss the institutions that could potentially enforce the concession: regulation, legal redress, and the federal executive's concern with Argentina's reputation in international capital markets.

Enforcement by the regulator and judiciary

The politicization of the regulator and the weaknesses of Argentina's judiciary made it less credible that the government would adhere to its side of the regulatory bargain if political circumstances changed. The contract also provides for international arbitration, but this is useful only as a last resort.

Politicization was built into ETOSS' structure. It is governed by a six-member board: two representatives of the presidency, two of the province of Buenos Aires and two of the municipality of Buenos Aires; the presidency rotates between them. Directors are appointed for six years with the possibility of serving two terms. Although they are supposed to be removed only for just cause, most directors have spent no more than two years in their positions; only two have been with the regulator since it was created in 1993. The different levels of government represented on the board have brought pressure on the regulator to take politically motivated decisions. For example, in 1994 the mayor of Buenos Aires wanted to build a highway and promised to resettle residents of a shanty town that was in the route of the highway. He pressured Aguas Argentinas to build the necessary water and sewerage connections, and when the company asked for a tariff increase to cover the cost, the ETOSS' directors representing the municipality pressured the regulator for a tariff increase. Some observers (FIEL, 1996) believe that this increase was too high and the change in responsibilities and tariff was in violation of the concession contract.

Not surprisingly, such a politicized board is subject to partisan disputes and paralysis. At the time of its creation the federal executive branch dominated the ETOSS board, since the presidency appointed the mayor of Buenos Aires. The mayor's position became directly elected in 1994 and was captured by the opposition party. Adding to the partisan divisions, the governor of the province was Menem's main opponent within the Justicialista party. This situation raised the risks that regulatory decisions would be based on partisan conflicts, and reduced the credibility of the concession.

In the event of a conflict between ETOSS and Aguas Argentinas, the company can appeal to the judiciary. The weakness of the judiciary and its vulnerability to political influence weaken the credibility of such commitments, however. For example, one of Menem's first initiatives as president was to increase the number of supreme court judges from 5 to 9 and to appoint a large number of federal judges and state prosecutors (De Michele and Manzetti, 1996).

Sustainability may be enhanced by a constitutional reform enacted a year after the concession was signed, which makes drastic revision in utility regulation less likely. This is because it changed electoral rules in ways that make it harder for one faction to capture the upper house and ignore the minority (Heller and McCubbins, 1996).[30] However, this change can protect the company against arbitrary action by the legislature, not from regulatory discretion.

Reputation effects

An important factor enhancing the credibility of government's commitment to the concession was the concern of the Argentine government with its reputation in global financial markets. Menem's market-oriented reforms had made the economy open and introduced a currency board that fixed the peso to the dollar. This openness to global economic influences meant that any future federal executive would be concerned about how foreign private investors might react should government renege on the regulatory promises it had made to Aguas Argentinas as part of a large and visible transaction. Reputation can be a powerful tool for contract enforcement, but also somewhat ephemeral. In this case, the force of reputation would depend on how salient investors perceived the concession contract to be when compared to other reputation factors, as well as whether the firm appears to have reneged on its part of the bargain. As we shall show, the executive branch has intervened repeatedly to support Aguas Argentinas in conflicts with the regulator, and to bypass the regulator entirely in renegotiating the contract.

Reputation concerns of the company also have an effect on the sustainability of the contract. The concession is one of the largest and its success is important to the reputation of the consortium partners that are competing in global markets for water contracts. This gives them an incentive to comply with the contract even if enforcement is weak. The strength of this incentive depends on the cost of compliance and whether government appears to have reneged on its commitments.

Access and Affordability

A final characteristic of the concession with important implications for performance is affordability of access. As we have seen, the cost to expand the primary network was shared by all users as part of Aguas Argentinas' general tariff, while the cost of expansion of the secondary network was charged to new customers, along with the cost of connection. The total access charge is given in Table 3.7 (to this must be added any cost to modify household plumbing and appliances).

All consumers were required to connect to piped water and sewerage when it was made available. Since most of those not connected to the municipal water system relied on wells and cesspools, we can compare these construction and operation costs with the cost of connection. Abdala (1996) estimates that the cost to construct a well and

[30] Prior to 1994, senators were elected by provincial legislatures from two-member districts by plurality rule. When the 1994 electoral changes took effect in 2001, senators became elected in three-member districts with the first two seats going to the party with the most votes and the third seat to the second-place party. In the 1995 and 1998 elections, senators up for election (one third every two years) were still elected by the provincial legislatures, but respecting the 1994 rules assuring minority representation. The new rules have created more pressure for compromise with the opposition.

Table 3.7

Buenos Aires access charges in 1995 (US dollars)

	Minimum	*Maximum*
Water:		
Connection fee	208	297
Infrastructure charge	43	340
Total water access charge	251	637
Sewerage:		
Connection fee	284	319
Infrastructure charge	572	572
Total sewerage access charge	856	891
Total access charge	1107	1528

Source: Abdala (1996). The charge varied depending on the total property area (ST), and a factor R that takes into account the type of soil and level of repair needed to streets and sidewalks. (The formula is ST × R × $0.825/m^3.) Abdala (1996) assumed an average soil and repair coefficient of 1.03 and land areas of 50 m^2 for the minimum and 400 m^2 for the maximum.

cesspool in 1995 (US $3506) was considerably more than the total access charges that year (maximum US $1528), but for most users these costs were already sunk. The annual cost to operate an electric pump and evacuate a cesspool (US $189) was more than the annual cost of water and sewerage usage (US $109). However, many of the unconnected households were poor and probably many used hand pumps and other less expensive forms of water and sewerage. Plus, these numbers do not include the considerable costs to modify plumbing inside the home.

Affordability of the connection was a problem for poorer households. Aguas Argentinas was required by the concession to help new customers finance the infrastructure charge. The repayment period had to be at least two years. Although Aguas Argentinas could have allowed a longer repayment period, it opted for two years, which made the average monthly charge $44. Although the community could build the connection itself, rather than pay Aguas Argentinas, consumers would still have to cover the cost. Some 85% of the unconnected consumers had low or low to middle incomes. Many of these were in the poorest sections of the city, where average monthly household income was US $200 to 245 a month, such that the infrastructure charge alone was 18% of their income.

Even new consumers who could afford the access charges resented the requirement to connect, partly because they had alternative water and sanitation systems that were now outlawed. These alternatives were risky — unconnected consumers had higher rates of waterborne diseases because their well water was contaminated by their cesspools — but many were not well informed about their risks and undervalued water quality. As for sewerage, since it has large externalities consumers place a lower value on it than society (Munasinghe, 1992). Another source of resentment was the fact that customers who had been connected before the concession had not had to pay an infrastructure fee;

instead the cost of expanding the secondary network had been spread over all customers. The infrastructure charge was a new feature introduced by the concession. As we report below, the access fees were adjusted several times but affordability problems combined with resentment led many customers to refuse to pay the access charges and helped precipitate the renegotiation of the contract in 1997.

Revisions and Renegotiation

Since the contract was signed, several major changes were implemented and the concession was renegotiated. These illustrate the weaknesses inherent in the initial design of the agreement and in the underlying institutions that supported enforcement.

Revisions to tariffs and other changes

The first revision occurred in 1994, when the *K* factor was increased by 13.5%. Part of this was a 9.5% cost pass-through increase because Aguas Argentinas' cost index had increased by more than 7% during the year. The rest was due to changes in the contract requested by the government and agreed to by the company.[31]

In addition, the minimum water connection fee was increased by 84% and the water infrastructure fee by 38%, while the minimum sewerage connection fee was increased by 42% and the sewerage infrastructure fee by 46%. These increases aggravated the affordability and resentment problems we described earlier. In May of 1995 ETOSS decided that it should modify this abrupt increase in the access charge somewhat. The minimum connection fee for water was reduced by 16.1%, still far less than the earlier increase, and the maximum by 48%. In addition the infrastructure fee for water was changed from a flat fee to one that varied depending on the size and characteristics of the property. These two changes combined to reduce the average infrastructure fee for a property of 100 m^3 by 81%. (One effect of these changes was that sewerage access became substantially more costly than water.) In September 1995 the infrastructure fees for both water and sewerage were reduced again, this time by 15%, when by contractual agreement the firm passed on to consumers a drop in its labor taxes.

Notwithstanding these reductions, the infrastructure fee combined with the connection fee was still a problem for many poor consumers. An increasing number of people refused to pay the fee, and Aguas Argentinas became concerned about these growing receivables, which reached US $30 million at the end of 1996. The Menem government was also concerned because opposition politicians were exploiting the issue, which affected one of Menem's core bases of support. The growing arrears led to the renegotiation of the contract, which we consider next.

[31] One increase was to cover the municipality's request asking the company to extend service to a shanty town moved because of a highway (at an estimated cost of US $9.8 million). A second change was to accelerate the investment in the upgrade of the Belgrano water plant and the expansion of the distribution system to 30,000 people (US $80.4 million), and an acceleration in the program to substitute nitrate contaminated wells with clean ones (US $31.8 million). (Aguas Argentinas failed to deliver on all of these new commitments and was fined by ETOSS, something that became an issue in the renegotiation described below.)

Renegotiation of the contract

In February 1997 the company asked to renegotiate the contract. Aguas Argentinas argued that its revenues after three years of the concession were US $217 million lower than anticipated, of which US $143 million resulted from factors beyond the control of the company. The company also requested suspension of fines that ETOSS had imposed because of various delays, including Aguas Argentinas' failure to meet the accelerated investment targets agreed to as part of the 1994 tariff increase.[32] Regardless of the merit of these claims, there was clearly a problem with the collection of the infrastructure charge, and that convinced the government that renegotiation was necessary. The administration also wanted to use the renegotiations to address environmental concerns, including measures to reduce contamination of the Matanza and Riachuelo rivers.

Following the firm's request for renegotiations, ETOSS and Aguas Argentinas presented a series of widely differing proposals. There was little ground for compromise, and the negotiations bogged down, partly because of the political polarization of ETOSS' board. Ultimately, ETOSS staff were bypassed and two federal ministries — the Public Works Secretariat and the Natural Resources and Human Development Secretariat — reached an agreement directly with the company. The final agreement was passed in August 1997 and had bipartisan approval, including by the Office of the President, a Bicameral Congressional Commission and representatives of the province and city of Buenos Aires.

The renegotiation introduced the following main changes.

(1) Replaced the infrastructure charge to new users with a Universal Service and Environmental Improvement Fee (SUMA) imposed on all customers. Part of the SUMA was to cover the loss of the infrastructure charge, including past receivables, and part would fund environmental improvements that were not included in the original contract. Thus, in the case of water, a charge of US $2.01 was imposed to cover the cost of expanding the system to new users, and another US $0.99 was imposed to finance the environmental improvements. The change was required to be revenue neutral, and a methodology was specified to try to assure this.

(2) Created a connection charge (CIS) for new users with an average value of about US $120 for water. The CIS could be paid off in installments averaging about US $4 every two months.

(3) Reduced Aguas Argentinas' contractual obligations by: (i) cutting expansion targets for the first five-year plan by about 15% for water, and about 13% for sewerage;[33] (ii) changing the date for completion of the first plan from the end of April to the end of December 1998; (iii) eliminating the fines imposed by ETOSS for the failure to complete the goals of the investment program.[34]

---

[32] By 1996, the promised works on the Belgrano treatment plant and the distribution network were only about 30% completed, the work on the shanty town was about 60% completed.

[33] The fines are supposed to result in lower SUMA charges and increased investment in a new neighborhood (Puerto Madero).

[34] The targets for sewerage connection in the second plan (1998–2003) were increased by 240 thousand people.

Table 3.8

Comparison of average bimonthly charges in Buenos Aires, before and after contract renegotiation in Argentinean pesos (equivalent to US dollars)

	Before	*After*
Average charges for already connected residential customers:		
Water and sewerage services	30.00	30.00
Regulatory fee (2.67%)	0.80	0.80
SUMA charge	–	6.00
VAT (21%)	5.46	7.72
Total	37.26	44.52
Average charge for a new customer (water only):		
Water services	6.00	6.00
Regulatory fee (2.67%)	0.16	0.16
SUMA charge	–	3.00
CIS charge	–	4.00
Infrastructure charge	44.00 [a]	–
VAT (21%)	10.53	2.76
Total	60.69	15.92

[a] Average monthly payment for the first two years.
Source: *Diario La Nación*, February 24, 1998.

As Table 3.8 shows, as a result of these changes the average bimonthly bill for already connected residential consumers increased by 19% from US $37.26 to US $44.52, while the average bill for new consumers decreased by 74%, from US $60.69 to US $15.92.

The renegotiation made two fundamental changes in the nature of the concession contract. First, whereas under the original contract the firm received payment for expanding the secondary network only when the work was done, the introduction of SUMA meant that Aguas Argentinas would be paid beforehand. This reduced the government's leverage to assure that the works were promptly completed. At the same time, the decision to delay the timing of planned investment without penalizing the firm may have been a recognition that the original targets were unrealistic, but it also established a precedent that the company would not be held accountable for delays. Second, the regulations associated with the renegotiation shifted the tariff regime from a form of price cap cum cost plus, towards something more akin to rate of return regulation. The new rules require the regulator to evaluate the impact of regulatory changes on the company's level of indebtedness. (These criteria were under review by ETOSS in 2001.)

The 19% increase in bimonthly bills did not excite much public comment, perhaps because the absolute amount was small relative to the average income of connected households. Nevertheless, various consumer associations filed suit to stop the SUMA from going into effect, and the ombudsman (Defensor del Pueblo) for Buenos Aires also requested that the two decrees authorizing the renegotiation be annulled. The judge initially suspended the universal service decree, but it was reinstated upon appeal by the Natural Resources and Human Development Secretariat, the government of the province of Buenos Aires and a group of 16 mayors of those districts where expansion works were planned.

More recent tariff changes did excite public opposition, however, since they came on top of the earlier increases and at a time when economic growth was slowing. In 1998 Aguas Argentinas requested an 11.7% cost pass-through increase, since its cost index had gone up by more than 7%. The proposed hike stimulated strenuous opposition from the two ETOSS directors appointed by the mayor of Buenos Aires, whose constituency had been the most affected by the SUMA. The ETOSS board voted on the increases cost by cost instead of on the entire cost index, and granted only a 1.6% increase. The two Buenos Aires directors voted against any increase. The Federal Executive Branch again intervened: the Secretary of Natural Resources granted Aguas Argentinas another 3% on top of the 1.6% awarded by ETOSS, for a total of 4.6%, still less than the 11.7 originally requested.

The adjustments and renegotiation contributed to a highly partisan debate in the press and some public disillusionment with the concession.[35] Some commentators (FIEL, Delfino) argue that the adjustments show the company acted opportunistically from the beginning, underbidding and accepting patently unreliable information in the expectation that tariffs would be renegotiated later. The fact that Aguas Argentinas signed a loan agreement with IFC in 1994 that required it to meet a debt to net worth ratio that it could only achieve if prices were increased or its investment targets reduced, seems to suggest an expectation of renegotiation.[36] If this is true, then at least one other bidder would have had to be equally strategic, since the second offer from the consortium headed by Thames Water was so close to the winning bid (26.1% versus 26.9%). Regardless of the truth of this assertion, the subsequent adjustments of the contract illustrate the risks of auctioning and regulating with such poor information, the problems inherent in a politicized regulator and weak judiciary, and the difficulty of winning public trust under such circumstances.

Effects on Performance and Welfare

Notwithstanding the flaws in the concession and the institutional weaknesses we have described above, the concession led to major gains in performance and in consumer welfare. Furthermore, the revisions and renegotiation did not have substantial negative effects on these achievements.

The effects of the reform on performance

The most dramatic effect of the concession was the increase in investment (Fig. 3.2). Aguas Argentinas estimated that to achieve the original targets under the (pre-renegoti-

[35] Public support for privatization in general has decreased from 38–39% in 1990–1991 to 16–18% in 1996–1997 according to polls by the Centros de Estudios Unión parra la Nueve Mayoria. This opposition is due to a sense that privatized enterprises are reaping profits at public expense because of weak regulation (Field interview with Manuel Mora y Araujo, public opinion expert, February 1998).

[36] Based on the assumption that all profits are fully reinvested; see Abdala (1996).

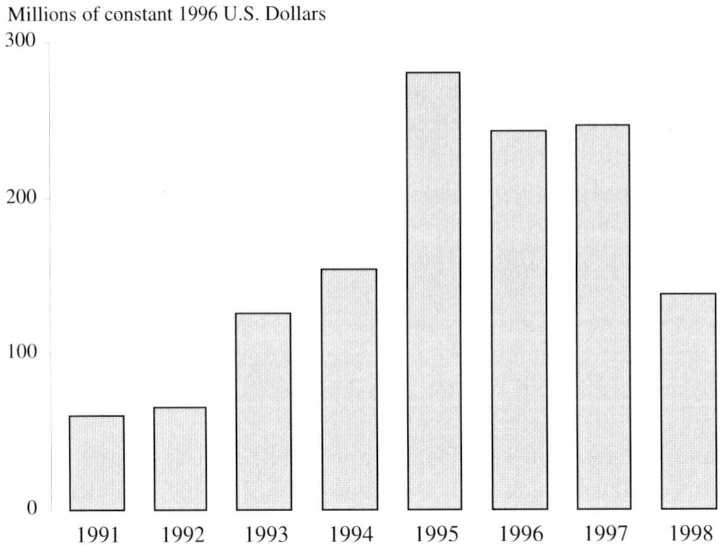

Source: Authors' calculations based on annual reports of OSN and Aguas Argentinas.

Figure 3.2. Investment by OSN and Aguas Argentinas, 1991–1998.

ation) concession for the first five-year period it would have to invest US $1.2 billion. Actual investment was less, about US $1.05 billion from 1993 to 1997. As a result of this investment new connections increased by 11% over the five years and coverage increased from 70% of customers in the service area in 1992 to 83% by 1997 (Fig. 3.3).

This new investment was financed largely through internal funds and debt (Fig. 3.4). In contrast to OSN, which had been allowed to borrow only twice from 1980 to 1993, Aguas Argentinas raised its debt : equity ratio from 1.17 in 1993 to 2.37 in 1996 by issuing short-term commercial paper and long-term borrowing from IFC.

Internal savings were an important source of capital thanks to the fact that operating revenues expanded despite the initial price reduction. Prior to 1997, most of the increase in operating revenues was due to improved billing and collection. Judging from changes in accounts receivable, collections went from 90% of amount billed in 1992 to 97.8% in 1993, then dropped to 94.4% by 1997.

Overall trends in real consumer prices for water services are hard to judge because the initial drop in the *K* factor was partly offset by changes in access fees, the introduction of SUMA and the reclassifications of consumers to higher rate categories. The trend in real operating revenues per cubic meter of water delivered to the consumer (i.e., minus UFW), is shown in Fig. 3.5. It suggests that unit prices have risen since the initial drop, but are not yet back to 1992 levels. Controversy has arisen over Aguas Argentinas' prices because some observers, adding up revenue neutral increases and increases due to cost inflation, have concluded that the company bid strategically with a plan to increase prices later. However, taking into account all adjustments since 1992 including SUMA, an already connected customer was paying about 23.4% more for water services in 1998 than in 1992; during that same period the consumer price index increased by 21.2%.

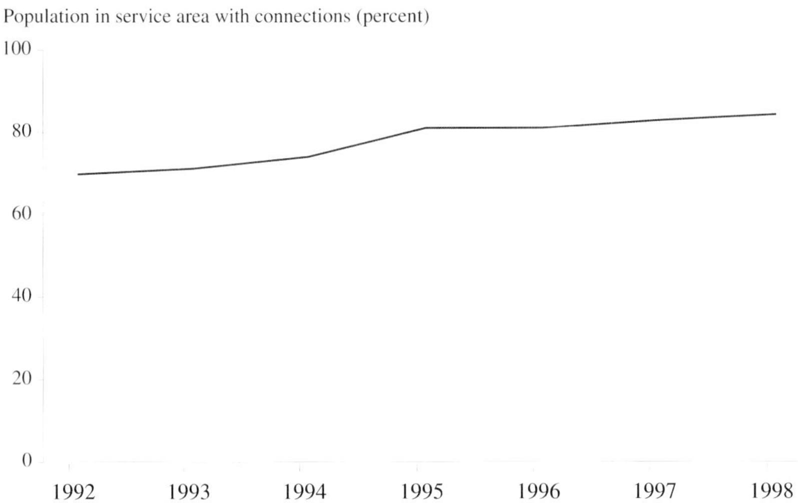

Source: OSN and Aguas Argentinas annual reports..

Figure 3.3. Water coverage in Buenos Aires.

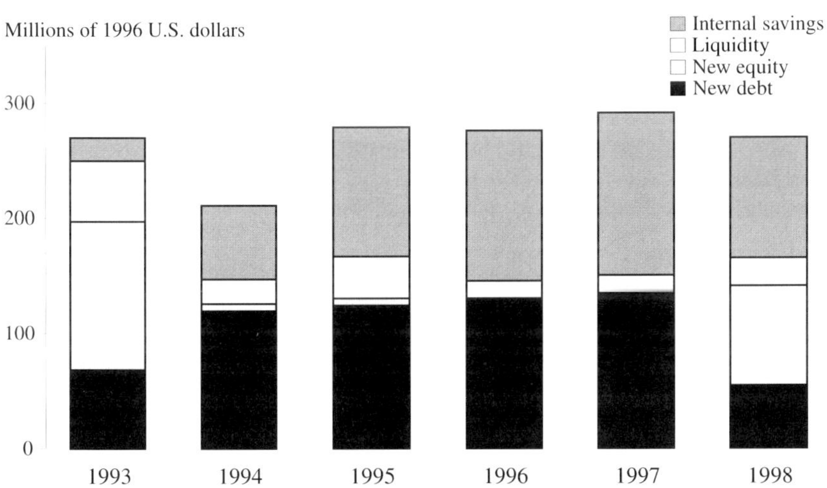

Source: Aguas Argentinas annual reports.

Figure 3.4. Sources of funds, Aguas Argentinas.

The margin of profits to sales has increased (Fig. 3.6) but is not exceptionally high. Unfortunately return on assets is hard to calculate since the company does not own most of the assets of the system. Estimates, which simulate fictional assets including assets owned by the government but operated by the firm, suggest a return on assets of about 3% in 1998.

The increase in operating profits and in the generation of internal savings for investment were largely a result of reductions in Aguas Argentinas' operating costs, which

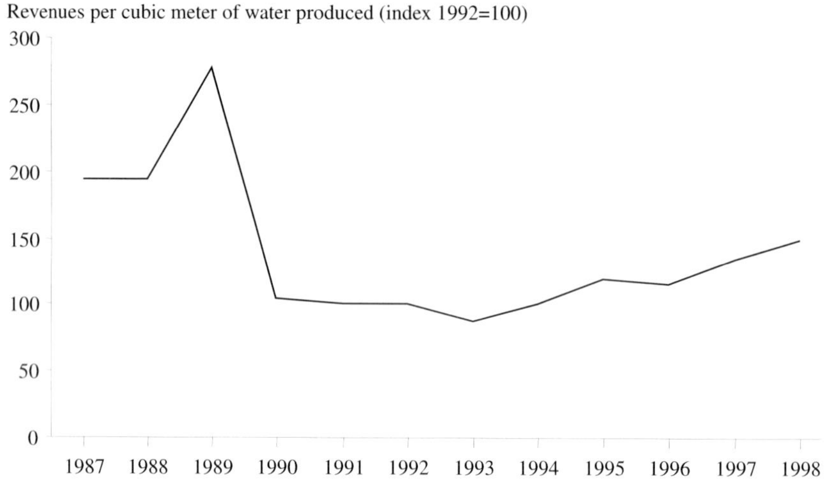

Source: Aguas Argentinas annual reports.

Figure 3.5. Real revenues, Aguas Argentinas.

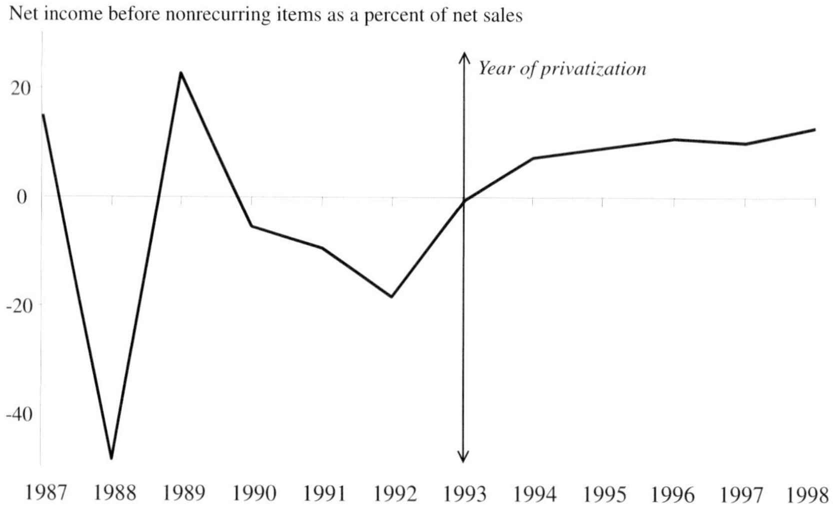

Source: Authors' calculations based on Aguas Argentinas annual reports.

Figure 3.6. Net profit margin, Aguas Argentinas.

fell by more than 14% in real terms from 1992 to 1998. This decline was partly due to a sharp reduction in labor costs thanks to the retirement programs described earlier. The number of employees went from 7666 in 1992 to 3829 by 1996, then rose to 4494 by 1998, which brought labor per thousand connections down from 3.3 to 1.5 and back to 1.7 by 1998, similar to other efficient companies such as EMOS in Santiago or Thames Water in the United Kingdom. The real cost of intermediate inputs also fell

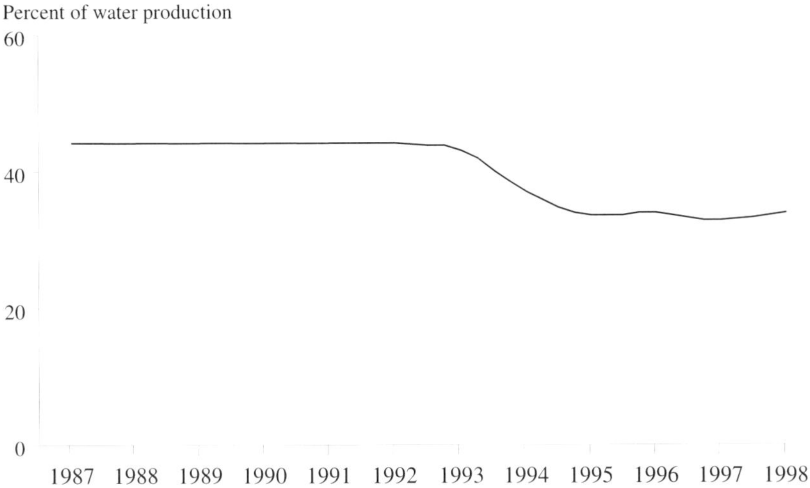

Source: Authors' calculations based on OSN and Aguas Argentinas annual reports.

Figure 3.7. Unaccounted-for water in Buenos Aires.

as the company improved its use of chemical products for water treatment and brought down its real electricity prices by bargaining for a term contract as a large customer in Argentina's wholesale electricity market. At the same time, production of water increased only moderately, to 1.4 billion cubic meters a year, or a total of 4% from 1992 to 1998. UFW was curtailed from 0.44 in 1992 to 0.34 by 1998 (Fig. 3.7). However, consumption of water by final users continued to be high and metering remained low, increasing from 1% before the concession to 12% by 1998.

As a result of these changes and the increases in revenues, operating expenses dropped from about 99% of operating revenues in 1992 to 61% by 1998. Again this is not much higher than other low cost systems, such as EMOS in Santiago where the ratio was 47% in 1997, or Thames Water in London (58%), and low given the massive rehabilitation program.

Total factor productivity increased dramatically the first three years of the concession (Fig. 3.8). Service quality also improved markedly after the concession (see Table 3.9). Although the number of complaints increased this was probably because channels of communication improved. Response time to complaints dropped sharply despite the larger number. And water pressure, which was always good in the Federal District, improved in Greater Buenos Aires as well.

Welfare effects of the concession

We measured the welfare effects of the concession by comparing the performance of Aguas Argentinas after the concession with what the expected performance would have been had the company not been privatized. To do this we projected the flows generated by the firm to the different parties involved — consumers, workers, government, and

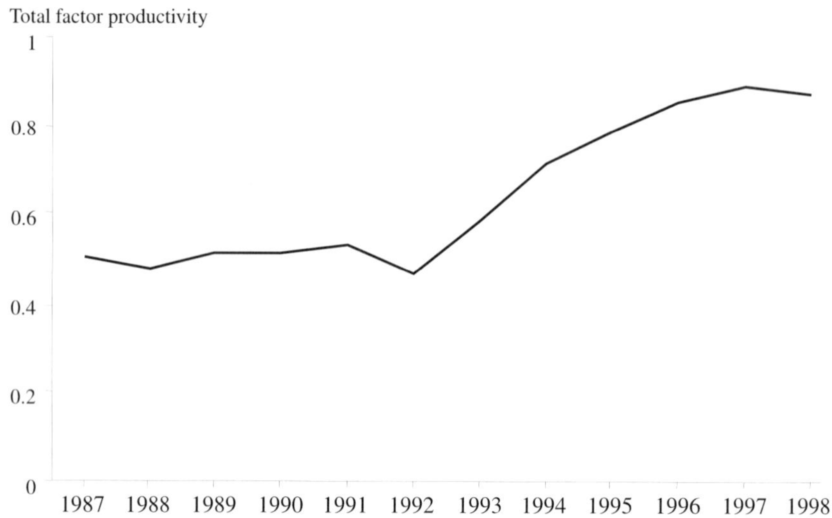

Source: Authors' calculations.

Figure 3.8. Total factor productivity in Buenos Aires.

Table 3.9

Service indicators for the Buenos Aires water system

	Before concession (1992)	*After concession (1995)*
Water complaints (number)	43,800	143,739
Response time (hour)	144	48
Sewerage complaints (number)	99,400	164,911
Response time (hour)	240	30
Customers with pressure greater than 8 wcm in Greater BA (%) [a]	17 [b]	54 [c]

[a] wcm = water column meters where 10 wcm = 1 atmosphere.
[b] May, 1993.
[c] April 1996.
Source: Data from ETOSS and Aguas Argentinas.

buyers — and compared them with the counterfactual flows under continued public ownership. The methodology is based on Jones et al. (1990) and Galal et al. (1994) and details on the analysis can be found in Abdala (1996). The data cover 1987 to 1998. We use the period of actual public operation to construct a counterfactual public Aguas Argentinas and project its operation for ten years after privatization to 2002. We take the actual private operation and project the private Aguas Argentinas to the same year.[37] We assumed that Aguas Argentinas would fulfill the terms of the concession with some exceptions, and we incorporated the changes introduced by the 1997 renegotiation in

[37] Of course there would be gains beyond this period but they would be reduced by discounting.

our calculations.[38] Thus, the actual and projected years are the following:

1987	....	1993	....	1998	....	2002
			Welfare calculation period			
	Actual public		Counterfactual Actual private		Projected public Projected private	

To construct the counterfactual we made three assumptions about the effects of the concession. First we assumed that the concession removed an investment constraint. Thus, if OSN had not been privatized, we assume that investment would have risen only enough to permit connections to expand at the same low rate observed during the actual period of public operation — about 1.5% a year in both water and sewerage. This assumption seems reasonable since the Menem government's policy was not to prop up investment in SOEs that were not being privatized. Indeed, following the 1994 so-called Tequila crisis, resources for public investment projects were cut. For example, funds for the Yacyretá hydroelectric project were cut from $300 million to $66 million in 1996 and to zero thereafter. A lower rate of increase in its customer base would result in less revenues for the counterfactual company than Aguas Argentinas captured, and we therefore assume it would not have had funds to reduce UFW or increase metering. The counterfactual company would also have had less funds than Aguas Argentinas to improve service and product quality, and collection rates would have been lower because of lower investments in information systems and management.

Second, we assumed that much of the gains in labor productivity and intermediate inputs would not have occurred without the concession. Instead, we assume that the counterfactual company would continue OSN's historic trend in labor productivity, which improved by an annual rate of 6.5% a year from 1987 to 1992, largely through attrition. Aguas Argentina's retirement program would have been difficult for OSN to finance in the absence of a serious reform program and might not have been credible to workers without privatization. Aguas Argentinas' real reductions in operating expenses were not likely to have occurred without privatization either. OSN had shown no efforts to minimize input purchases and it seems unlikely that this would have changed.

Finally, we assumed that the reduction in prices in 1993 would not have occurred without the auction of the concession, and therefore, the counterfactual's nominal prices would have remained the same. Other state enterprises in Argentina did not change prices in the absence of privatization. Moreover, nominal prices would not have increased since the Convertibility Law fixing the parity between the peso and the dollar prohibited indexation clauses, nor would prices have gone down since that would trigger the need for funds from the Treasury to finance the company's operating deficit. Average prices in the counterfactual and the actual scenarios differ for other reasons as well. As we have seen the net effect of the SUMA was to raise prices for some consumers. In

[38] Our assumption is that Aguas Argentinas would not have met the investment targets even if the renegotiation had not occurred. We made this assumption because the company could not meet the targets without violating its loan agreement with the IFC that the debt : equity ratio would be less than 1.9 from 1998 onwards.

Table 3.10

Welfare effects from the Aguas Argentinas concession

	NPV (1992 US dollar millions)	*Percent of 1992 sales*	*Annual welfare per group (1992 US dollars)*
Government	−137.97	−4.60	
Domestic investors	66.76	2.23	
Foreign investors	349.60	11.66	
Workers	49.52	1.65	$11,602 per employee
Competitors	2.27	0.08	
Consumers	1,326.58	44.25	$519.13 per connection
Total domestic	1,656.77	55.27	$175.80 per capita

Source: Authors' calculations based on ETOSS and Aguas Argentinas data (see Abdala, 1996).

addition, Aguas Argentinas was able to reclassify about 11% of residential customers who should have been charged as nonresidential (still not as many as the company wanted, as we have seen), and to meter some customers. The company's photographic and on-site inspections also enabled it to correct the reported size of built areas, which led to increases for some 425,000 customers.

Table 3.10 shows that the welfare gains from the concession over continued public operation total US $1.66 billion. Since these gains are the net present value of a flow in perpetuity, the final column of the table expresses the annual equivalent of the net welfare gains as a percent of OSN's sales revenue in the last year before sale. The big winners are consumers, with 80% of the increased gains; 79% of the gains are reaped by domestic actors.

Government loses slightly as a result of the concession. Although Treasury gets more in taxes, we must deduct from that the levy government collects to pay ETOSS' costs. Government receives fewer quasi-rents as a shareholder of Aguas Argentinas than as the owner of OSN, plus it has to pay for the transaction and OSN's debts.

Employees do better because they own 10% of Aguas Argentinas' shares. We assumed that workers who took voluntary retirement did so because the payments were sufficient compensation for being laid off. Buyers are better off by US $416 million, thanks to the quasi-rents they earn on their investment in the company. Competitors in the effluent business lose slightly because Aguas Argentinas takes some of their market.

Consumers are the big winners, better off by almost US $1.33 billion. The gain in consumer surplus is calculated using a demand elasticity based on studies of willingness-to-pay in similar cities.[39] This is a conservative estimate of their gains since it does not take into account improvements in product and service quality or the effects of improved public health and reduction of sewer overflows. It does take into account the benefits from reduced interruptions.

The gain in consumer surplus disproportionately benefited high- and middle-income

[39] See Abdala (1996). The elasticity in the base case was −0.35 for residential users and −0.85 for nonresidential.

Table 3.11

Sensitivity analysis (annualized net present value of welfare gains in millions of 1992 US dollars and as a percentage of 1992 sales)

	Government		*Domestic investors*		*Workers*		*Consumers*		*Total domestic*	
	NPV	*1992 sales*	*NPV*	*1992 sales*	*NPV*	*1992 sales*	*NPV*	*1992 sales*	*NPV*	*1992 sales*
Base	−138	−4.60	67	2.23	50	1.65	1327	44	1657	55
Random rationing	−138	−4.60	67	2.23	50	1.65	604	20	934	31
Residential elasticity −0.6	−138	−4.60	67	2.23	50	1.65	1207	40	1537	51
Residential elasticity −0.1	−138	−4.60	67	2.23	50	1.65	2045	68	2376	79
Nonresidential elasticity −0.95	−138	−4.60	67	2.23	50	1.65	1226	41	1556	52
Nonresidential elasticity −0.75	−138	−4.60	67	2.23	50	1.65	1454	49	1784	60
Projected annual price increase 1%	−127	−4.23	77	2.56	56	1.86	1297	43	1702	57
Projected dividend payout 100%	−183	−6.11	191	6.36	128	4.27	1327	44	2396	80

Source: Authors' calculations.

users. A large part of this welfare gain results from the drop in tariffs following the concession tender, despite subsequent tariff increases prior to the renegotiation. Most of this gain accrued to predominately higher income consumers who were already connected at the time of the concession. The other major source of gains in consumer surplus is reduced rationing of demand thanks to Aguas Argentinas' higher pace of building new connections. As we saw in Table 3.5, almost two thirds of the expected beneficiaries from new connections were lower middle to upper income. Moreover, as we have seen the infrastructure charge proved unaffordable for the very poor. In addition during the first two years after privatization the company did not try to reach low-income households. Hence, it seems likely that the percentage of new connections going to poor households, at least until 1998, was lower than the 36% that Table 3.5 suggests.

We tested the sensitivity of the results to our assumptions about rationing policy, elasticity of demand, prices, discount rate, investment costs, labor productivity and the like. Changes in most of these parameters had little impact, so we focused on the few that did (Table 3.11). The direction of the results is robust to these changes, which only have an effect on magnitudes. The most important effect on consumer welfare comes if we change the rationing policy. We assumed that demand is rationed by willingness-to-pay. If instead demand is rationed randomly, the gains to consumers would be much less. Random rationing, however, is not a realistic assumption given what we know about the city. Clearly people with connections or getting connections first were wealthier than those without connections which suggests that rationing was by willingness-to-pay.

We also tested changes from our basic assumptions of ± 0.25 in elasticity for residential and nonresidential consumers. As we would expect, a lower residential price elasticity raises consumer surplus substantially.

Gains to investors are most sensitive to changes in output prices and dividend pay out ratios. If prices increase by 1% a year in real terms, buyers are slightly better off (by about US $10 million) and consumers are slightly worse off (by about US $30 million). If 100% of dividends are paid off, buyers' gains are almost tripled, while government is worse off.

We also compared the gains to what they would have been had the 1997 renegotiation not occurred. The effect was negligible. While the SUMA reduced consumer surplus for those already connected it raised it for new customers, so the net effect was slight. The reduction in Aguas Argentinas' investment targets does not cause much difference, since we already had assumed that the company would have to curtail investment to meet the conditions of the IFC loan. We also tested the effects of lifting the constraints on Aguas Argentinas' investment, but the gains in consumer surplus are too small to offset the loss to shareholders and the net effect on welfare is negative.

CONCLUSION

This chapter finds that the Buenos Aires water system concession had large benefits for consumer welfare, along with smaller gains for buyers and employees. Performance improved dramatically. The average annual investment of Aguas Argentinas in the first six years of the concession was 2.4 times that of OSN in the last decade of its operation, operating efficiency improved markedly, and product and service quality were much higher. There were also important social and external benefits from the increase in coverage. Some 1.46 million people who now have access to piped water and 583 thousand with access to sewerage will no longer be consuming contaminated well water or polluting groundwater or rivers.

What features of the concession account for these gains? First was competition. Despite the flaws in information, the competitive auction of the contract produced a major price reduction and made it clear that the market was contestable. Second was appropriate alignment of incentives. The company's revenues increase when access expands, since average prices allow the company to cover the cost of service and earn a return on its investment, plus the company can be penalized if it fails to meet its targets. The strength of the incentive has been diminished somewhat with the introduction of SUMA and the credibility of regulatory penalties was reduced when ETOSS fines were waved during the 1997 renegotiation. The company also has an incentive to curb UFW and wasteful consumption, since this reduces the amount it must invest in new production. Third, the currency board, the privatization program more broadly and the general openness of Argentina's economy gave the concession credibility with investors.

Even though it is not unusual for an economic crisis and sector problems to bring about circumstances where efficient change is possible, the political circumstances that made the concession politically feasible, politically desirable and credible in Argentina are not common. Radical changes are not typical in democracies, and it is unusual for an administration to exploit the opportunity for consensus as effectively as the Menem government did. It is very rare for water to have such a low cost that government can attract an investor to reverse years of disinvestment and neglect — and also offer a lower consumer tariff. Low cost raw water was not the only reason this was possible. It was also important that the concession was credible to private investors, thanks to the government's prior history of privatization, the transparent auction, the broad base of support built through the privatization process, and favorable public attitudes.

The sustainability of the concession was and is less secure, however. Although the 1994 constitutional change will make it harder to overturn the concession in Congress, the weakness of the regulatory framework has put sustainability in doubt. ETOSS is vulnerable to politically motivated interventions in its decisions, as evidenced by the 1998 price decision, and can be paralyzed by partisan disputes in its board, as during the renegotiation of the contract. It suffers from serious information asymmetries created by poor historical data and the newness and weakness of its staff. The provision for cost pass-through when Aguas Argentinas' cost go up by more than 7% adds to ETOSS' information burden, since it is ambiguous and requires more judgment than a simple inflation adjustment. The lack of metering and the obscure nature of the tariff formula also make it harder for the regulator to judge the revenue effects of price increases. It gives the company an incentive to negotiate individual tariff components, such as the classification of users or the location coefficient, rather than just the *K* factor, further complicating the regulator's task.

ETOSS' weakness feeds consumers' doubts that their interests are not being protected, and that in turn increase political pressure to renege on agreed price increases.

The lumpy nature of the cost plus increases adds to consumer concerns. The poor information base, lack of transparency in regulatory decisions and ad hoc nature of executive branch interventions make it harder to reassure consumers that their welfare is being protected.

Better information on the system and a rationalized pricing policy could have reduced many of these problems. However, it is hard to know whether the necessary delay would have ultimately halted the concession, as the administration feared. Delays in other cities, such as Lima, Mexico City, Santiago, allowed political changes that ultimately derailed private operation of their water systems.

The concession agreement states that the operator assumes the information risks and that "claims based on defects of information provided will not be allowed". Yet Aguas Argentinas' claim for redress of costs beyond its control stated that about half these costs were due to wrong information from OSN. Nasser, FIEL (1993, 1996) and other observers have argued that the winning bid was aggressive given the poor information base, suggesting that the company was assuming that it could shift the burden of the information risk to the government. However, the second bid was very close to the winning bid, which suggests that more than one party was acting aggressively, possibly in response to some reasonable expectation that renegotiation would be feasible. The company was also responsible for collecting all usage and access fees under the contract, another risk that was lifted during negotiation. Here again, part of the problem was lack of information — in this case, the government's lack of information on affordability.

There are several lessons from the Aguas Argentinas' concession. First the experience reinforces conventional wisdom that risks and responsibilities, as well as the grounds for renegotiation, should be spelled out as clearly as possible in the contract. It also suggests that contracts that shift all information risk to bidders may not be credible.

Second, safeguards are needed to reduce the risk of regulatory capture by the company or politicians. One such safeguard is to reduce regulatory discretion by enshrining in law less ad hoc procedures for setting tariffs. Since the 1994 constitutional change made laws harder to reverse in Argentina, detailing tariff regulations in the law would likely enhance regulatory stability, as it has in Chile. It might also reduce consumer opposition. Experience in other countries suggests that consumers are more likely to react to tariff increases that seem unfair or unjustified than to high tariffs per se. Cities such as Santiago, for example, have higher tariffs and lower per capita incomes than Buenos Aires, yet there have been no protests when water tariffs were increased for the private and public operators.[40] This is partly because the laws detail a technical process that occurs automatically no more than once every five years, with inflation adjustments in the interim. The cost plus provision in the Buenos Aires concession makes interim tariff increases ad hoc and lumpy. We have argued that these increases, the additional tariff adjustments to cover new tasks, the renegotiation and the perceived weakness of the regulator have contributed to the public's sense of being poorly protected from arbitrary change, and this increases the risk of politically motivated manipulation of price policy.

[40] Although most of Santiago is served by a public company, about 5% of consumers are served by a privately owned water company. Santiago's means tested subsidy for water services also reduces opposition to increases.

Transparent, rule-based decision making, such as that used in Santiago, for example, is important to maintaining public confidence in regulated industries. Ad hoc intervention by the federal executive in regulatory decisions is inconsistent with rule-based regulation, which can only be overturned by a pre-established process of appeal.

Another safeguard against political intervention is to avoid a politicized board; ETOSS' board currently pits partisan interests against one another, and allows short term, sectarian interests to control decision making. ETOSS' board could be replaced or enlarged with technical appointees (such as economists and hydraulic engineers) to reduce the dominance of political appointees. U.S. experience suggests that better public information on regulatory decisions and an open public comment period would also protect against capture by narrow political interests or by the regulated company.

Third, the Buenos Aires experience raises questions about the practice of bidding on the basis of consumer price in systems where there are large benefits from expanded coverage and those without access are poor. The concession could have been bid on the basis of the cost of access, in order to maximize new connections. The welfare gains of already connected consumers would have been lower, but the gains to newly connected households and from reduced health externalities would have been greater than they were, along with much more equitable redistributive effects. As it was, the price reduction was based on unrealistic assumptions about the willingness and ability of poor new consumers to pay large fees for access. The introduction of SUMA has created another unsustainable contract. By virtually eliminating the incentive of the company to expand services to the poor (who pay below marginal cost), SUMA opens the way for future problems. Buenos Aires' experience suggests that the poor benefit more from subsidized financing for access than from subsidized usage charges, which benefit the rich disproportionately.

Finally, the Aguas Argentinas experience suggests that the information benefits from metering need to be considered more carefully in designing regulatory policy. Typically, the cost of metering is measured against the benefits of reduced consumption, which may be low in a system where raw water is cheap and capacity is not constrained, as in Buenos Aires. But metering also reduces the information burden for regulators and makes it possible for consumers to monitor billing and reduce their bills by controlling consumption and these benefits, albeit intangible, may be high in weak regulatory settings.

REFERENCES

Abdala, Manuel A., 1996. *Welfare Effects of Buenos Aires Water and Sewerage Services Privatization.* Washington, DC: World Bank. Unpublished manuscript.

Alcázar, Lorena and Penelope Brook-Cowen, 1996. *Institutions, Politics, and Contracts: Private Sector Participation in Urban Water Supply.* Washington, DC: World Bank. Unpublished research proposal.

Artana, Daniel, Fernando Navajas and Santiago Urbiztondo, 1997. "La Regulaciòn Economica en las Concesiones de Agua Potable y Desagues Cloacales in Buenos Aires y Corrientes, Argentina". InterAmerican Development Bank Working Paper R-312.

De Michele, Roberto and Luigi Manzetti, 1996. "Legal security and market reforms: the case of Argentina". Judicial Security and Markets. Unpublished paper prepared for the University of Maryland at College Park.

ETOSS. *Memorias.* Various issues. Buenos Aires: ETOSS.

FIEL, 1993. *Inversión en Infraestructura 1992–2000*. Buenos Aires: FIEL.

FIEL, 1996. *Regulación, Organización e Incentivos, La Economía Política de los Servicios de Agua Potable: El Caso Argentino*. Washington, DC: Inter-American Development Bank.

Galal, Ahmed, Leroy Jones, Pankaj Tandon and Ingo Vogelsang, 1994. *Welfare Consequences of Selling Public Enterprises: An Empirical Analysis*. New York, NY: Oxford University Press.

Heller, William and Mathew D. McCubbins, 1996. "Politics, institutions, and outcomes: electricity regulation in Argentina and Chile". *Journal of Political Reform* 1: 357–387.

Hill, Alice and Manuel A. Abdala, 1994. *Regulation, Institutions, and Commitment: Privatization and Regulation in the Argentine Telecommunications Sector*. Policy Research Working Paper 1216, Washington, DC: World Bank.

Idelovitch, Emanuel, 1994. *Private Sector Participation in Water Supply and Sanitation: The Transition from Public to Private Operation: The Buenos Aires Case*. Washington, DC: World Bank.

Idelovitch Emanuel and Klaus Ringskog, 1995. *Private Sector Participation in Water Supply and Sanitation in Latin America*. Direction in Development Series, Washington, DC: World Bank.

INDEC, 1985/86. *El Consumo y el Ingreso de los Hogares Frente a una Caída de la Inflación*. Buenos Aires: INDEC.

Jones, Leroy P., Pankaj Tandon and Ingo Vogelsang, 1990. *Selling Public Enterprises: A Cost–Benefit Methodology*. Cambridge, MA: MIT Press.

Laffont, Jean-Jacques and Jean Tirole, 1993. *A Theory of Incentives in Procurement and Regulation*. Cambridge, MA: MIT Press.

Levy, Brian and Pablo Spiller, 1994. "Regulations, institutions, and commitment in telecommunications: a comparative analysis of five country studies". In: *Proceedings of the Annual Conference on Development Economics*. Washington, DC: World Bank.

McCubbins, Mathew D., 1993. The Prospects for Stability in Argentine Politics, Unpublished manuscript.

Munasinghe, Mohan, 1992. *Water Supply and Environmental Management. Developing World Applications*. Boulder, CO: Westview Press.

Nasser, Thomas-Olivier, 1997. "Water Concession in Buenos Aires: Issues and Propositions". Processed.

Republic of Argentina, 1949. *OSN Ley Organica*. Buenos Aires: Government of Argentina.

Republic of Argentina, 1993. *Concession Contract with Aguas Argentinas*. Buenos Aires: Government of Argentina. Processed.

Shaikh, Hafeez, 1996. *Argentine Privatization Program*. CFS Discussion Paper Series, 117, Washington, DC: World Bank.

Shirley, Mary M. and Lixin Colin Xu, 1998. "Information, incentives and commitment: an empirical analysis of contracts between government and state enterprises". *Journal of Law, Economic and Organization* 14(2).

Shirley, Mary M. and Lixin Colin Xu, 2001. "The empirical effects of performance contracts. *Journal of Law, Economics and Organization* 17(1): 168–200.

Snow, Peter, 1996. "Argentina: politics in a conflict society". In: Wiarda and Kline, eds., *Latin America Politics and Development*. Westview Press.

Williamson, Oliver, 1985. *The Economic Institutions of Capitalism*. New York: Free Press.

World Bank, 1994a. *Argentina: Water Supply and Sewerage Sector Note*. Unpublished report, Washington, DC: World Bank.

World Bank, 1994b. *World Development Report*. Washington, DC: Oxford University Press.

World Bank, 1995. *Bureaucrats in Business: The Economics and Politics of Government Ownership*. Washington, DC: Oxford University Press.

Thirsting for Efficiency: The Economics and
Politics of Urban Water System Reform
Mary M. Shirley (Ed.)

CHAPTER 4

INSTITUTIONS, POLITICS, AND CONTRACTS: THE PRIVATIZATION ATTEMPT OF THE WATER AND SANITATION UTILITY OF LIMA, PERU

LORENA ALCÁZAR, LIXIN COLIN XU AND ANA M. ZULUAGA

INTRODUCTION

In 1989–1990 Peru underwent a severe economic crisis with a 5.4% decline in GDP and inflation peaking at an annual rate of over 7600%. The newly elected President, Alberto Fujimori, embarked on a program of sweeping structural reforms that turned Peru from a state-directed to a market-led economy. The reforms changed the role of the state through the sale of more than 200 state owned enterprises, sharply reduced state subsidies and price controls, and introduced tax, wage and other measures designed to improve the efficiency and financial soundness of government. Included in the program was a proposal to privatize the operation of Lima's water and sewerage utility, by means of a concession contract.

As we shall show, Lima's water and sewage system was in a state of near collapse by the time of Fujimori's election in 1990. Water is scarce in the arid costal region where Lima is located, and leakage, waste, and contamination of water sources compounded this natural scarcity. The result was severe rationing and frequent interruptions for the 75% of the population connected to the system, and high costs and inconvenience for the 25% without direct access to piped water. In addition, almost a third of Lima's 1991 population of 6.5 million were not connected to the sewage system. Sewerage treatment was almost nonexistent and Lima dumped raw sewage into the ocean at a rate of 17 to 18 m^3 per second. Low tariffs for water services, lack of metering and low rates of bill collection had left SEDAPAL with neither the incentive nor the resources to expand or maintain the system, and gave consumers little incentive to curb water consumption except during rationing.[1] The social and external costs of the system were large: high rates of illness and death from water-borne diseases, pollution of rivers used for irrigation and of the ocean used for fishing, depletion of the groundwater, and losses of exports and tourism during the 1991 cholera epidemic.

Conditions thus seemed ripe for Lima's water and sewerage concession. The water and sewerage system was in crisis and the political circumstances favored privatization. Yet the concession did not happen. Even though a number of reforms short of privatization were implemented, we shall show that the welfare gains to Peru from these were well below the potential gains had the concession gone ahead as planned. The objective of this chapter is to analyze the reforms that SEDAPAL underwent instead of the concession, to explore the consequences of that change for water services in Lima, and to explain why the concession was not signed.

The next section of this chapter discusses the circumstances in the water sector leading to reform, and the following section discusses the reforms that were implemented. The chapter then compares the welfare effects of reform with the concession. Next it explores the political circumstances that explain why the operation of the water system was not privatized. We conclude with a discussion of the remaining problems and the implications of the Lima case for policy.

CIRCUMSTANCES IN THE WATER SECTOR LEADING TO REFORM

Cost and Scarcity of Raw Water

Lima's sources of raw water are scarce and variable. Lima is located in the coastal region of Peru, a region that receives less than 15 mm of rainfall a year. The river flow through this region is strongly seasonal: 25 of the 53 main rivers dry up entirely during the dry season from May to December. As a result, the average availability of surface water in the coastal area is estimated at 2885 m^3 per capita, compared to an estimated world average of 8500 m^3 (Macroconsult, 1996).

Pollution aggravates the scarcity problem. The major source of Lima's water supply is the Rimac River, which is contaminated by heavy metals from nearby mines as well as untreated sewage. About a third of Lima's water comes from wells, which depend on a shrinking supply of groundwater. The aquifer is increasingly polluted by salinity when the water table near the ocean drops because of increased pumping in the dry season (World Bank, 1994).

Demand and Water Tariffs

The potential demand for water was increasing rapidly in the period before reform. Lima was growing by an average of about 2.7% a year from 1981 to 1992 as terrorism and economic factors drove rural populations to move to low-income communities on the outskirts of the city. These so-called *pueblos jóvenes* accounted for 58% of Lima's population of 6.5 million by 1991.

---

[1] See Appendix A, Table 4A.2 for the rate of metering in 1988.

Throughout the 1980s and 1990s water and sewerage tariffs, which are combined in Lima, were well below the opportunity cost of supply in an arid area, and they did not reflect the social costs of pollution. Under the populist policies of President Alan García (1985–1990), the water and sewerage tariff was allowed to decline in real terms to the point that by 1989 the average tariff was about US \$0.17 m^3, less than half of the 1985 price in constant (1989) soles. Even if tariffs had been higher they would have done little to curb consumption, since only about a third of connections were metered in 1991 and only about 10% of users were billed according to a meter reading. Unmetered consumers paid a flat rate regardless of what they consume and hence had no incentive to conserve. Moreover, much water was free — only about 43% of billed amounts were collected.

The main constraint on demand was not the price but low coverage and rationing. Only about 75% of the population was connected, 48% of the connected population received water service for less than 12 h a day, and 28% for less than 6 h.[2] Nevertheless, water consumption in the city was 236 l per capita per day in 1987 during the wet season when rationing was low, which is high compared to European averages (150–200 l.p.d.). Consumption varied widely across the city, however; for example in 1993 it ranged from highs of 567 in wealthier neighborhoods with high rates of connection to lows of 105 in poorer zones (World Bank, 1994, annex 8). About a third of those without connections relied on standpipes or group taps, another third on water vendors and the rest on other sources such as wells. Vendor water was expensive, about US \$2.50–2.75 per cubic meter (m^3) and persons who relied on vendors consumed only about 30 l.p.d. (World Bank, 1994).

Management

Mismanagement added to the water shortages and serious sanitation problems. Massive amounts of water were wasted: unaccounted-for water was 43% of production in 1991. Some two thirds of these losses were due to leakage; the rest was consumed but not billed.

The utility, SEDAPAL (Servicio de Agua Potable y Alcantarillado de Lima) had little incentive to reduce UFW in order to sell more water. SEDAPAL's tariff did not cover even its operating costs and it was unable to cut off and punish non-payers. More importantly SEDAPAL did not have the money to replace leaking pipes or increase metering in order to reduce waste. Not only was it making losses, but also it could not borrow after President García unilaterally reduced foreign debt payments to 10% of the owed amount in 1988. As its liquidity situation worsened, the company also stopped paying its debts and by 1993 had US \$45 million of external debt in arrears. SEDAPAL depended on government transfers to meet its current expenditures and these were also declining as the central government's fiscal crisis worsened under President García. As

[2] Rationing costs have been estimated at US \$0.27 m^3, assuming the cost of a household water storage reservoir is US \$1000; operation and maintenance costs are 5% of the investment costs; the economic life of the investment is 15 years with a discount rate of 10%. If physical losses are 30%, than the cost per m^3 consumed is about US \$0.38 (World Bank, 1994).

a result, in the late 1980s the company reduced its investment in expansion and even maintenance investment fell: investment per 1000 connections fell from US $28.6 in 1987 to a low of US $16.6 in 1988. From 1987 to 1989 only about 3 km of water pipe were replaced in a system with more than 6700 km (World Bank, 1994b).

SEDAPAL's lack of resources also meant that close to two million people were not connected to the sewerage system. In addition, the sewerage system was suffering from lack of maintenance: close to 85% of the pipes needed to be replaced in older parts of the city. Only about 5% of sewage was treated.

Externalities

The scarcity of water and the pollution problems had major social costs. Because of supply interruptions, even people with connections were storing water under unsanitary conditions. People without connections spent hours queuing at public standpipes.[3] Lack of water also meant that personal hygiene was often substandard. Inadequate sewage disposal added to health problems; in particular, foodstuffs were contaminated by the dumping of raw sewage into rivers used for irrigation and into the ocean used for fishing. As a consequence, water-borne and water-related diseases were a major cause of morbidity and mortality, especially in poorer neighborhoods of Lima.[4] The medical costs and lost wages from such diseases were a high part of household income for the poor, 27% by one estimate.[5] The situation contributed to a cholera epidemic centered in Lima in 1991 during which almost 3000 people died throughout Peru.

REFORM OF LIMA'S WATER AND SEWERAGE SYSTEM

Partly in preparation for privatization and partly in compliance with the World Bank project, the government instituted a number of important reforms in water regulation and in the operations of the utility. This section presents and discusses the consequences of the main changes.

Tariff Reform

As Fig. 4.1 shows, the government reversed the decline in the real price of water and sewerage (SEDAPAL does not charge separately), raising real tariffs from an average of US $0.17 per m^3 in 1989 to US $0.41 in 1995 and 1996 (1996 prices). Notwithstanding these increases, average tariffs were still below the 1994 estimated marginal cost of water and sewerage of US $0.45 (World Bank, 1994).

Furthermore, tariffs continued to be badly distorted. By the end of 1998, SEDAPAL was still charging a small fixed charge (US $0.90) and five different variable charges: (i) social (for hospitals, charities and public standpipes), (ii) residential, (iii) government,

[3] One person reported a total time of 7 h a day queuing during the dry season (Webb and Associates, 1992).
[4] The pueblos jóvenes reported eight episodes of diarrhea per person a year (World Bank, 1994b).
[5] Webb and Associates estimated the medical cost of water-borne diseases at US $13 per capita a year in 1992. Average family size is five persons and household income is US $240.

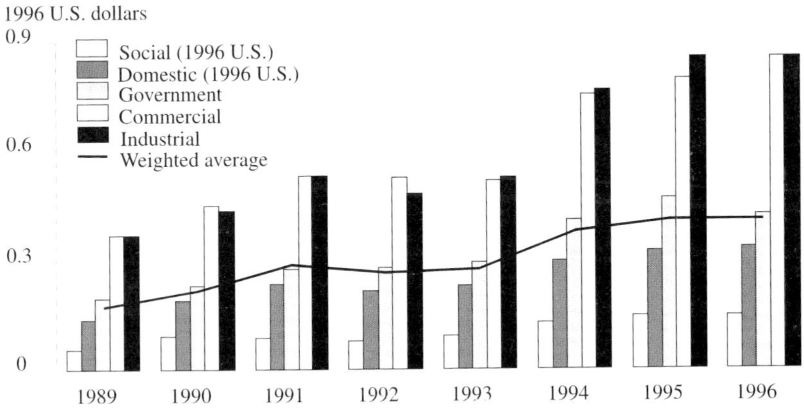

Source: Authors' calculations based on data from SEDAPAL Annual Reports

Figure 4.1. Average water tariff.

(iv) commercial, and (v) industrial.[6] Initially, the rate of increase for social and residential tariffs was somewhat higher than for other uses, which reduced cross subsidies slightly, but this has been reversed somewhat since 1993 (Fig. 4.1). For example, in 1989 industrial rates were 170% higher than residential; these fell to 130% in 1993, then rose to 160% in 1996.[7] Although rates are higher for industrial users, these fall mostly on smaller industries, since large industries can afford to drill their own boreholes.

Only 39% of connections were metered by 1996, and only 15% of connections had meters that were read (see Appendix A, Table 4A.3); an unknown but even smaller number were actually charged according to their meters.[8] Metering varied by type of user, with a greater proportion of industrial users charged according to a meter.[9] Consumers who were not charged according to their meter reading were charged on the basis of estimated consumption volumes that varied by type of consumer and type of neighborhood (see Appendix A, Table 4A.4) as well as by number of interruptions. SEDAPAL's consumption estimates were often higher than actual consumption because interruptions in service were not fully taken into account, a source of consumer complaints.

Notwithstanding the increase in real tariffs, Lima's pricing regime provided no incentive for unmetered consumers to conserve water. Indeed the price increases had the opposite effect, encouraging consumers to try to increase consumption at least to the amount for which they are billed and encouraging large industrial users to drill their own wells. Currently SEDAPAL estimates that average demand in Lima has increased to the equivalent of 460 l per capita per day, above that of cities with ample water

[6] The connection charge in 1998 for water was US $224 and for sewerage US $296.

[7] The residential rates were US $0.18 per m^3 compared to industrial tariffs of US $0.49 in 1989; by 1996 the rates were US $0.33 and US $0.85, respectively.

[8] In 1992 only 8% were charged according to a meter (LMG Consultants). These data are not strictly comparable, however, because SEDAPAL has contracted out metering and meter reading.

[9] In 1988 about a third of industrial consumers were charged according to a meter versus 16% of residential users (see Appendix A, Table 4A.3).

sources, although rationing keeps per capita sales much lower. On the other hand, higher prices gave SEDAPAL the incentive and wherewithal to reduce physical water losses and pursue bill collection more rigorously, as we discuss in more detail in the section on SEDAPAL's operations below.

A weakness of the reform was the failure to implement an objective price setting formula. Tariff regulation was defined in the water and sanitation law (Ley 26338, 1994) and its Reglamento (DS 09-95, 1995). However, SUNASS determined that the application of the new tariff regime would be gradual. The implementation of the new tariff system has three stages: preparatory, improvement, and definite. During the preparatory stage, the financial feasibility of the water utility will be stressed so that the enterprises will cover their operation costs. During the improvement stage, the enterprises should index, in a term of five years, their tariffs to the variation of wholesale price index — every time 3% is surpassed. Finally, at the definite stage, tariffs should reflect the long-term marginal cost and they would simply be adjusted according to the inflation. However, the schedule established to advance through the system stages has not been fulfilled and tariffs are still basically adjusted to cover operating costs. In mid-1997 the regulator ordered SEDAPAL to bring estimated consumption closer to actual hours of water supply and to reduce the dispersion in tariffs, the number of categories for commercial and industrial users and the number of categories of residential consumers from 10 to 4 (SUNASS, Boletin Informativo, No. 3, 1997). Even with these changes, however, the price regime continued to send distorted signals to consumers and the utility.

Regulatory Framework

In preparation for private participation several laws were passed from 1991 to 1994 that, among other things, established financial viability as one of the principles for tariff setting. They also consolidated water regulation under MIPRE and created a new regulatory body for water services, the National Superintendence of Sanitary Services, SUNASS (Superintendencia Nacional de Servicios de Saneamiento).[10] The new framework had several serious weaknesses, all described below, that most notably included the following: (i) it lacked a clear delineation of the function and purpose of the regulator and other actors in the sector; (ii) the regulator was not sufficiently autonomous from government; (iii) the regulator was organizationally weak; (iv) there were no viable mechanisms for conflict resolution; and (v) the regulation contained provisions not fully supportive of private participation in water. As a result, the framework would not support reforms of the type envisioned in the concession. The regulation failed to provide credible commitment to future private investors that their returns would not

[10] Specifically, the private investment law (D.L. No. 697) of November 1991 incorporated SEDAPAL into the privatization process; D.L. No. 25738 consolidated regulation of the sector under the Infrastructure Vice-Ministry and PRONAP (Programa Nacional de Agua Potable y Alcantarillado). D.L. No. 25965, in December 1992, created SUNASS with responsibility to supervise water and sewerage enterprises; and the general law of sanitation service, D.L. No. 26338 in July 1994, defined new institutional responsibilities for the sector and established transparency, financial viability and social equity as the basic principles to govern tariff setting and called for autonomous and commercially viable water companies.

be threatened by politically motivated under-pricing or other forms of confiscation and credible assurance to consumers that the regulator could not be captured.

The first problem, a lack of clarity in roles, arose because SUNASS was not only supposed to regulate the water service enterprises, but also to promote and strengthen them. As other countries with this mix of roles have found, Chile for example, the task of promotion harms the arms length relationship a regulator must maintain to assure objectivity and neutrality. This role confusion was illustrated by the remarks of the former Superintendent, Ms. Lidia Oblitas, who stated that SUNASS could not regulate very weak water service enterprises that lack resources. She explained that SUNASS instead tries to strengthen these companies first, and only then demand that they comply with the rules (*El Comercio*, issue of November 9, 1994, confirmed in field interviews in 1997).

Another confusion arose because of the conflict between the role of SUNASS and PRONAP. Some of PRONAP's responsibilities overlapped with SUNASS' regulatory powers. For example, PRONAP was empowered to propose a subsidy policy similar to Chile's income support for water bills (Ochoa, 1997), and to develop accounting techniques and financial models for water enterprises. Conflicts between the two entities were regarded as a major problem by officials of both SUNASS and PRONAP as well as sector authorities in MIPRE and experts consulted in the World Bank and Inter-American Development Bank (field interviews).[11]

Autonomy was the second regulatory problem. SUNASS received 2% of all tariff revenues, making it financially self-sufficient, yet it was still vulnerable to political intervention. Unlike other regulatory bodies in Peru, such as the regulator for telecommunications, SUNASS did not have a board or governing council in which government was only one of several stakeholders, as a way to insulate it from direct intervention by the ministry. Instead, all power was vested in the Superintendent, who was a political appointee and until recently reported to MIPRE and could be removed by the Ministry at any time. This was a problem for relations with SEDAPAL, which unlike the other water companies that were attached to their municipality or province, reported directly to MIPRE. The new law (No. 26922, February 1998) partly solved this problem by putting all regulatory agencies under the Presidential Council of Ministers.

A third problem was SUNASS' organizational weaknesses. Although SUNASS, like other regulatory bodies, could pay as much as other regulators and above civil service levels, it had proportionately fewer technically qualified staff — less than 5% of professionals have graduate degrees — than regulators for telecommunications, electricity or consumer- and market-loyal competition protection. Moreover, over one fourth of SUNASS staff came from the former supervisory body for water, and brought with then an approach that focussed more on control than on regulation (field interviews).

Fourth, there were no adequate mechanisms for enforcement or conflict resolution. The law allowed SUNASS to assess a penalty of up to 30% of an enterprise's revenues for failure to comply with the regulation, but at least until the end of 1998 this power had never been used. According to SUNASS' tariff supervisor, the regulator

[11] PRONAP, Programa Nacional de Agua Potable y Alcantarillado (National Program of Water and Sanitation), was designed as temporary and scheduled to disappear but its existence has been expanded.

had never assessed a penalty because there was no provision for what would happen if the enterprise failed to comply with the penalty (field interview). Under the present circumstances with a weak regulator and a publicly owned utility, the absence of a neutral mechanism for conflict resolution has not been an issue, but it might become one should private participation be used in the future. The law provides for a water company to appeal SUNASS' decisions to the courts, but the weakness of the judiciary reduces the credibility of the regulatory framework. In interviews, foreign investors cite the corruption and political nature of the judiciary as a particularly negative aspect of Peru's current investment environment (Barthelemy, 1996).

Finally, the water system regulation did not fully support private participation. For example, the general law of sanitary services included a provision stating that "tariffs can be modified in the event that the interest of the population is affected by the privatization process" (law No. 26338, July 1994, authors' translation). None of the legislation specifies how private enterprise tariffs will be set and is ambiguous on details of the concession process.

Reforms in SEDAPAL

In preparation for the concession and to comply with the requirements of the World Bank project, SEDAPAL began a program of restructuring in 1992, that continued after the concession was postponed. The project funds combined with the increased revenues resulting from tariff increases allowed the company to improve its financial performance and increase investment. These changes also helped ease the pressure for the concession.

Through voluntary retirement the number of SEDAPAL workers was reduced from 3769 in 1988 to 1359 in 1996; at the same time the number of connections was expanding, so workers per 1000 connections fell from over 6 to 2. This reduction in the labor force was accomplished partly by removing redundant workers and partly by outsourcing activities that were previously done in-house, especially bill collection and maintenance, installation and repair of connections. As a result, labor costs fell sharply. Even though outsourcing raised administrative and maintenance expenses and electricity costs increased, overall operating expenses per cubic meter of water sold stayed fairly constant (Fig. 4.2).

SEDAPAL also introduced a number of reforms designed to change management and personnel incentives in 1996. This included a performance-based pay system with specific targets for each employee and quarterly bonuses of up to 10% of pay.[12] Partly as a result, the average annual salary of SEDAPAL's personnel began to increase in real terms, as can be seen in Fig. 4.3.

The reduction in labor expenses and the rise in tariffs allowed SEDAPAL to make a profit in 1993 for the first time in more than a decade. Further price increases, as well as improvements in billing and collection have kept SEDAPAL solvent (Fig. 4.4).

The increase in SEDAPAL's liquidity and the World Bank project allowed a surge in investment in expansion and maintenance. As Fig. 4.5 shows, investment per con-

[12] To implement this change management won agreement of the union for the 55% of staff that is unionized.

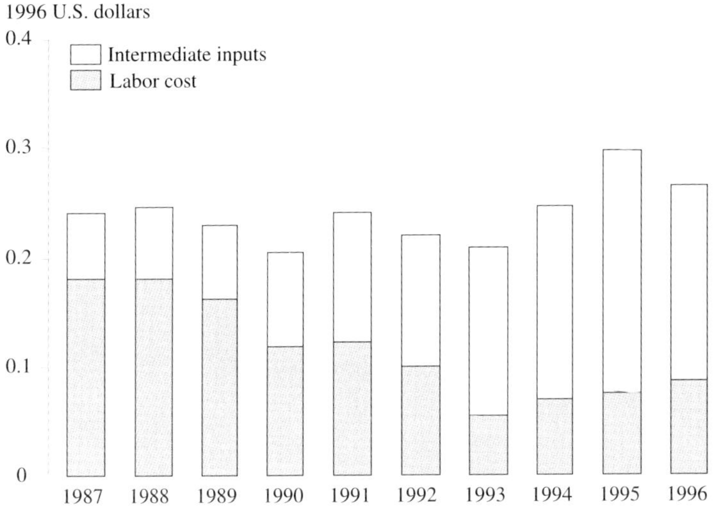

Source: Authors' calculations based on data from SEDAPAL annual reports

Figure 4.2. Operating cost per cubic meter of water sold.

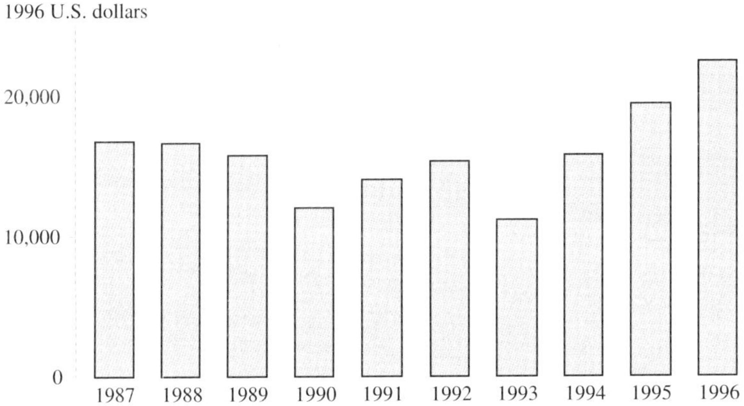

Source: Authors' calculations based on data from SEDAPAL annual reports

Figure 4.3. Average annual salary.

nection rose from US $26 in 1990 to US $80 by 1996. Most of the investment has been earmarked for expansion and there has been less emphasis on rehabilitation and maintenance of the existing network.

THE EFFECTS OF THE REFORM ON PERFORMANCE AND WELFARE

This section shows the impact of the reforms described in the previous section on several performance measures and then compares their welfare effects with those that might have been achieved under the concession.

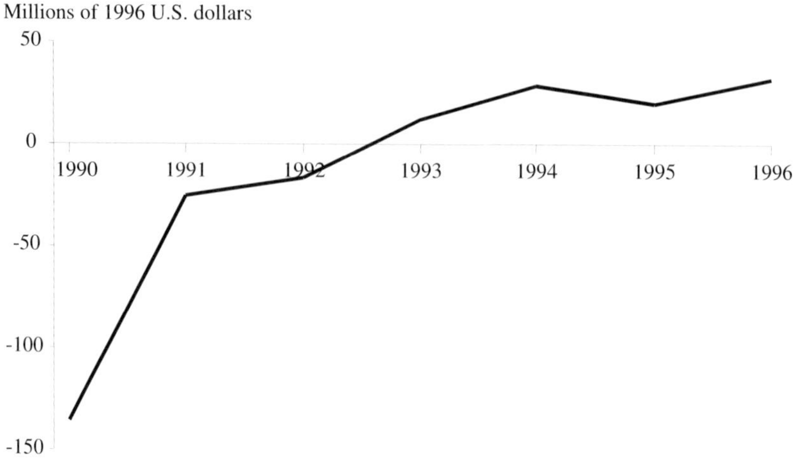

Source: SEDAPAL annual reports

Figure 4.4. SEDAPAL profits.

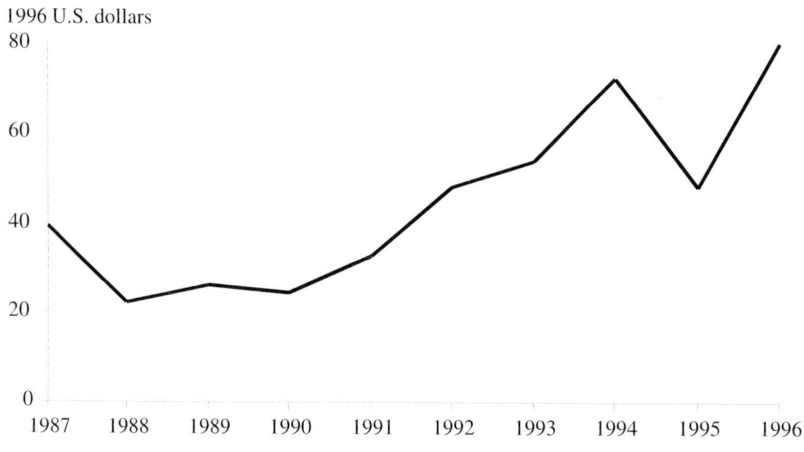

Source: Authors' calculations based on data from SEDAPAL annual reports.

Figure 4.5. Annual investment per thousand connection.

The Effects of the Reform on Performance

Additional funding from the World Bank project and higher water tariffs enabled SEDAPAL to add close to 200,000 additional water connections between 1989 and 1996 (Fig. 4.6). This growth in connections only allowed the company to keep pace with Lima's rising population, however, and the coverage ratio of 70% barely changed (Fig. 4.7). Moreover, the quality of service does not seem to have improved. For example, field interviews suggest that some districts of Lima received no water service for more than two months in 1997.

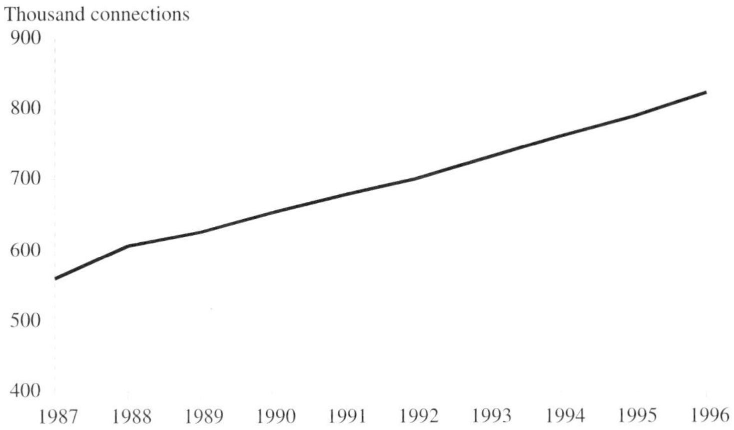

Source: SEDAPAL annual reports.

Figure 4.6. Thousands of water connections.

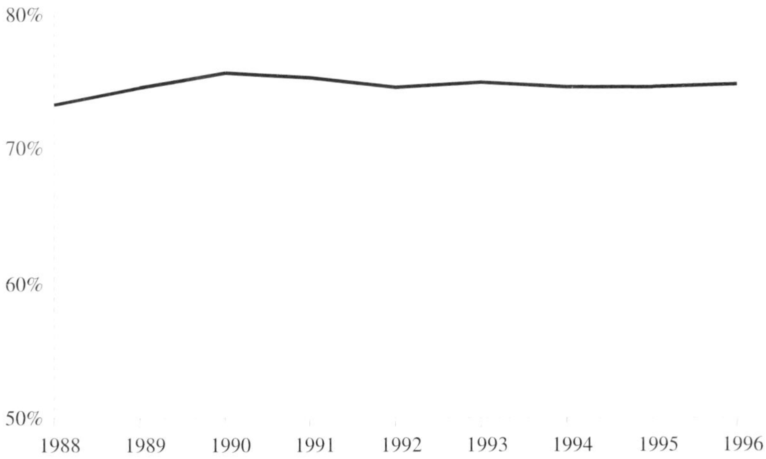

Source: SEDAPAL annual reports.

Figure 4.7. Percentage of water coverage.

The efficiency improvements in SEDAPAL did not have a sustained effect on unaccounted-for water. UFW fell from close to 50% in 1987 to a low of about 27% in 1992 but then began to increase again (Fig. 4.8). By 1996 it was 36%.

Furthermore, the changes in management and incentives did not lead to any measurable improvement in total factor productivity (Fig. 4.9).

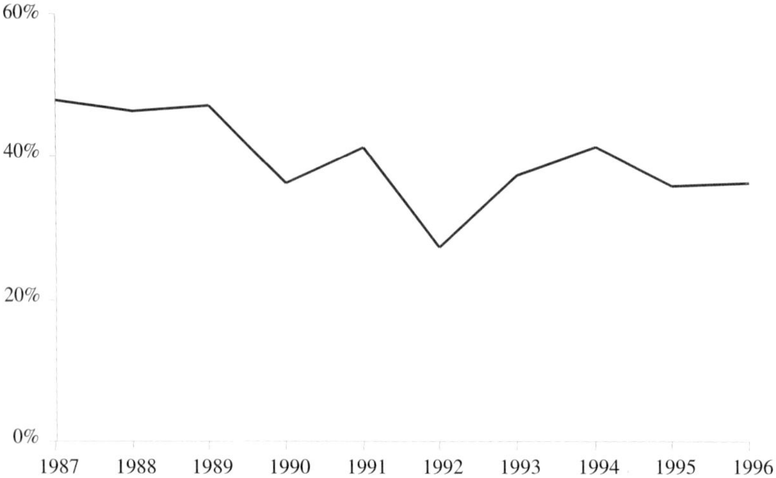

Source: SEDAPAL annual reports.

Figure 4.8. Unaccounted-for water.

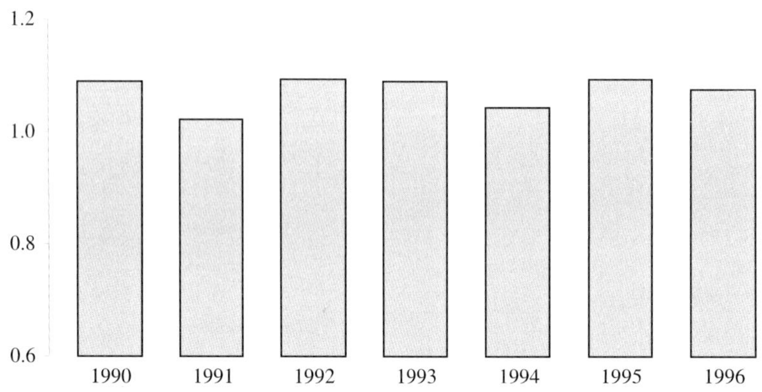

Source: Authors' calculations based on SEDAPAL annual reports.

Figure 4.9. Total factor productivity.

Methodology for Calculating the Welfare Effects

What would have been the net benefits to Peru had the concession with a private operator been signed and fully implemented? Using a partial equilibrium cost–benefit methodology developed by Jones et al. (1990), we compared the actual reform outcome with a counterfactual projected on the basis of the draft concession contract (CEPRI-SEDEPAL 1994).[13] We assumed that the private operator would invest to meet all of the minimum

[13] We projected performance under the concession for ten years (from 1995 to 2004), which required us also to project the actual case for eight years beyond our data, and compared the results under the concession with what actually occurred.

requirements in the concession, using retained earnings and borrowing. According to the concession document, the winning bidder would be responsible for all financial risk and capital expenditures as well as take over all the company's liabilities, in return for all profits from operating the concession for 30 years. The concession also required the company to increase the company's paid in capital by US $100 million in the first five years, increase water service from 15 to 24 h per day in five years, expand market coverage from 75% to 95% in ten years, and raise metered connections from 38% to 95% of total connections in ten years. Unaccounted-for water was required to drop from 36 to 25%.[14]

It is doubtless unrealistic to assume that the winner bidder will meet all the terms of the concession. However, without any reasonable basis on which to adjust the concession terms, we chose to project the concession as drafted and then consider the magnitude of the gain if we discounted it by 50%.

We assumed that a private operator would be able to improve further SEDAPAL's operating efficiency, hence we projected rather conservatively that the concession operator would reduce the use of intermediate inputs by 5%. However, we expected that any further improvement in labor over and above the cuts already introduced would be unlikely. Finally, we assumed that the opportunity cost of capital would not change.

The concession was to be awarded to the bidder who offered the lowest average water tariff. The maximum allowable increase for the first year was 40% and afterwards tariffs would be increased according to a factor *K* plus inflation. *K* was set to zero for the first five years, after which *K* would vary depending on investment and efficiency improvements. For our base projections we assumed that all bidders would increase prices to cover marginal costs (using the World Bank estimate of US $0.45) or by about 38%, with no further real increases during the projection period. A bidder, if offered lower price increases, would be more likely to renegotiate the concession to reduce the investment requirements, thus reducing the net effect on our projected consumer surplus.

We compared the concession projections with SEDAPAL's actual performance, projected to 2004 based on linear trends from 1988 to 1996. Residential and nonresidential demand was estimated for both the actual and counterfactual using the following methodology. We assume that total demand (TQ) is a function of population (N) times the individual demand curve ($a - bP$), where a is the demand when the marginal price (p) is zero, such that:

$$TQ_t = N_t(a - bP)$$

The elasticity of demand is assumed to be −0.30 for metered residential customers and −0.20 for nonresidential; unmetered customers are assumed to have an elasticity of zero since their volume of consumption is unaffected by price.[15] We assume that

[14] According to the concession document the qualified international operator must retain at least 25% of newly issued shares. Employees must receive at least 5% of new shares, and at least 20% of the new shares must be issued in the stock exchange. In the base scenario, we assumed that the international corporation retains 75% of the stocks.

[15] To our knowledge, there is no study of the demand elasticity for Lima. We thus arbitrarily assigned a plausible elasticity based on the literature and the elasticity in other Latin American cities. Later we shall check the sensitivity of the welfare results to these assumptions.

an individual's demand curve remains the same under the actual and the counterfactual and that the total demand consists of all the residents of the SEDAPAL service area, in other words that the unconnected population constitutes excess demand. Given the relative low price of water and sewerage when the reforms began, this assumption seems justified. We also assume that, when facing excess demand, rationing is based on willingness to pay.[16] From these assumptions we obtain two equations and two unknowns, *a* and *b*:

$$-\varepsilon_{\mathrm{d}} = \frac{\mathrm{d}\ln TQ_{\mathrm{metered}}}{\mathrm{d}\ln P_{\mathrm{metered}}} = -bN_{\mathrm{metered}}\frac{P_{\mathrm{metered}}}{TQ_{\mathrm{metered}}}$$

$$\frac{Q_{\mathrm{metered}}}{\mathrm{coverage} \times \mathrm{continuity}} = N_{\mathrm{metered}}(a - bP)$$

Coverage is the percentage of population with access to connections and continuity is the hours that a connected resident has access to water divided by 24 h. We assume that for half of the interrupted hours the residents still have access to water from personal water storage facilities. We arbitrarily take the last year of our data set, 1996, to calculate our demand function in order to estimate the consumer surplus associated with any price and quantity pair. The demand function for unmetered individuals is then assumed to have the same *a* and *b* parameters as for metered consumers, although we treated the marginal prices differently. We then estimated consumer surplus using the methodology described in Appendix B. In calculating consumer surplus, we assume that excess demand is allocated by willingness to pay.

Welfare Effects

The total welfare effects of the concession are the sum of the differences between the actual and counterfactual projections of consumer surplus, employee welfare, government revenue, and investors' revenue. This calculation suggests that the concession would have yielded significant gains totaling US $864.03 million (net present value in 1994 dollars) for the period 1995–2004. Many of these gains would have gone to foreign investors, but if we deduct their gains, the total improvement in domestic welfare is still substantial, US $557.80 million. Foreign gains are large because we assumed that they would own 50% of the stock of the private operator, although the concession draft only requires the strategic investor to own 25%. If we relax that assumption, domestic gains would be much higher, as we show in the next section. We can get a sense of the magnitude of the gains by calculating the annual gain as a percentage of the annual sales of SEDAPAL in the last pre-privatization year (1994). Thus, the domestic gain from the concession would have been almost 40% of annual sales, in perpetuity (see Table 4.1).

Despite price increases, consumers are better off by US $251 million or about US $33 per connection annually, thanks to the increased coverage and more continuous service projected under the concession. Employees are better off because they own 5% of shares; their gains amount to about US $1579 per employee per year. The new owners

[16] Later we shall present a sensitivity test with respect to random rationing.

Table 4.1
The distribution of welfare changes: base scenario

	NPV of cumulative welfare gain 1995–2004 (1994 US dollars, millions)	*Annual welfare gain/1994 sales (%)*	*Annual welfare gain per group (1994 US dollars)*
Government	12.45	0.9	–
Employees	30.02	2.1	1579.32 per employee
Domestic investors	264.20	18.6	–
Foreign investors	306.23	21.5	–
Consumers	251.12	17.7	32.92 per connection
Total domestic welfare	557.80	39.2	8.46 per capita
Total welfare	864.03	60.7	–

Source: authors' calculations.

of the company, foreign and domestic, are the biggest winners, while government is only better off by about US $12.45 million. This is because under the concession it would lose the quasi-rents that it received as owner of SEDAPAL, which would now accrue to the investors as the return on their capital. This reflects the fact that to encourage large new investment under private operation government can no longer confiscate these quasi-rents.

This calculation underestimates the benefits because it is not possible for us to include externalities or the direct social benefits to consumers. If indeed the concessions brought water and sewerage to 95% of Lima in 10 years, the health improvements and savings in time queuing would be considerable. The rapid increase in prices and metering would have reduced demand and the UFW targets would have brought down losses, conserving a scarce resource. Better sewerage coverage would allow the city to safely recharge the aquifer with rainwater.

Sensitivity Tests

To test the sensitivity of the results to alternative assumptions, we experimented with changing one parameter and kept the rest of the base scenario the same. In this way we tested our assumptions about how water demand was satisfied, price elasticity, price increases, foreign ownership shares and efficiency of private operation. As reported in Table 4.2, the total gain changes very little except when we assume that excess demand is rationed randomly instead of by willingness to pay. If the pool of consumers who were without a connection at the time of the reform included many people who were willing to pay a lot to be connected, then consumer surplus and total welfare gains would be much higher. Random rationing is not a plausible assumption for Lima, however, since we know that most of the unconnected population was poor and that wealthy people who were not connected had their own sources of water.

As Table 2 shows, changing assumptions about the initial price increase and elasticities change the consumer surplus only slightly. Changing assumptions about the

Table 4.2

Sensitivity analysis (annualized net present value of welfare gains in millions of 1994 US dollars as percentage of 1994 sales)

	Government	*Investors*		*Workers*	*Consumers*	*Total domestic welfare*	*Total welfare*
		Foreign	*Domestic*				
Base	0.88	21.53	18.57	2.11	17.65	39.21	60.74
Random rationing	0.88	21.53	18.57	2.11	31.53	53.09	74.62
Elasticity of 5% more	0.88	21.53	18.57	2.11	16.73	38.29	59.82
Elasticity of 5% less	0.87	21.53	18.57	2.11	19.02	40.58	62.11
Price increase 40%	2.10	21.92	18.91	2.15	17.04	40.19	62.11
25% foreign ownership	0.88	10.55	29.55	2.11	17.65	50.19	60.74
75% foreign ownership	0.88	31.66	8.44	2.11	17.65	29.08	60.74
Equal cost of international inputs	−1.99	20.64	17.81	2.02	17.65	35.50	56.14

Source: authors' calculations.

ownership of the company has large effects only on the distribution of gains between foreign and domestic owners of the private operating company. Not surprisingly, if the private operator is not more efficient in using intermediate inputs than the publicly operated SEDAPAL, then welfare will decrease, although the magnitude is not large because our base assumption was conservative.

Welfare Effects of Partial Reforms

Using the same methodology, we estimated the welfare gains from the reforms that actually occurred compared to no reform at all. Since the quality of the data for 1990–1991 was poor because of hyperinflation, we relied largely on data from 1987 to 1989 to predict the behavior of SEDAPAL without reform to 1996, our last year of observations for the factual. We assumed that SEDAPAL would have followed the trend from 1987 to 1989 without the reforms.

Table 4.3 reports the welfare gains of reforms as a percent of SEDAPAL's sales in 1992, the base year in which reforms started, and in 1994, to allow comparison with the gains from the concession shown in Table 4.1. The reforms led to total welfare gains of about US $40 million (net present value in 1994 US dollars), or an annual gain of 7% of SEDAPAL's 1994 sales in perpetuity. Government gains by some US $38 million, thanks to the increase in quasi-rents with higher prices. Consumers gained by US $54 million, which reflects the increase in connections net of the effect of higher prices. Workers were the only losers, worse off by about US $52 million, because of the effects of forced early retirements.

These calculations suggest that the welfare effects from the partial reforms are much lower than the gains under the concession. Consumer surplus is almost four and a half time larger with the concession. This suggests that even if only half the concession's goals had been attained, consumers would still be two and a half time better off than under the actual reforms.

How unrealistic were the concession's goals? We have raised some doubts about Peru's regulatory credibility, but that did not prevent investment in other infrastructure.

Table 4.3

Welfare effects of the 1992 reforms (net present values in millions of 1994 US dollars and as a percent of 1994 sales)

1992 Reform	*Government*	*Investors*		*Workers*	*Consumers*	*Total domestic welfare*	*Total welfare* [a]
		Foreign	*Domestic*				
Cumulative welfare gain 1993–96 (millions of 1994 dollars)	38.1	0.0	0.0	−52.2	54.5	40.5	40.5
Annual welfare gain/ 1994 sales (percent)	6.7	0.0	0.0	−9.1	9.5	7.1	7.1

[a] Total domestic welfare is the same as total welfare since there were no foreign players.
Source: authors' calculations.

Moreover, Lima's concession tender attracted three bidders who invested an estimated US $1 million each to prequalify. One serious problem was affordability. The fact that the price of a house in the slums would increase fourfold if it had a water connection suggests that people would have been willing to pay a great deal more for piped water if they could finance the connection. Connection charges, however, would probably not have been affordable to many poor people without a better form of financing (better than the efforts made under the FONAVI program). If the experience of Buenos Aires is any indication, the operator might have pushed government to make new connections affordable to more people through cross subsidies or through the provision of subsidized credit to the poor. Another serious problem was scarcity. Even assuming reduced consumption through metering and better maintenance, and investment in storage and sources of supply, the water scarcity is such that the provision of water without interruption to all Lima's population in the medium term is likely to be very high.

The partial reforms are better for government than a concession because government can retain the quasi-rents from higher prices, but the consumers lose because of much lower investment. In other words, under a concession the higher prices give the private investors a return on their capital which they invest to expand and maintain the system, while under the partial reforms that actually took place, prices go up less quickly and government retains some of the quasi-rents and invests less.

POLITICAL CIRCUMSTANCES LEADING TO REFORM

Although the water system was in a state of collapse by 1990, that crisis was not a sufficient condition for reform. The circumstances that precipitated change were political. Politics also explain why the utility was not privatized. As we discuss below, reform occurs only when three necessary conditions are met: (i) the reform is politically desirable, i.e., the political benefits to the president or other reformers outweigh the political costs; (ii) the reform is politically feasible, i.e., those favoring reform control the levels of power (such as the legislature and the judiciary) and opposition to the reform cannot prevent its implementation; (iii) the reform is credible, i.e., the promises the government makes to implement and sustain the reform, to reward the winners and compensate the losers, are credible to investors, opponents and other groups who might otherwise derail the reform by refusing to go along (World Bank, 1995a). As we shall show, the SEDAPAL concession initially met all three conditions, but a change in circumstances changed the political equation.

The Failure of Traditional Politics

To understand the politics of water privatization, we must first understand the circumstances that brought President Fujimori to power in 1990 and made reform of Lima's water system a possibility. Throughout the 1960s and 70s Peru was governed by a succession of authoritarian and democratic governments pursuing generally populist policies. A 1968 coup installed a military government that greatly increased the state's role in the economy and the number of state enterprises (from 18 to 174). As the eco-

Table 4.4

Attitudes towards market reforms in Lima (percentage in agreement)

	April 1989	*July 1990*	*August 1991*	*April 1993*
Foreign investment should be promoted	71	87	84	76
Private firms are favorable for the country	63	70	80	73
The market-oriented economy is the most convenient for the country	54	58	72	58
The state should give productive activities to the private sector	51	56	59	57
Most state-owned enterprises should be privatized	49	48	58	51

Source: Opinion Reports — Apoyo.

nomic situation deteriorated under their rule, the military decided to restore elections, and introduced a new constitution in 1978.

Although the two democratically elected presidents who followed came from different political backgrounds, both regimes were characterized by increased state spending, rule by decree, economic decline and corruption. President Alan García, who ruled from 1985 to 1990, froze the prices of many commodities including water, unilaterally reduced payments on Peru's foreign debt to 10% of obligations, and increased deficit spending, contributing to a deep financial crisis and hyperinflation.[17] In addition, rising terrorism contributed to a sense of widespread insecurity and growing disillusionment with government. By March 1989 only 23% of those polled in Lima reported that they trusted the President, 20% trusted the legislature or the judiciary and 17% trusted the political parties, compared to 42% who reported they trusted the military (see Table 4.5).

The economic crisis and hyperinflation also led most people to prefer a more market-oriented economy to the state-directed economy of the last three decades (Table 4.4). A poll in August 1989 found that 50% of the population in Lima agreed that state enterprises should be privatized (84% of those in the highest income group compared to only 32% in the lowest; Alvarez Rodrich, 1991).

The 1990 election was held in this context of crisis and distrust of traditional political organizations. Initially the independent candidate, Mario Vargas Llosa, with strong backing from the business community and right wing parties, won public support for his program of radical economic reform and privatization. However, he was ultimately defeated in a runoff by a relatively unknown independent, Alberto Fujimori, who had few ties to the political elite and promised vague, heterodox policies. Fujimori won the support of the lower middle and lowest income groups, partly because they feared that Vargas Llosa would institute an economic austerity program that would injure them, but also because they identified Vargas Llosa with the traditional 'rich' and 'white' political forces.

[17] Between 1987 and 1990 GDP decreased 23.4% in real terms while inflation reached 2775% in 1989.

Lima was an important source of Fujimori's support. In the first round of the election he won 33.5% of the votes in Lima, compared to 29.1% in the entire country; in the runoff he won 65.2% of Lima compared to 62.5% in all of Peru. Polls suggest that within Lima, poorer voters were his strongest base. A poll in July 1990, shortly after the election in April, found that among those with average monthly family incomes between US $200 and US $240 who comprise about 43% of Lima's households, Fujimori's rate of approval was 72% compared to only 32% approval among those whose average monthly family income was between US $3200 and US $5000, who make up the top 4.5% of households.[18]

Despite this support base, Fujimori proceeded to implement a program of radical reforms very similar to that of his opponent for reasons we will explain in the next section.

Political Conditions Favoring Reform

The economic crisis, hyperinflation and terrorism had reached a point where the new President had little choice but to institute change. By moving rapidly and decisively, he might be able to turn around the economic situation, defeat the terrorists, and win back his support group. Fujimori's decision to implement a reform program that was not favored by his constituent base was also based on the possibility that radical reforms could win the support of three important veto players. These veto players who could help him retain power and provide financing that he could use to regain support of his core constituency later were: the military, Peruvian private investors, foreign investors and financial institutions, and both private and bilateral and international aid agencies.

Shortly after taking office, President Fujimori radically reduced tariffs, subsidies and price controls. For example, the price of gasoline was increased by a factor of 30 in one day. He also drastically cut government spending and introduced restrictive fiscal and monetary policies. He began to redefine the role of the state in the economy by privatizing state enterprises, reducing regulation of financial and labor markets, and introducing monetary and tax adjustments. He also moved to reduce terrorism and capture the terrorist leaders.[19] He instituted these changes by decree, ignoring legislative and judicial obstacles and bypassing the traditional ministries and strengthening the Ministry of the Presidency (MIPRE). Tensions with the legislative and judiciary rose throughout this period and these two branches formed the principal opposition to the reform program.

This approach paid off both economically and politically. Economically the most important change was a rapid drop in inflation, from a monthly rate of 398% in August 1990 to 4% by the end of 1992. The decline in GDP was halted and then reversed; GDP grew by over 6% in 1993 and over 13% in 1994.[20] The reforms had negative effects on

---

[18] The Fujimori administration's approval rate for the two household income groups between these two extremes was about 50% (from *Apoyo Opinion Reports*, several issues).

[19] Culminating in the capture of Abimael Guzman, top leader of the 'Shining Path' or *Sendero Luminoso*, in September 1992.

[20] The GDP growth rates were: 1987 8.5%; 1988 −8.4%; 1989 −11.7%; 1990 −5.4%; 1991 7.0%; 1992 −1.8%; 1993 6.4%; 1994 13.1%.

Table 4.5

Do you trust the following institutions? (percentage of trust in Lima)

Institution	*1989 (March)*	*1990 (Sept.)*	*1991 (Sept.)*	*1992 (Sept.)*	*1993 (Sept.)*	*1994 (Sept.)*	*1995 (Sept.)*	*1996 (Sept.)*
Private firms	–	63	–	53	63	65	64	49
Military forces	42	58	47	57	66	59	68	55
President	23	57	26	54	60	63	72	56
Council of Ministers	–	–	24	37	43	44	48	43
Senate	20	47	19	–	44	42	52	41
Unions	–	35	–	20	32	32	35	34
Judiciary	20	23	22	28	34	33	36	39
Political parties	17	21	13	13	11	14	19	18

Source: Opinion Reports — Apoyo.

the poor, however. From 1990 to 1992 the number of people living in poverty increased from 9 to 14 million and real wages dropped by 40%.

Notwithstanding the adverse consequences for his poorer supporters, the political effects were strongly positive. After an initial drop in popularity, public approval of President Fujimori rose – from 30% shortly after the election to 60% in November of 1991 (Apoyo Opinion Reports). He won the support of the military, which in April 1992 collaborated when he dismissed the Congress and the Judiciary in a so-called *autogolpe*, or self-inflicted coup d'état. The *autogolpe* consolidated the President's power and added to his popularity. It was supported by 80% of the population according to polls (Apoyo Opinion Reports). The capture of the leader of the terrorist organization 'Shining Path' (Abimael Guzman) and a decline in terrorist activities further enhanced the new regime's political strength. Trust in the presidency, as an institution was very high by 1992, as was trust in Fujimori's allies: private firms and the military. Trust in the President's main opponent organizations, the traditional political parties, the judiciary, unions and, to a lesser extent, the Senate, continued being low according to surveys in Lima; see Table 4.5.[21]

Investors' perception also began to change; for example, studies comparing Peru's business environment with that of 30 other developing countries ranked Peru last in 1989; it had risen to third rank by 1991.[22] Another sign of growing confidence in Peru's economy was an improvement in the face value of Peru's external debt trading in world capital markets. It rose from 4% of face value in July 1990 to over 17% by July 1992 and 69% in July 1993. The government also won the support of bilateral aid agencies

[21] Other signs of weakness of the unions were the drop in union membership from one third of the labor force in the seventies to 12% at the beginning of the nineties, and a drop in strikes (man hours lost to strikes fell from 38 million in 1983 to 15 million in 1990).

[22] Study by the Futures Group updated for 1991 by Apoyo. The ranking was based on trade liberalization, exchange rate policies, liberalization of financial markets, prices and salaries, fiscal and monetary policies, privatization, attitude towards foreign investment, information, property rights, government management, legal and accounting systems, and infrastructure.

(notably Japan) which provided funds to restore Peru's credit standing with the World Bank and the IMF.

The Political Conditions for the Privatization of SEDAPAL

Fujimori's strategy, as we just described, was to implement a very rapid and decisive reform that would concentrate the negative effects on the poorer population in the early part of the new administration's term in office ('honeymoon period'), and later in his period in office restore growth and support. Thus the administration began quickly to implement its plan to privatize more than 200 state-owned enterprises (SOEs) in two and a half years, i.e., by mid-1994. Included in this plan was a concession to operate the state-owned water utility, SEDAPAL.[23]

Privatization was critical to the success of the entire reform program. It would improve the fiscal situation by generating capital, including much needed foreign exchange, and reducing transfers to SOEs that were hemorrhaging funds — the SOE deficit was 0.5% of GDP in 1991. It would also attract funds for badly needed investments in key infrastructure services that the government could not afford to finance. Finally, privatization would help restore Peru's credibility with investors and foreign financial institutions. Because privatization is harder to reverse than less structural changes, such as reductions in price controls or tariffs, it would help signal investors that the Fujimori government was indeed committed to sustained reform (World Bank, 1995a). The government's strong commitment to privatization was evident through this period. The government's plan included more than 200 SOEs. By the end of 1994 over US $3 billion in revenues had been raised from privatization, over US $7 billion by April 1997 (see Appendix A, Table 4A.5).

For all these reasons, private participation in SEDAPAL was politically desirable at that time, at least to the market-oriented technocrats who were responsible for designing the concession transaction. The concession was a logical component of a sweeping privatization program designed to restore business confidence and remove bottlenecks to modernization. Additional funds would allow SEDAPAL to improve service, which would especially benefit the lower- and middle-income consumers who lived in communities that suffered the most frequent interruptions, and who could least afford to build private cisterns for water storage. Private operation would also remove a fiscal drain; government transfers to SEDAPAL in 1991 were US $12.2 million. However, political desirability was less strong in the top ranks of the administration, for reasons we describe below.

The reform was also politically feasible, in the sense that the government would

[23] The privatization process officially began in September 1991 with decree No. 674 (law for promotion of private investment) which created the body in charge of privatization, COPRI (Comisión de Promoción de la Inversión Privada or Commission for Promotion of Private Investment). Also in 1991 decree No. 662 guaranteed equal treatment for domestic and foreign investment and the right of foreign investors to repatriate their profits and dividends. In June of 1992 the government officially announced the inclusion of SEDAPAL in the privatization program (Resolución Suprema No. 349-92-PCM) and created a special committee, or CEPRI, to oversee the process (CEPRI-SEDAPAL).

be able to overcome opposition and implement the concession. The *autogolpe* had left Fujimori in complete control of the apparatus of the state with the backing of the military and most of the public. The main political opponents of SEDAPAL's privatization, the workers, were not veto players. High unemployment and the growth of the service and informal sectors had reduced union membership from one third of the workforce in the 1970's to only 12% by the early 1990's. As we saw in Table 4.5 unions had low public support, in part because they were associated with the discredited traditional political parties. Moreover, they were not constituents of Fujimori. Polls also suggest strong public support for personnel reductions in SOEs: in January 1991, 75% of those surveyed considered a reduction in the SOE labor force necessary (82% of high-income respondents and 71% of low-income; Alvarez Rodrich, 1991). Furthermore, compensation packages from US $5000 to US $6000 were promised to redundant personnel in SEDAPAL, which helped reduce opposition to layoffs.[24]

The concession was also credible, in the sense that potential investors had reason to believe that government would keep its promises to abide by the agreement. Government had carried through the rest of its reform program as promised, including selling large and politically important SOEs such as Hierro Perú, Compañía Peruana de Teléfonos and Edelnor and Edegel. One sign of the government's credibility was that international investors had been willing to sink large amounts in capital in sales of, for example, telecommunications and electricity. Another sign of the administration's credibility was the further drop in the discount on its foreign debt instruments; by the end of 1995 Peru's external debt was trading at 71% of its nominal value. It is possible that investors might not have seen government's commitment to raise prices as fully credible in a sector as politically sensitive as water, especially given adverse public reaction to tariff increases in other services. However, they probably expected that they could renegotiate their investment obligations under the concession down to levels that would allow adequate returns (as has occurred in concessions in Argentina and elsewhere).

Accordingly, based on recommendations of a consultant's study, the government decided on a thirty-year concession with a required minimum investment program of US $3 billion. The concession would be awarded to the bidder offering to meet all the quality and coverage targets in the contract at the lowest retail tariff. In November 1994 three international consortiums prequalified to bid on the concession.[25] The successful groups were headed by Canal de Isabel II from Spain, Compagnie Générale des Eaux from France and Lyonnaise des Eaux from France.[26] However, the ultimate bid for the concession itself was postponed until after the 1995 elections. It was then further

[24] The average annual wage in 1991 was US $4704.

[25] Bidders had to have experience operating systems with minimum sales of US $180 million and assets with a net value above US $750 million, in urban centers of more than 4 million inhabitants.

[26] The Spanish partnership also included Argentaria (financial company) and Aguas de Barcelona (which is also a member of the Aguas Argentinas partnership). The Générale des Eaux partnership also included Thames Water (which operates much of the system for London), Mitsui from Japan, and three Peruvian companies, Banco de Crédito, Cosapi and Gremco. Lyonnaise leads the Aguas Argentinas partnership, which also includes England Company Water, Mitsubishi from Japan, Graña y Montero from Peru and Aguas de Barcelona.

delayed, and finally in November 1997 President Fujimori declared the concession postponed indefinitely.[27] (It was probably not cancelled outright only because of concerns about litigation by the foreign investors, who are reported to have spent about US $1 million each in preparing their bids.) In the next section we explain why the concession was cancelled.

Why Was the Operation of SEDAPAL not Privatized?

There are three main reasons why SEDAPAL's concession was ultimately not implemented. First and foremost, it became politically undesirable, second the window of opportunity for the sale closed, and third the most serious problems in the sector were partly alleviated. Another important factor to mention that made the concession difficult from the beginning, contributing to its postponement and to lose the window of opportunity was the lack of a 'blueprint' or model to follow. In contrast to the other cases, such as telecommunications, there did not exist an international model or 'best practice' to help the team in charge in the design of the concession and regulatory drafts.

As mentioned before, the desirability of water privatization was always weak. Water had high political saliency for Fujimori's core base of support, Lima's poor.[28] The concession would deprive the government of this ability to direct investments to areas with high political payoffs, a factor that has delayed private participation in the management of water elsewhere, in Mexico City for example.

The benefits of the concession to Fujimori's poor constituents in Lima would be further reduced by an increase in prices. The concession allowed up to a 40% price increase, which would raise the average residential tariff for water and sewerage from US $0.30 in 1994 to US $0.42 per m^3. According to World Bank calculations, the average incremental cost for a smaller expansion in the system was US $0.45 m^3 (World Bank, 1994b).[29] If this estimate is correct, any private bidder would have been strongly motivated to renegotiate the agreement to win a higher price or lower investment targets than allowed under the concessions. Although there may have been operational inefficiencies assumed in the World Bank estimate, marginal costs were indeed high in Lima because water was in scarce supply, the required expansion was large and so much of the system had deteriorated and needed replacement. New, deeper wells would have to be drilled to tap the reduced aquifer. From one third to 85% of the sewerage system

[27] In April 1996 Jorge González Izquierdo, the President of the privatization commission, COPRI, stated that government had decided to postpone the privatization of three controversial and complicated SOEs — SEDAPAL, the petroleum corporation and a large mining company — until the end of the whole process (*El Comercio*, April 1996). Later he stated that the government had decided to keep both SEDAPAL's property and operation public because " . . . we think that water is something vital and there are some problems in this activity that the government must solve . . . " (*Semana Económica*, No. 578, June 29, 1997).

[28] It is also noteworthy that Lima was not designated a region in the 1993 Constitution enacted by the Fujimori administration, which kept SEDAPAL under the direction of MIPRE rather than transferring it to the municipalities or provinces when responsibility for the other water companies was moved.

[29] Based on a project to bring new connections to over 300,000 persons and improve service by another 300,000 while increasing water supply by 5.2 m^3 by reducing water losses through rehabilitation and consumption (through metering).

needed to be replaced and half of the over 300 operating wells had diminished yields because of aquifer depletion or pumps beyond their design life.

A private operator would probably have been able to reduce SEDAPAL's operating costs by improving efficiency, but the effects on tariffs would be limited since by far the biggest part of costs would be investment. For example, SEDAPAL's operating costs before interest, depreciation and taxes averaged US $70 million a year from 1990 to 1993, average annual projected investment during the first three years of the concession would have been US $476 millions.[30]

The higher tariffs combined with connection charges would make water unaffordable to many unconnected poor consumers, even compared to water from vendors.[31] Poorer and middle-income consumers who were already connected would have faced higher bills, not only because of rate increases but also with the introduction of metering and better bill collection. SEDAPAL charged unmetered, low-income consumers on the assumption that they used 22 m^3 a month, but a very large, poor household could be using more.

None of this means that poor consumers would not have been willing to pay more for a connection; a SEDAPAL survey in 1993 found that the average property value of houses in the *pueblos jóvenes* quadrupled after being connected to piped water (from US $1542 to US $6375). Moreover, poor consumers already paid high prices for water from vendors; a survey in December 1993 found that the average price was US $2.50 per m^3.

The cost to the poor could have been made affordable through subsidies, by including the connection costs in the variable water tariff charged to all consumers, or by extending longer-term, lower-interest loans for connections.[32] However, none of these arrangements were contemplated in the concession agreement, and would not have been easily added. Any efficient provider, public or private, would have wanted to reduce the already large cross subsidies in Lima's tariffs. Industrial tariffs were 130% larger than residential. An efficient operator would have also wanted to raise metering and change per meter prices closer to marginal cost, not only to assure a reasonable return on investment, but also to keep the demand for water down so as to reduce the

---

[30] US $100 million a year or US $3 billion over 30 years to increase coverage and US $59 million a year or US $1.8 billion in 30 years to improve the continuity of service (Black and Veatch, 1992).

[31] SEDAPAL assumed a minimum water consumption level of 22 m^3 a month for low-income households for billing purposes. At 1993 tariffs (US $0.22 per m^3 for domestic users) this amounted to $4.92 a month or 2.42.1 to 2.92.5% of the household income for the 43% of Lima's households in the lowest-income category (US $200–240 a month). Connection costs in poor-income neighborhoods were about US $850 and could be paid for by a loan, with a five year repayment period and monthly interest rate of 1.2%, which meant that water and sewerage charges would be 16% of income for a new connection. Raising tariffs to cover marginal cost would more than double rates. Without any comparable increase in income or some sort of subsidy, the poorest households would pay around 5% of their income for water and sewerage, or over 18% including the connection charge.

[32] Loans were in fact offered to water companies and consumers to finance construction of primary networks and to cover investments in secondary networks and costs of connection through a government fund financed by a social tax, FONAVI. FONAVI channeled about US $1000 million to the water sector at moderately favorable interest rates, out of which 75% went to customers. However, customers failed to repay the loans. By the end of 1998, the government recognized the big problem and lunched a plan to 'recover' the FONAVI debts offering the water companies the option to issue shares in favor of the central government.

need for further costly expansion. It was therefore inevitable that poorer consumers would face substantially higher prices, as indeed occurred under public ownership later, reducing the political benefits of the concession.

Not only would price increases reduce the net benefits from the concession for lower income consumers, the size of the projected benefits from investment was open to question. Even if prices could be increased to cover marginal costs, an investor might well be wary of sinking large sums in nontransferable assets given Peru's weak institutional environment. Barthelemy (1996) reports that foreign investors had concerns about the sustainability of Peru's reforms given the concentration of power in the hands of the President and the weakness of the judiciary and other institutions. Furthermore, should demand for new connections be less than expected once prices go up, the operator might well argue for renegotiation of the investment targets. In Buenos Aires, for example, the operator renegotiated the concession contract to win larger price increases and lower investment commitments than planned partly because high connection charges proved uncollectable. The possibility that renegotiation could happen in Lima was probably not lost on the President and his advisors.

The second reason why the concession for SEDAPAL was postponed and then abandoned was a problem of timing. Privatization of water had lower priority than other sectors. Although the proposed SEDAPAL concession would remove a fiscal drain and generate new investment, it would not bring any sales proceeds or other revenues to the government. In contrast, the sale of electricity, telecommunications or mining would attract investment and also generate substantial funds for the Treasury. For example, the sale of 35% of the telecommunications company in February 1994 raised US $1.4 billion for the Treasury and US $600 million of new investment. This initial lower priority of SEDAPAL's privatization worsened by the difficulties faced by the privatization process in general and by the SEDAPAL's bid process in particular. In addition, as was mentioned above, there were significant technical difficulties. It was more difficult to prepare the privatization of SEDAPAL than the others because of the lack of international experiences and of an appropriate model to follow.

Notwithstanding the government's best efforts, it proved impossible to privatize even the top priority SOEs as fast as the timetable required. By the end of 1994 only about 45% of the total value of transactions that ultimately would occur (by April 1997), had occurred (Table 4.6). These other higher-priority privatizations delayed work on the SEDAPAL concession which only began in 1993. Also, its relatively weaker political desirability was reflected in a weak team responsible for the SEDAPAL concession (CEPRI-SEDAPAL). This group had lower incentives to privatize than the others and faced strong opposition from some politicians close to the President and to SEDAPAL. Because the concession would affect an important constituency of the President, considerable attention was paid to the design of the legal and regulatory framework by top officials in the Ministry of the Presidency (MIPRE) who were close to Fujimori. This further delayed the process, and, when the MIPRE officials refused to approve the bidding and concession documents, ultimately led to the resignation of the CEPRI-SEDAPAL team responsible for the privatization in April 1995. As a result of the delays in the privatization of SEDAPAL's operation fell perilously close to the 1995

Table 4.6

Proceeds and investment from privatization transactions in Peru (in million US dollars)

Year	*Sale of shares*	*Concession*	*Transfer option*	*Other revenue*	*Total proceeds*	*Projected investment by buyer*
1991	2.59	–	–	–	2.59	–
1992	207.53	–	–	1.39	208.92	706.00
1993	316.70	20.70	–	6.49	323.89	589.34
1994	2648.57	–	2.21	2.52	2653.30	3137.55
1995	1089.01	48.76	1.18	6.45	1145.40	216.07
1996	2291.81	344.20	1.00	5.79	2642.80	2852.20
1997 [a]	67.13	–	–	1.32	68.45	218.19
Total	6623.34	413.66	4.39	23.96	7065.35	7719.35

[a] Till April.
Source: COPRI data.

Table 4.7

Rate of approval of privatization by socioeconomic level in Lima (percent)

Level (average monthly household income)	*Sept. 1991*	*Feb. 1992*	*Dec. 1994*	*Dec. 1993*	*Dec. 1995*	*Dec. 1996*	*Dec. 1997*
Total	52	59		48	49	42	28
High (US $3200–5000 [a])	87	83	84	80	77	65	67
High middle (US $750–1150 [b])	65	81	73	64	66	46	42
Low middle (US $330–470 [c])	55	63	58	45	43	39	25
Low (US $200–240 [d])	37	42	49	36	42	40	19

[a] Approximately 4.5% of total households in Lima.
[b] Approximately 19.5% of total households in Lima.
[c] Approximately 33% of total households in Lima.
[d] Approximately 43% of total households in Lima.
Source: Opinion Reports — Apoyo (income ranges from July 1997).

election. This was politically risky; a concession would sharply increase prices, an effect that would ran counter to Fujimori's need to win the support of his core constituency in Lima. Hence the concession lender was postponed.

After the elections the concession became less politically desirable because of growing public disillusionment with privatization (Table 4.7). Although a new SEDAPAL privatization team was appointed in December 1995, government's interest in the concession was much weaker after the elections (field interviews). As Table 4.4 showed, support for privatization in general peaked among all income groups in 1993 and fell off gradually, then sharply in 1997. The decline in support stemmed from a perception that the benefits were concentrated in a small group of foreign investors and wealthy Peruvians, plus sharp increases in the prices of services, such as telephone and electricity rates. The SEDAPAL concession would also have combined large rate increases and higher returns to investors in the short run, with most consumer benefits coming in the medium term as major investments are completed. Support for privatization of SEDAPAL was never as strong as for other SOEs and fell sharply over time. In 1989, 47% of those surveyed in Lima supported transferring SEDAPAL to the private sector; by 1997 this was down to 22% (Apoyo Opinion Reports).

Furthermore, support for Fujimori was sharply down by late 1997 (Appendix A, Table 4A.1) and it became even riskier to chance alienating his core supporters, especially when he was trying to change the constitution so he could run for President again in 2000.

Since 1997, although several privatization transactions were implemented, they severely lost public approval (63% of the population was against them in October of 1999) and partly government support, making even more difficult to complete the concession of SEDAPAL. However, the results of the privatization reform proved a success. Among many other benefits from the process, more than US $8000 million were obtained from privatization transactions (1991–1999); two thirds of foreign investment came from privatizations; the government received US $6391; privatized firms substantially increased internal production and exports; the Peruvian economy became more competitive (4th in Latin America); and 410,000 Peruvian citizens and 16,000 workers became shareholders of privatized firms (Apoyo Comunicaciones, 1999). Particularly noteworthy are the results regarding the coverage, quality, and prices of utilities. A 100% coverage was achieved in Lima with the privatization of the electric utilities and almost 50% in the case of telephones. Furthermore, from 1990 to 1998, water tariffs increased in real terms more than their privatized counterparts: electricity and telecommunications (17.6% increment of water tariffs versus −34.6% and −23.1 in the cases of electricity and telephones, respectively).

The third reason why the SEDAPAL concession was not implemented was that most of the quick gains from reducing the water sector problems were achieved without it. The World Bank and other aid agencies put together a US $600 million financing package to rehabilitate the distribution network for water and sewerage in Lima, expand connections in the poorest areas, install metering in high-consumption areas, and improve SEDAPAL's efficiency and the legal and regulatory framework for the water system, all within five years. Although the package was conditioned to the implementation of many changes to make privatization possible and contemplated the concession of SEDAPAL, this condition was revised to adjust for the changes in the political climate.

The aid-financed program of investment and reforms reduced the pressure for privatization because it alleviated some of the worse problems of the system. It was combined with a program of layoffs and management improvements in SEDAPAL that eliminated SEDAPAL's losses and improved operations. The reforms expanded service in the *pueblos jóvenes* for lower price increases than might have occurred with a concession since recovery of capital costs was not necessary. Although the net benefits from a concession would have been much greater, according to our above calculations, they would only come in the medium term, after major investment works were completed.

CONCLUSION

Although Lima's water system was in near crisis, that was not a sufficient condition for radical change. Partial reforms, carried out under public management partly due to the credible threat of privatization, reduced many of the problems but left a quarter of citizens without access to water or sewerage connections, supplied a service subject to frequent and extended interruptions, and wasted more than a third of the water in circumstances of very scarce supply.

The fundamental reason for Lima's failure to implement the concession was geographical — the scarcity of water sources meant that marginal costs were high, requiring pumping water from deep wells and building adequate storage for dry periods. High extraction costs were compounded by years of neglect, so that much of the system needed to be replaced. To attract a private investor, prices would have to be set to recover these high costs, including a reasonable return on capital. Since government had kept prices well below cost recovery for years, a concession would have required a sharp and sudden price increase to cover marginal costs. Not only that, but any forward looking investor would want to reduce the pace of future investment, by curbing demand through metering and more effective bill collection based on meters. This logic would also argue for reduced cross subsidies, since these reduce the incentive to conserve water.

Although the ultimate cause of the concession's failure was geographical, the proximate cause was political. Privatization of any utility is politically tricky if it involves higher prices and controversy over ceding monopoly powers to private parties, especially foreigners. Private participation in water suffers from all these problems, magnified by the social importance of water as an essential good and by the lack of international experience and thus technical difficulties to design a privatization reform in the sector. At the same time water offers fewer offsetting benefits compared to other utilities because it raises less or no revenues that can be used to reward supporters or compensate losers. In Lima's case the benefits were further reduced by the likelihood that the price increases would especially affect the urban poor who were important to the President's support base. Thus, despite initially favorable conditions, the political equation shifted against private operation of SEDAPAL as elections neared.

The failure to introduce the concession was costly for Lima, and not just because the welfare gains from the concession, even if the access goals were not fully met, would have been larger than the gains from the reform. The cost also included the adverse effects on health in a city where over 1.7 million people are not connected to water or sewerage and those who are connected suffer interruptions in water service that last for months at a time. In addition, the failure to curb consumption through metering has also meant that the aquifer continued to be depleted much faster than it was replenished, especially as the number of connections grew, and hence was increasingly contaminated by salt from the ocean.

Although the reforms introduced some important improvements in the operation of SEDAPAL, especially in reducing labor costs, these too were far short of what a private operator might be expected to implement. Our assumptions about efficiency gains under the concession projections were very conservative; it would be more realistic to assume that a profit maximizing operator facing price cap regulation would sharply curb unaccounted-for water, and try to minimize the consumption of electricity and other inputs.[33]

Our assumptions about the benefits from the concession do not take into account Peru's regulatory weaknesses, however. The improvements introduced since 1992 have been very partial and left the regulator vulnerable to either political manipulation or capture. Had the concession been signed, the government might have had greater motivation to protect the regulator from capture and the investor would had an incentive to push for safeguards against politically motivated regulatory changes. Peru's institutional weaknesses, and in particular its lack of an autonomous judiciary, would limit how much could have been achieved. Nevertheless, considering the net gain from private operation in much weaker institutional settings in Africa, it seems safe to conclude that Lima would have been better off with a concession.

ANNEX 4A: STATISTICAL APPENDIX

Table 4A.1.

Rates of approval in Lima of Fujimori as president of the country (percentages)

Month/Year	*1990*	*1991*	*1992*	*1993*	*1994*	*1995*	*1996*	*1997*
January		43	65	60	66	64	75	47
February		35	64	62	58	74	66	46
March		38	53	61	61	66	64	67 [a]
April		49	81	63	60	75	62	48
May		45	76	59	61	80	59	47
June		35	76	66	64	76	65	34
July		31	65	61	65	68	58	23 [b]
August	46	39	62	65	68	77	62	34
September	51	32	74	63	68	78	60	39
October	53	54	68	70	65	76	55	34
November	59	58	65	67	63	70	52	37
December	61	60	64	64	67	73	45	34

[a] After the release of the hostages of the Japanese Embassy.
[b] After accusations of telephone conversation interventions.
Source: Apoyo Opinion Reports.

[33] That there was scope for this is supported by the fact that the cost of intermediate inputs for Santiago's water and sewerage company were half those of SEDAPAL in 1996 (when Santiago had about 1 million connections compared to 823 thousand in Lima).

Table 4A.2

Metering of connections in 1988 (percentage distribution)

Type of Connection	*Social*	*Residential*	*Commercial*	*Industrial*	*State*	*Total*
Unmetered	72.4	71.2	62.3	51.6	74.8	70.5
Metered:	27.6	28.8	37.7	48.4	25.2	29.5
Charged according to meter-reading	9.4	16.2	13.6	29.3	12.2	16.1
Total	100.0	100.0	100.0	100.0	100.0	100.0

Source: SEDAPAL Gerencia de Planeacion y Presupuesto, "Analysis of Demand in Lima and El Callao."

Table 4A.3

Trends in connections, metered connections and meter reading

Year	*Connections (1000s)*	*Metered connections (%)*	*Connections read (%)*
1987	559.44	56	Not applicable
1988	607.40	60	Not applicable
1989	626.59	59	Not applicable
1990	654.14	57	11
1991	678.54	54	9
1992	700.16	52	7
1993	732.26	50	4
1994	762.93	47	5
1995	790.69	45	6
1996	823.16	39	8
1997	839.34	39	16
1998	871.72	40	32

Source: SEDAPAL data.

Table 4A.4

Estimated consumption per capita in Lima (liters per day)

District	*Consumption*	*District*	*Consumption*
SMP	172	Lima Cercado	372
Rimac	229	La Victoria	231
La Molina	560	San Miguel	235
San Luis	239	Brena	261
Chorrillos	243	Pueblo Libre	272
Santiago de Surco	405	Jesus Maria	264
San Borja	369	Magdalena	325
Surquillo	234	Callao Cercado	243
Miraflores	567	Bellavista	231
Linca	324	La Perla	254
San Isidro	465	C. de la Legua	252
Barranco	307	La Punta	312

Source: World Bank, 1994b.

Table 4A.5

Principal privatized enterprises (in millions of US dollars)

Enterprise	*Date of sale*	*Shares sold (%)*	*Basis price*	*Sale price*	*Investment commitment*
Sogewiese Leasing	10/06/91	15.0	1.00	1.08	–
Hierroperú	30/10/92	100.0	22.00	120.00	150.00
Cerro Verde	10/11/93	100.0	30.00	35.44	485.30
ENTEL/CPT	28/02/94	35.0 both	535.00	2002.00	1800.00
Edelnor	12/07/94	60.0	127.72	176.49	150.00
Edelsur	12/07/94	60.0	129.42	212.12	120.00
Banco Continental	18/04/95	60.0	–	195.70	–
Egenor	25/06/96	60.0	228.20	42.00	–

Source: COPRI data.

NOTES ON THE METHODOLOGY FOR MEASURING THE WELFARE GAINS OF CONCESSION

Actual Scenario. Since SEDAPAL did not charge separately for sewerage and water, they were treated as a composite product. To distinguish the differences in behavior and pricing, we decomposed the revenue from water services into domestic, non-domestic, and social sources. For domestic and non-domestic sources, we further classified them as metered and unmetered. In 1988 only 16% of the domestic consumers were charged based upon meter-reading in 1988 (Table 4A.2); and in 1994, 7%. Although the company reported that 29% of the connections were metered, that number did not represent accurately the number of connections billed based on meter-reading. Most metered connections were actually charged either the company's estimated consumption or the actual reading, whichever was higher.

We calculated water revenues based on the quantity and the average price reported by SEDAPAL. For the period before 1994, due to data limitation, industrial discharges ware calculated based on the quantity reported by the company and the total revenue recorded by the World Bank.[34]. From 1994 on, the data used were from SEDAPAL reports. Connections and 'other revenues' were the residual, and its price was assumed to increase at the same rate as the consumer price index.[35]

When a connection was metered, the demand elasticity was assumed to be −0.30 for domestic usage, −0.20 for non-domestic usage.[36] When a connection was not metered, the elasticity was assumed to be zero. This assumption aimed to capture the idea that the prices underlying unmetered connections were represented by a fixed charge and a marginal price of zero.

[34] "Analysis of the Financial Situation of SEDAPAL (1987–2003)". Luz Marina Gonzáles.

[35] The issue of connection fee is not treated explicitly in this analysis.

[36] To our knowledge, there is no study of the demand elasticity for Lima. We thus had to somewhat arbitrarily assign a plausible from the literature about water demand. We used sensitivity analysis to check the robustness of the welfare results.

Excess demand for water estimated based on service coverage and water continuity. The percentage of people not served and the hours of interruption of service were translated into cubic meters of excess demand for water.[37] There were no specific records of discontinuity of service in Lima except by the Living Standards Measurement Survey done by the World Bank in 1991. This suggested that 48% of Peru's population received water service for less than 12 hours a day, and, another 28%, less than 6 hours a day. We also know that some districts of Lima in 1997 received no service for more than two months. To be consistent with these observations, we used 15 hours a day, which was the amount estimated by the Consortia for SEDAPAL Privatization.

Counterfactual Scenario. The bidding variable for the concession contract was the tariff. The operator could increase prices at the beginning of the contract and then adjust them every year by inflation plus *K*%. *K* must be equal to zero for the first five years and from then on, it could vary depending on investment and efficiency improvements. The bidding documents mandated that the maximum price increase for the first year would be 40%, and the percentage of price increase we assumed was 38.24%. This increase, well within the ceiling of price increase of 40%, allowed the price to rise to equal to the estimated long-term marginal cost (World Bank estimated it to be $0.45 or 0.99 soles in 1994). After the first-time increase, the price increase was assumed to be zero per year afterwards.

The bidding document also provided some minimum targets, which were used for the assumptions about the counterfactual. The service coverage was required to increase from 15 to 24 hours a day in 5 years, and the market coverage, from 75% to 95% in ten years. These conditions were directly quantified in output quantity and excess demand under the counterfactual. In addition, unaccounted-for water was required to drop from 36% to 25%, and metered connections to rise from 38% to 95% of total connections in 10 years. Finally, in terms of ownership of new issues, the qualified international operator was required to retain at least 25% of newly issued shares. Employees must have at least 5% of new shares. And at least 20% of the new shares must be issued in the stock exchange. In the base scenario, we assumed that the international corporation retains 50% of the stocks.

Although the contract would be extended for 30 years, we projected only up to the year 2004, 8 years after the end of the actual series, and 10 years after the beginning of the presumed concession. Simulation about further time series would entail too much prediction error; moreover, large discounting factors beyond the year 2002 would render this omission unimportant.

Since there was no evidence that SEDAPAL would have improved its performance at a faster pace than it had in the past years under public ownership, we projected the future for the factual scenario based on the linear trends from 1988 to 1996. For prices,

---

[37] Note that the rising tariffs did not bring about the reduction of monthly consumption per connection. While the tariff rose between 1992 and 1996, the average monthly consumption per connection increased for two years and then decreased for another two years. This suggests that the unmet demand should be treated as excess demand.

Table 4B.1

Assumptions used in the counterfactual scenario

Key elements	*Assumptions for counterfactual time series (i.e., the private enterprise scenario)*
Price of water	Increase of 38.24% in first year prices to cover long-term marginal costs. In all the following years, prices only keep up with inflation.
Quantity of water	*Continuity:* increase continuity to 24 hours a day in 5 years. *Coverage:* domestic water coverage increases to 85% in 5 years and 95% in 10 years. All other categories are assumed to increase based on the historical trend.
Investment	Under both the factual and the counterfactual, retained earnings are used for investment. Additional funding for investment is financed through loans. Capital stock was also adjusted by inflation. Paid-in capital increases by US $100 million in the first 5 years. At the beginning of the contract, it increases by US $25 million and the following 5 years by US $15 million.
Unit cost of intermediate inputs	The unit cost of intermediate inputs is assumed to drop by 5% in 1995 and 1996, then stay constant after 1996 (i.e., the projection period). This assumption is somewhat arbitrary, but in sensibility tests turns out not to be important for the welfare outcomes.
Unit cost of labor	The unit cost of labor is assumed to be the same as the actual. Since the factual already witnessed a very large reduction in labor force, there is no reason to assume that private ownership could slash the labor force further.
Opportunity cost of working capital	The same as the actual. There is no reason to assume that the market rate would change with the reform of SEDAPAL.

we used the data after 1990 when the World Bank project required that tariffs rise to cover marginal costs. Unit costs of labor and intermediate inputs for the projection periods were assumed to be the same as the actual in 1996.

The deposit and the loan rate were assumed to be equal under the public and private scenarios. The loan rate was set to the average between 7.32% for multilateral organization loans and 11% for other sources. The Consortia estimated this rate for the privatization of SEDAPAL, using the last 50 Latin American bonds issued plus a country risk of 3.5% and an industry risk of 4.5%. A summary of the main assumptions used for the counterfactual is presented in Table 4B.1.

REFERENCES

Alcázar, Lorena and Penelope Brook-Cowen, 1996. *Institutions, Politics, and Contracts: Private Sector Participation in Urban Water Supply.* Research Proposal, Washington, DC: World Bank, Processed.

Alvarez Rodrich, Augusto, 1991. *Empresas Estatales y Privatización.* Lima: Apoyo.

Apoyo Opinion Reports (monthly) several issues. Apoyo, Lima.

Barthelemy, K, 1996. "Improvement of Peru's investment climate." In: *Peru: Beyond the Reforms*. Promperu, Lima.

Bates, Robert and Anne Krueger, 1993. *Political and Economic Interaction in Economic Policy Reform*. Cambridge, MA: Blackwell.

Black and Veatch International, Binnie and partners, Information for Investment Decisions Inc., 1994. *Evaluación Técnica de los Servicios de Agua Potable y Alcantarillado para Lima Metropolitana*. Lima.

Brook-Cowen, Penelope, 1994. *The Rationale for Private Sector Participation in Water and Sanitation*. Workshop on Private Sector Participation in Water Supply, Washington, DC: World Bank, Processed.

Cotler, Julio, 1995. "Political parties and the problems of democratic consolidation in Peru". In: Scott Mainwaring and Timothy Scully, eds., *Building Democratic Institutions*. Stanford, CA: Stanford University Press.

Cox, Gary and Mathew D. McCubbins, 2000. "Structure and policy: the institutional determinants of policy outcomes." In: Stephan Haggard and Mathew McCubbins, eds., *Presidents, Politics and Policy*. Cambridge: Cambridge University Press.

Fernandez Baca, Jorge and Lorena Alcázar, 1997. *Definición de Mecanismos de Capacitación de Fondos para un Futuro Programa de Inversiones en el Sector de Saneamiento*. Lima, Processed.

Galal, Ahmed, 1994. *Regulation and Commitment in the Development of Telecommunications in Chile*. Policy Research Working Paper 1278, Washington, DC: World Bank.

Galal, Ahmed, Leroy Jones, Pankaj Tandon and Ingo Vogelsang, 1994. *Welfare Consequences of Selling Public Enterprises: An Empirical Analysis*. New York, NY: Oxford University Press.

Graham, Carol, 1996. *Raising the Stakes in the Social Services, Social Security, and Privatization in Peru*. Processed.

Graham, Carol and Moises Naim, 1998. "The political economy of institutional reform in Latin America". In: Nancy Birdsall, Carol Graham and Richard Sabot, eds., *Beyond Trade-offs: Market Reforms and Equitable Growth in Latin America*. Washington, DC: Brookings Institution and IDB.

Haggard, Stephan and Steven Webb, eds., 1994. *Voting for Reform: Democracy, Political Liberalization, and Economic Adjustment*. New York, NY: Oxford University Press.

Heller, William and Mathew D. McCubbins, 2000. "Political institutions, utility regulation, and economic development in Argentina and Chile: the case of electricity." In: Stephan Haggard and Mathew McCubbins, eds., *Presidents, Politics and Policy*. Cambridge: Cambridge University Press.

Hill, Alice and Manuel A. Abdala, 1994. *Regulation, Institutions, and Commitment: Privatization and Regulation in the Argentine Telecommunications Sector*. Policy Research Working Paper 1216, Washington, DC: World Bank.

Idelovitch, Emanuel and Klaus Ringskog, 1995. *Private Sector Participation in Water Supply and Sanitation in Latin America*. Washington, DC: World Bank, Processed.

Jones, Leroy P. Pankaj Tandon and Ingo Vogelsang, 1990. *Selling Public Enterprises: A Cost–Benefit Methodology*. Cambridge, MA: MIT Press.

Klein, Michael, 1996. *Economic Regulation of Water Companies*. Direction in Development Series, Washington, DC: World Bank. Processed.

Klein, Michael and Neil, Roger, 1995. *Back to the Future: The Potential in Infrastructure Privatization*. Silver Award Essay, Washington, DC: World Bank. Processed.

Levy, Brian and Pablo Spiller, 1994. "Regulations, institutions, and commitment in: telecommunications: a comparative analysis of five country studies." *Proceedings of the World Bank Annual Conference on Development Economics*. Washington, DC: World Bank.

Macroconsult, 1996. *Organización, Regulación e Incentivos de los Sistemas de Agua Potable y Saneamiento en el Perú*. Research Project, Washington, DC: InterAmerican Development Bank, Processed.

McCubbins, Mathew D., Roger Noll and Barry Weingast, 1989. "Structure and process, politics and policy: administrative arrangements and the political control of agencies". *Virginia Law Review* 75: 431–482.

Ministerio de la Presidencia, Vice-Ministerio de Infraestructura, 1997. *Informe del Sector de Saneamiento Básico, Período 1985–2000*. Lima.

Nelson, Joan, 1994. *A Precarious Balance: An Overview of Democracy and Economic Reforms in Eastern Europe and Latin America*. San Francisco, CA: International Center for Economic Growth and Overseas Development Council.

North, Douglass, 1990. *Institutions, Institutional Change, and Economic Performance*. New York, NY: Cambridge University Press.

Oblitas, Lidia, 1996. "Experiencia Peruana en la regulación de la prestación de servicios de Saneamiento." Paper presented at the forum *Estudio de Legislación Comparada de los Servicios de Saneamiento de Colombia, Chile y Perú*. Lima.

Ochoa, Fernando, 1997. *Política de Subsidios para los Servicios de Saneamiento en el Perú: Propuesta para Discusión*. Lima: PRONAP, Processed.

OXERA, 1996. *Assessment of the Institutional and Regulatory Framework for the Water and Sewerage Sector.* Processed.

PROMPERU, 1996. *Peru: Beyond the Reforms*. Lima.

Richard, Barbara and Thelma Triche, 1994. *Reducing Regulatory Barriers to Private Sector Participation in Latin America's Water and Sanitation Services*. Policy Research Working Paper 1322, Washington, DC: World Bank.

Stigler, George J., 1971. "The theory of economic regulation." *Bell Journal of Economics* 2: 3–21.

Superintendencia Nacional de Servicios de Saneamiento SUNASS. Various, Boletín Informativo, Lima: SUNASS.

Triche, Thelma, 1994. *Designing Water Supply and Sewerage Oversight Institutions to Fit the Local Context.* World Bank Water Week Report, Washington, DC: World Bank, Processed.

Webb, Sheila and Associates, 1992. "Waterborne diseases in Peru." Background paper for *World Development Report 1992*, The World Bank.

World Bank, 1994a. *World Development Report*. Washington, DC: World Bank.

World Bank, 1994b. *Staff Appraisal Report: Peru, Lima Water Rehabilitation and Management Project.* Washington, DC: World Bank, Processed.

World Bank, 1995a. *Bureaucrats in Business: The Economics and Politics of Government Ownership*. A World Bank Policy Research Report, Washington, DC: Oxford University Press.

World Bank, 1995b. *Peru, Water Utility Management: Beyond the Reform.* Washington, DC: World Bank, Processed.

Thirsting for Efficiency: The Economics and
Politics of Urban Water System Reform
Mary M. Shirley (Ed.)

CHAPTER 5

WATER SECTOR SERVICE CONTRACTS IN MEXICO CITY, MEXICO

LUKE HAGGARTY, PENELOPE BROOK AND ANA MARIA ZULUAGA

INTRODUCTION

In the early 1990s Mexico City — more precisely the Distrito Federal (D.F.) — seeking to reform and improve urban water services, initiated a set of service contracts with private sector partners. While service contracts are commonplace in the water and sanitation sectors, the Mexico City reform had two unusual features. First, it involved not one but four private sector operators. Second, the contract with each operator was to be implemented in three stages over a ten-year period. These features are of interest to the extent that the Mexico City experience can tell us something about the use of multiple private partners and progressive step processes to increase the benefits and sustainability of private sector reforms in water.

The decision to involve four operators introduces the possibility of competitive pressures for improved performance through comparisons and benchmarking. The decision to use a staged approach is of interest to the extent that it can shed light on the question of whether 'gradualist' approaches to private sector participation are more effective in reducing the kind of social and political opposition often faced by more radical forms of private involvement. In particular, a gradualist approach holds out the possibility of introducing the private sector over an extended period, thus building up confidence and trust within the government and among the general population. It may also allow the government to address pricing problems and introduce or strengthen regulatory arrangements on a gradual basis after securing private partners — reducing the risk of potentially politically costly price shocks, and allowing careful attention to the design of regulatory arrangements.

Incremental reforms also present risks, however. The long-term nature of such plans generally does not fit well with the shorter-term horizons of politicians, particularly in electoral government systems. The difficult decisions in the implementation of such plans are often left to the later years so as to push potentially unpopular actions onto future administrations while allowing the current government to claim credit for instituting reforms (Brook Cowen, 1997). The Mexico City case presents an opportunity for observing the advantages and disadvantages of gradualist reforms in practice.

More broadly, the case is of interest because the chosen reform tackled only tangentially, and in a limited way, what is perhaps the most serious long-term issue in the water and sewerage sector in the city: the improved management of a scarce and fragile water resource. An understanding of the constraints that produced this result and the implications it had on reform effectiveness may help give a better sense of the prerequisites for fully effective sectoral reforms.

This chapter seeks to address the following questions: What were the factors motivating and constraining reform, and why did the reform take the form it did? Once implemented, to what extent did the reform follow the intended path, and produce expected outcomes? To the extent that there was divergence from the intended path, why did this occur?

The chapter is set out as follows: first we describe the conditions in the sector and the city prior to reform, next we discuss political factors motivating and constraining reform, then we describe the service contract approach adopted in Mexico City, the implementation experience and outcomes of the reform, next we examine the extent to which the reform had any impact on the broader policy issues of water resource management, pollution control and improved services to the poor and finally we set out some conclusions that can be drawn from the case.

CONDITIONS IN THE WATER SECTOR PRIOR TO REFORM

Historical and Macroeconomic Circumstances Leading to Reform

We generally expect reforms involving substantial, long-term private sector participation (PSP) to be easier to implement when motivated by a strong or long-term crisis than by a mild or temporary crisis. In Mexico, while a macroeconomic crisis did help to precipitate reform of the Federal District's water system, the resulting reform was years in gestation and ended up being a rather limited one, that deferred difficult political decisions.

The precipitating crisis

In 1982 the 'Mexican miracle' of over three decades of strong economic growth ended abruptly.[1] Easy foreign credit, fueled by vast quantities of petro-dollars, had been used to prop up the overextended and inefficient Mexican state. The sudden disappearance of these propelled Mexico into deep economic crisis: GDP contracted by over 4% and

[1] During the decade before the crisis (1972–1981), GDP had grown at an annual rate of over 7% and inflation had generally been low by Latin American standards (World Bank, 1997).

inflation rose to over 100% in 1983 (World Bank, 1997). GDP contracted again in 1986 and inflation continued at levels between 50 and 130%. The government of President de la Madrid (1982–1988) took a number of steps to curb the decline, including a large privatization program.

In 1987, towards the end of de la Madrid's six-year term, inflation was soaring to over 130% annually despite multiple austerity programs and growth was still elusive. The ruling Partido Revolucionario Institucional (PRI) was suffering defections on both the left and the right and facing the most vigorous electoral opposition in its history. To stave off a renewed round of capital flight and a loss of confidence by investors, President de la Madrid ignored dissenters within the PRI and chose his technocratic Minister of Budget and Planning (SPP), Carlos Salinas de Gortari, as the PRI presidential candidate. This increased the internal divisions within the party leading to the defection of a number of high level party members who formed a left-wing opposition coalition, the Frente Democratico Nacional (FDN). The outcome of the presidential elections in 1988 shocked the PRI. For the first time parties on both the left and the right hotly contested a presidential election. President Salinas was elected with only 50.7% of the vote, although many believe that the counting process was so plagued by fraud that Salinas' margin of victory would have been much smaller had the election been fully free and fair.[2]

In the face of continued sluggish economic growth of slightly more than 1% and inflation of over 100% in 1988, President Salinas decided to push reforms further and faster. In 1989 his economic cabinet approved the divestiture of all public enterprises in sectors not expressly classified as 'strategic' in the constitution (Teichman, 1995). This opened up a number of sectors to private participation where it had not previously been considered, including infrastructure such as roads, telecommunications, energy and water.

The water sector was targeted for early reform (Buras, 1996). In 1989, a new federal water policy was promulgated which aimed at increasing the efficiency of water usage by decentralizing management to state level water commissions. The administration also passed a new water law in 1992 that (a) established the concept of water as a commodity (rather than a 'social good'), (b) endorsed the use of market mechanisms to manage water resources, and (c) promoted local participation in water management (ibid). State water commissions were established with control over and responsibility for water services in their areas. And reform of service providers was decentralized, thus providing a framework for private participation.

The Water Sector in Mexico City

Mexico City had for a long time faced acute problems in the water and sanitation sectors. Some of these problems — such as weak and fragmented arrangements for the management of water and sanitation services, inadequate maintenance, poor billing and collections performance and entrenched underpricing relative to service costs — were,

[2] Centeno, 1994; Purcell, 1995. There are also a minority of scholars who claim Salinas would have lost the election entirely had it been fair. See Reding (1989).

at least in theory, solvable through a well-designed program to engage the private sector in service delivery. Other problems — most notably involving the protection and sustainable management of water resources in the Valley of Mexico and surrounding states — were likely to be less amenable to private sector based solutions.

This section focuses on three key factors: the scarcity of water resources and therefore the high opportunity cost of meeting the city's needs, the high demand for services, and the fragmentation of responsibilities for the provision of water (and sewerage) services in the city. It then provides data on the performance of service providers in the city prior to reform.

Scarcity of water resources

The availability of raw water resources to supply the D.F. is seriously constrained by geographic factors. Mexico City is built on the floor of a drained lake — the site of the former Aztec city Tenochtitlan — high in a mountain valley. The city has a long and precarious hydrological history, combining severe water shortage with severe flooding. Both have been combated by heroic engineering projects to mine the aquifer underlying the city, to bring water from ever more distant river valleys which are one kilometer, in altitude, below the city, and to provide drainage away from the city for wastewater and floodwaters.

At the time of the reforms (1993) total water production for the D.F. was around 35.3 m^3/s (in the Valley of Mexico as a whole, production was around 60 m^3/s) (Joint Academies Committee on the Mexico City Water Supply, 1995). The greater part of Mexico City's water was drawn from the aquifers underlying the city, from wells with depths of 70 to 200 m. Most of the rest of the water came from two distant sources: the well fields in the Lerma Basin 60 km distant, and the Cutzamala River 127 km distant, and some 1200 m in altitude below the city (Joint Academies Committee on the Mexico City Water Supply, 1995). Table 5.1 summarizes the main water sources for the Federal District.

The extraction rate from the aquifer (for the metropolitan area as a whole) was around 42 m^3/s, while the replenishment rate was only around 24 m^3/s. From 1986 to 1994 alone the aquifer beneath the central sections of the city dropped by 6–14 m (DGCOH, 1994). Because Mexico City is built on a series of drained lakebeds, the

Table 5.1

Raw water sources for the Federal District (1993)

Raw water sources	*Total supply (m^3/s)*	*Share of total supply (%)*
Internal (basin of Mexico)	23.4	66.3
Imported sources	11.9	33.7
Total water supply	35.3	100.0

Source: Joint Academies Committee on the Mexico City Water Supply (1995).

land is soft, and prone to subsidence as the aquifer is depleted. Since the beginning of groundwater exploitation in the late 19th century, the downtown area of the D.F. has sunk by an average of 7.5 m. Several areas of the city center have sunk by over 2 m in the last decade alone (ibid). By virtue of its location and elevation, the city's alternative water sources are expensive ones. For example, in 1998 plans were underway to bring an additional 5 m^3/s to the D.F. and the State of Mexico from a new source some 140 km distance and 1000 m lower in elevation, at a cost of around US $500 million.

From a water resource perspective, therefore, the need for change would seem stark. However, the power to undertake reforms that could tackle the wider water resource management problems of the city and surrounding region lies outside the jurisdiction of the D.F. Therefore, any reforms that the D.F. could feasibly implement would be, at most, a partial contribution to resolving these problems. Similarly, responsibility for developing and funding new water sources, outside of the district does not fall directly on the D.F. but rather on the Federal Government. Such external funding of major supply projects means that the incentive to reform, so as to conserve on the very large costs associated with new supplies, is weakened.

Demand

Demand for water services was large and growing. Mexico City has one of the largest populations in the world. The D.F. has been growing rapidly: from a population of 4.8 million in 1959 it had grown to 8.5 million by the early 1990s (Gamboa de Buen, 1994). The Metropolitan Zone of the Valley of Mexico, in which the city is located, supported a population of 15 million (Joint Academies Committee on the Mexico City Water Supply, 1995).

At the time of the reforms, Mexico City had a relatively high rate of existing connections by developing country standards. As can be seen from Table 5.2, 95% of residents in 1990 had access to potable water in their houses or on their lots through the

Table 5.2

Water and sewerage coverage in the D.F. in 1990 and 1995 (percentages)

	1990	*1995*
Water services		
Inside the house	71.5	76.76
Inside the lot	23.5	20.41
Total direct water coverage	95.0	97.17
Standpipes	1.3	0.71
Total water coverage	96.3	97.88
Sewerage services		
Direct sewerage coverage	85.7	91.3
Septic tank	6.8	4.8
Total sewerage coverage	92.5	96.1

Source: Census 1990 and 1995, INEGI.

Table 5.3

Administrative arrangements in the water sector pre-reform (1991)

Agency	*Responsibility*	*Reports to*
DGCOH	Construction and maintenance of primary water and sewer networks	Secretary of Works and Services (D.F.)
Delegaciones	Construction and maintenance and operation of secondary water and sewerage networks	Secretary of Government (D.F.)
Treasury	Meter reading, billing and collections	Secretary of Finance (D.F.)

formal distribution network, and over 85% had direct sewerage connections.[3] Such high levels of connection generally reduces one of the possible motivations for attracting private capital since it does not require large new investment for the expansion of the network. While new investment was potentially needed to tap new sources of bulk supply, both to satisfy growing household needs and to take pressure off the aquifer, to expand wastewater treatment and to rehabilitate the ageing network, none but the last was contemplated as part of this reform.

Fragmentation of sectoral responsibilities

The management of the city's water distribution and sewerage system at the time of reforms in 1992 was fragmented and weak. There was no water utility, in the conventional sense; rather, responsibility for service delivery was spread across three distinct sets of institutions each answering to a different part of the District government (see Table 5.3).

The primary network was the responsibility of the *Direccion General de Construcion y Operacion Hydraulica* (DGCOH) under the Secretary of Works and Services. While its principal mandate was to construct and operate the primary network and water treatment plants, it also did a significant amount of construction and repair of the secondary network. Some observers describe it as a slush fund for the *delegaciones* (see below) because it frequently supplied them with material, engineering and labor to carry out repairs or extensions of the secondary network (field interviews in DGCOH and several *delegaciones*). Such an arrangement was politically useful as the mayor could redistribute resources without indicating this in formal budget documents. DGCOH's staff was generally perceived as highly professional, but strongly oriented toward construction rather than operation and maintenance of the existing system, since DGCOH had no revenue collection role this may not be surprising.

[3] There is some evidence that the official census overstates coverage rates because it tends to undercount low-income, self-housing (see Joint Academies Committee on the Mexico City Water Supply, 1995, p. 270); however, it is difficult to estimate the significance of the variation.

The operation and maintenance of the secondary distribution networks was the responsibility of the city's 16 *delegaciones*, political sub-units who report to the secretary of government and the mayor. Their operational arms are thus subject to strong political influences as is evidenced by the fact that they were put under the direction of the politically oriented secretary of government rather than a technical Secretariat such as Works and Services. One of the roles of the *delegacion* water department was to ration water because there was not enough supply within all of the *delegaciones* to meet demand, particularly in the poorer areas of the city. Such a situation was ripe for political exploitation and there have long been allegations that water shortages are created and then resolved by party officials to ensure voter loyalty (Coulomb, 1989).

Historical data on labor within the *delegacion* water departments was not available; however, it is estimated that in 1998 the average *delegacion* water and sewerage operations employed about 300 people.[4] Field interviews reveal that overstaffing was common and was estimated at about 50%, largely the result of years of political patronage. Workers in the water sector, like those of the rest of the D.F. Government, were highly unionized and labor rules made it nearly impossible to reduce the work force.

Finally, prior to reform, responsibility for billing and collection rested with the D.F. Treasury. In part this was justified on the basis that the DGCOH and the *delegaciones* lacked the requisite financial expertise, but this arrangement also fit well with the centralized political structure of the D.F. By centralizing billing and collection within the Treasury, the mayor was able to control all flows of funds and thus ensure that all the *delegaciones* would be dependent on his office and unable to finance an independent power base.

Sectoral performance

Under these arrangements, the management of operations was weak, creating scope for gains from private participation. The pressure on bulk water resources could have been significantly lessened by reducing consumption and waste, but little was done to effect this. Aggregate water use in the D.F. was high by both industrialized and developing country standards. Total water consumed by the city, measured by water produced over population covered, was about 379 liters per person per day in 1993. This was high in comparison with consumption levels in other Latin American and European cities which average between 200 and 300 liters per person per day (Fig. 5.1). In part, these high 'consumption' figures reflect high physical losses. However, they are also likely to result in part from lax billing, and from tariffs set well below the cost of delivering water to households.

The ratio of unaccounted-for water (physical plus commercial losses over production) was high even by developing country standards (see Table 5.4). Given the cost of new

[4] Obtaining detailed information on the water operations of the various delegations has been difficult since they vary tremendously in size and institutional arrangements; for example, in 1998 the central *delegacion* of Benito Juarez had 130 people working on the water and drainage system whereas the southeastern *delegacion* of Iztapalapa had 1000 (field interviews).

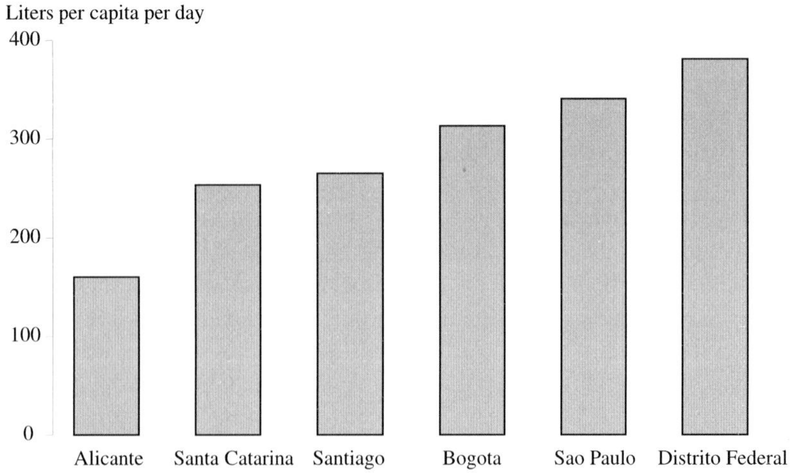

Source: Mexico DGCOH, other countries Yepes and Dianderas, 1996. All figures 1990 except Sao Paulo which is 1988.

Figure 5.1. Per capita water production in selected cities, 1990.

Table 5.4
Comparative statistics on water loss

City/Country	*Year*	*Physical losses (%)*	*Total UFW (%)*
Bogota, Colombia	1991	14	40
San Jose, Costa Rica	1990	21	46
Dakar, Senegal	1993	n.a.	29
Manila, Philippines	1988	n.a.	59
Mexico City	1992	37 [a]	51 [b]
Sao Paulo, Brazil	1992	n.a.	40

Sources: For all cases other than Mexico, Yepes and Dianderas (1996). For Mexico, [a] DGCOH (1997b) is the only estimate of losses available and is expected to have been the same in 1992. [b] Authors' lower bound estimates.

water sources, these losses imply a substantial financial burden — if not for the D.F. itself, then for the country as a whole. Such high physical losses reflected poor maintenance. At the time of reform there was no program to detect and repair nonvisible leaks in the secondary network and leak detection in the primary network used outmoded technology.[5] Leakage problems were aggravated by land subsidence since water and drainage pipes are buckled and broken as the city sinks.

[5] This was a very serious problem is some sections of the city, particularly the south and southeast where large caverns have been detected in the late 1990s from long undetected leaks which were spraying water down (field interviews with delegations).

Table 5.5

Labor productivity in selected water systems

Country/City	*Year*[a]	*Employees per 1000 connections*
Conakry, Guinea	1989	20.0
Ivory Coast (average)	1987	8.3
Mexico City[b]	1992	13.4
Lima	1991	4.6
USA (average)	1990	2.7

Sources: Other case studies in this volume; for USA, Yepes and Dianderas (1996), p. 17.
[a] For all cases except the USA, year chosen is that preceding a reform attempt.
[b] Employment figures for Mexico City have been calculated based on actual figures for the DGCOH, reported figures for Treasury employees involved in billing and collections for water, and estimates of *delegacion* employment for water services.

As for commercial losses, in 1992 over 22% of all connections were not registered. Even those customers who were registered and did receive a water bill often did not pay. While 53% of registered customers had meters installed most of them were not being read regularly. According to Ochoa (1998), prior to reforms the Treasury received regular payments from only 500,000 customers out of an average of 1,200,000 bills sent out (Ochoa, 1998) Using a combination of data from the census and the *Comision de Aguas del Distrito Federal* (CADF) and the most favorable assumptions we estimated maximum collection efficiency (collections/total billed) in 1992 at 62%.

Not all of these usage problems could have been addressed by private participation alone. Tariff policy was also at fault, providing little incentive to those customers who did receive a metered bill to reduce their consumption. Between 1970 and 1990 water tariffs had been changed just five times and by 1992 tariffs covered only around 28% of operating expenses plus investments, excluding interest and debt. The combination of low tariffs and poor billing and collection practices resulted in low revenues that, in 1992, covered just 54% of the operating costs.

Labor productivity was also poor by international standards (see Table 5.5) For example, Mexico City had three times as many employees per thousand connections in the pre-reform period as did Lima, Peru.

In sum in 1992 the water system in Mexico City faced serious problems. Demand was high, the aquifer was declining, the city was sinking, and alternative sources of new water were expensive. Weak operations and maintenance lowered efficiency and increased costs while low tariffs and deficient cost recovery aggravated the water supply problem.

Yet many factors made a radical reform involving private participation, such as a concession, unlikely. For one thing this 'crisis' was not new. Over-extraction of the aquifer had been recognized as a problem since at least the 1930s and the city had already committed to two very expensive projects for importing water from distant sources. Moreover, ultimate responsibility for developing these sources — and carrying a heavy share of the costs — was largely borne by the Federal Government. The price charged to the D.F. for bulk water delivered to the city did not reflect the investment costs, and the D.F. had a long history of not charging the opportunity cost for water extracted from the aquifer (field interview with CNA). The problems that motivated the city to seek private participation were largely operational, poor billing and collection, lack of maintenance of the secondary network and overmanning. Addressing even these problems required some difficult political decisions, as we show in the next section.

POLITICAL CIRCUMSTANCES LEADING TO REFORM OF WATER

As we have seen the physical and organizational state of water services in Mexico City made them ripe for reform. Now we consider the political factors that created the possibility of reform, while simultaneously constraining the shape that reform could take.

Political Factors Favoring Water Reform and Private Sector Participation

Water reform and the electoral fortunes of the PRI

Rapid urbanization and economic austerity contributed to a crisis in Mexico's ruling party, the PRI. In response to this crisis the ruling administration launched programs to expand and improve urban services. While initially politically successful, at least in the D.F., these programs proved to be unsustainable.

Nationally the PRI had been steadily losing electoral ground for many years as can be seen from Fig. 5.2. In part this decline was due to the national electoral reforms of 1977 and 1986 which the PRI had implemented in response to popular demands for greater democracy. However, although these reforms allowed greater opposition representation in the legislature, they did not seriously endanger the PRI's national hegemony (see Appendix A). More important to the dramatic fall in support between 1982 and 1988 were the economic austerity programs instituted by the de la Madrid administration, which reduced the resources that could be distributed by corporate or local Party leaders (Baer, 1993; Gomez Tagle, 1993).

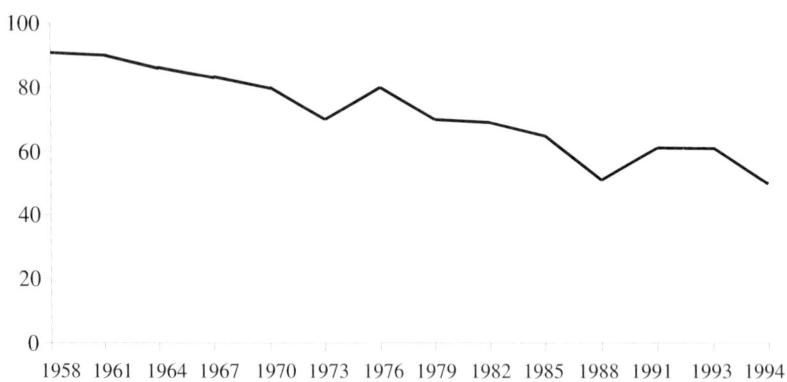

Source: Crespo, 1994.

Figure 5.2. Percent of the vote won by the PRI in congressional elections.

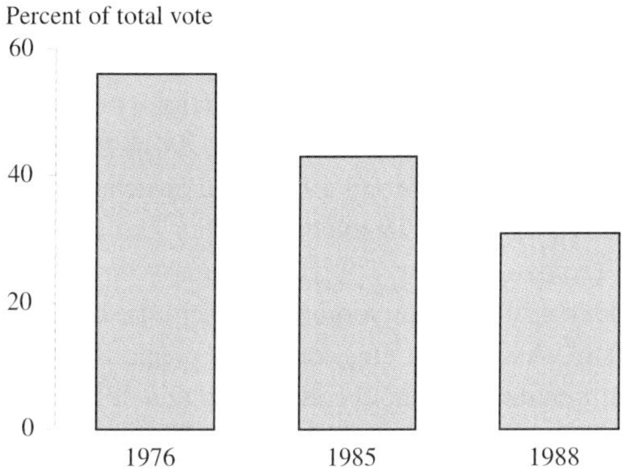

Source: Ward, 1989.

Figure 5.3. PRI share of votes in D.F. congressional elections, 1975–1988.

The other important factor in the PRI's declining support was the divergence between Mexico's growing urbanization and the Party's corporatist structure. The percentage of the population living in rural areas in Mexico fell from 78% in 1940 to only 28% by 1990 (Basáñez, 1993). However, the PRI had been designed for the political realities of the 1930s, with important representation for the rural sector and labor and was not well suited to incorporate the growing urban lower and middle classes. In particular the PRI had been ineffective at incorporating 'Popular Urban Movements', *Movimientos Urbanos Populares* (MUPs), that had come to represent an increasingly important segment of society since the 1970s. These autonomous associations were formed in an effort to secure services for the lower classes in many urban and peri-urban areas throughout Mexico. Because the PRI had failed to draw these organizations into its corporatist structures, the large numbers of migrants who moved to the city were largely uninvolved in the Party (Ramirez Saiz, 1992; Haber, 1993). Although not traditionally active in electoral politics, there is evidence that a number of these organizations and their members participated and heavily favored the leftist Frente Democratico Nacional (FDN) in the 1988 elections (Centeno, 1994; Haber, 1996; Klesner, 1996). As a result the PRI was losing ground even more rapidly in urban areas than in the nation as a whole, this was especially true in the D.F. (Fig. 5.3).

Despite winning the 1988 presidential elections, the PRI did poorly in most urban areas and in the D.F. in particular; winning just 27% of the vote, compared to 48% won by the FDN and 22% for the right-wing Partido Accion Nacional (PAN). In the D.F. the FDN dominance was near total taking 37 any of the 40 electoral districts (13 with over 50% of the vote) while Salinas did not win a single electoral district.[6]

After the narrowest presidential electoral victory in PRI history the Salinas admin-

[6] This section draws on data from Calderon Alzati and Cazes (1991). The FDN coalition also presented a unified list (single coalition candidates) in the D.F. Senate races and won both with 39% of the total vote. The coalition members could not agree on single candidates for congressional seats and so did not win a single district. The PRI won 26 congressional districts while the PAN won 14.

istration launched a concerted effort to recapture lost electoral ground, particularly in areas that were seen to have defected to the FDN. Aggressive new participatory programs were launched to raise the rates of delivery of public services to the poor with strong emphasis on potable water and sewerage. According to Lustig (1994) PRONASOL, the largest of these programs, devoted about 40% of total expenditure to water and sewerage, which has been described as the "heart of PRONASOL" (Bailey and Boone, 1994).[7,8]

The strategy was highly effective in reversing PRI's electoral fortunes in the capital. By the time the 1991 congressional (and D.F. Assembly) elections took place, the PRI had gained 46% of the D.F. votes in the senate and deputy elections and taken all *delegaciones* in the local assembly elections (Instituto Federal Electoral, 1991).

In the D.F., however, the strategy was not sustainable in the long run because it failed to address the underlying resource constraint or to reduce the high cost of the D.F. water system to the Federal Government.[9] For many years the D.F. had run large budgetary deficits which were routinely covered by the Federal Government (Salinas de Gortati, 1990; and Gamboa de Buen, 1994). Prior to 1989, the D.F. received some 75% of its total budget from federal funding through either revenue sharing or budgetary transfers (Beristain, 1995). The water system was one of the two largest expenditure categories in the D.F. budget (the other was the Metro). Transfers to the D.F. were unsustainable for two reasons: (i) low oil prices and greater difficulty in borrowing internationally had made federal funds scarce; and (ii) such large subsidies to the D.F. did not fit with Salinas' shift in focus to poorer communities. Therefore one of President Salinas' main concerns for the D.F. was to find a way to increase internally generated revenues, and yet continue to maintain the political gains among the poorer sections of the population.

Structure of the distrito federal and the rise of Camacho

Another important factor affecting the shape of the D.F. water reform was the dominant role of the presidency in local administration. Unlike every other state and municipality

---

[7] The actual targets of these programs has been the subject of some debate, with the administration claiming that they exclusively targeted the poor while numerous opposition figures, as well as political scientists, have claimed that the expenditures were politically driven. However, Molinar Horcasitas and Weldon (1994) clearly show that the programs were, at least in part, driven by electoral politics. They demonstrate that PRONASOL expenditures in 1990 were targeted precisely at areas that the Salinas administration wished to recover electorally in 1991. A significant portion of these funds was spent on areas that had voted for the FDN in 1988 (Molinar Horcasitas and Weldon, 1994).

[8] The underlying assumption of this strategy was that many of those who voted for the FDN in 1988 did so to protest the economic hardships they had undergone during the stabilization period of 1982–1988 and not because they were ideologically committed to the FDN (and later the Partido de la Revolución Democrática, PRD). This appears to have been the correct diagnosis as the urban poor in general, and MUPs in particular, were not attached to any political party but proved to be more pragmatic in their voting patterns, supporting whoever would meet their demands (Dresser, 1994).

[9] In fact these programs could exacerbate problems within an unreformed system by increasing water shortages and the percentage of water lost to leakage since the participatory program work was frequently of a lower standard (field interviews).

in the republic, the D.F. did not have a popularly elected mayor until 1997.[10] For most of the last 50 years the D.F. mayor was a direct presidential appointee who automatically held a high cabinet position in the Federal Government. The D.F. itself is broken up into 16 political units or municipalities known as *delegaciones* controlled by a local municipal head and the president also had the power to appoint or confirm the mayor's nominee for these positions.[11] The D.F. did (and does) elect deputies to the lower house of congress as well as two senators, just as in any other state, and during the early years of this reform (until 1993–1994) the national congress carried out all legislative duties for the Federal District.[12]

In 1988 President Salinas appointed Manuel Camacho Solis as mayor and part of his remit was to reduce the D.F. burden on the federal budget. Mayor Camacho saw the D.F. as a stepping stone to the presidency in 1994. To serve his goals and reduce the D.F. dependence on federal funds, the mayor had to build a strong support base within the D.F. (field interviews).

One way the Camacho administration built this support was by crushing a crime wave in 1987. Both Camacho and a former aide stated that the dramatic decline in crime rates after they took office was of critical importance in establishing the administration's credibility.[13] Whether or not the Camacho administration was actually responsible for the declines is less important than the political support generated from the perception that the new administration had addressed the crime situation effectively. The middle and upper classes, who felt themselves to be the targets of crime, were grateful for the reduction and therefore presented much less opposition when the administration raised property taxes and water rates in 1990. These extra funds, in turn, allowed the administration to increase spending in poorer areas of the city in their effort to reverse the PRIs political fortunes. In keeping with his national priorities, President Salinas did reduce transfers to the D.F. and, by 1993, the D.F. became a net contributor to the rest of the Federal Republic (field interviews).[14]

The PRI's strong showing in the 1991 mid-term elections set the stage for more ambitious reforms. The Salinas government launched the national water reforms described above, and the D.F. introduced contracts with the private sector.

[10] The Federal District of Mexico (D.F.) has existed as an entity since 1824 but was given its current form by a Presidential decree in 1928. This was in response to two basic problems; the inability of some of the local municipal governments to deliver services due to their poor financial state and the political infighting between factions in various municipalities which made coordinated city management impossible (Ward, 1989).

[11] This entire section of the structure of the D.F. Government draws heavily on Ward (1989).

[12] The Federal District Representative Assembly was created as a popularly elected body in 1986 but had limited authority during most of the reform period when it acted as a "consultative body of the mayor" (Gomez Tagle, 1993). The assembly would not become a truly integral part of the ruling of the D.F. and gain legislative authority until the mid-1990s.

[13] Separate interviews with Manuel Camacho and Ignacio Marvan in Mexico City, 1997.

[14] This is confirmed by the analysis in Beristain (1995). While transfers were still positive (though very small) the D.F. was contributing more in taxes than it was receiving from revenue sharing and so, by 1993 it had become a net contributor to the Republic.

Why private sector participation?

The question naturally arises as to why the D.F. chose private sector participation to reform their water sector. The main reason was disillusionment with prior efforts to reform public enterprises. Under President de la Madrid, the national government had implemented a series of performance contracts in an attempt to improve state-owned enterprise performance but by the early 1990s these efforts were perceived to be having little positive effect and the government had shifted the focus of the program to the privatization of state-owned enterprises. Similarly, in the D.F., the Camacho administration had initially put greater emphasis on the ongoing public sector reform attempt, the *Programa de Uso Eficiente del Agua* (PUEDA). This reform, begun in 1986, sought to improve the administration of water services by carrying out a user census, installing residential meters and promoting more efficient use of water, mainly through the improved technologies (toilets, sinks and faucets). However, by 1992 the program was judged to be progressing too slowly; inter-agency coordination problems, lack of funds, and general lack of interest on the part of users due to low tariff levels were all blamed for the program's lack of success.[15]

Political Factors Working Against Water Reform

Two major factors were working against a radical reform in the D.F. water system. First, some of the administration's core constituents stood to lose from any reform that passed substantial responsibilities to the private sector. Second, water had long been highly politicized in Mexico in general, and the D.F. in particular, resulting in tariffs set well below costs of service delivery. As we will discuss below, the government chose a reform that did not, at least in the initial stages, require it to confront either of these problems. However, while these factors did not prevent the initiation of private sector involvement, they did eventually prove important in bringing the reform to a standstill. In this section we describe what these factors were and how they affected the initial design of private sector participation; in the section on implementation we discuss how they helped to derail the later stages of reform.

Potential winners and losers from reform

The distribution of potential winners and losers of a reform depends on the exact reform to be carried out. It is essential therefore to examine how a reform that passed substantial responsibility for the water system to the private sector might affect different interest groups.

Those who might be expected to benefit from serious reform of the D.F. water system included: (a) private companies (domestic and foreign) who expected to win contracts to operate the system; (b) residents who were unconnected or had minimal service,

[15] Information on the reforms comes from Schteingart and Torres (1997), supplemented with an analysis of the data from the DGCOH *Compendio* (1997) and field interviews.

and might expect expanded and/or improved service and; (c) connected customers whose service might improve and who valued those improvements more than any price increases. National politicians would also benefit if the Federal Treasury made fewer transfers to the D.F. water system, and thereby freed resources they could use for their constituents.

Not all of these groups would weigh equally in the minds of the government. The private companies were clearly important for the PRI as it sought to raise funds for the upcoming 1994 campaign. The political pressure from those not connected to the system had diminished with the implementation of the participatory programs to bring water services to the poor. Both the level of the price rise necessary and the capacity to rapidly increase service quality were uncertain making benefits to connected customers more questionable.

Potential losers included: (a) current customers whose costs from increased prices might outweigh the value placed on improvements in service (e.g. better quality water, quicker connections, uninterrupted water supply, sufficient water pressure, rapid repair of leaks, etc.); (b) employees and heads of existing public sector agencies who might be laid off or lose stature or benefits; and (c) politicians who might lose the ability to reward supporters with jobs, investment contracts or increased water services. Again the groups varied in importance to the government and in their willingness and ability to affect reform design and implementation. As we saw above the currently connected consumers were a large portion of the public who might be expected to oppose any reform that entailed large price increases without dramatic improvements in quality. The public sector agencies and employees were of concern due to the high rates of unionization (which increased the likely incidence and effectiveness of broad public employee mobilization against reform) and the experience of problematic union opposition in the 1988 elections (Wiarda, 1988; Haber, 1996). Lastly a large number of local politicians from the local municipal heads to the District Secretaries and the mayor himself stood to lose significant patronage power if reforms went ahead as planned.

As we shall show in the discussion of the implementation experience below, the net political costs of implementing all the stages of the contracts as initially designed outweighed the net benefits to important government constituents within the D.F. The result was a partial and selective implementation of the planned reform.

The political importance of water in the D.F.

The second factor working against reform was the political sensitivity of water issues in the D.F. This sensitivity is perhaps reflected most strongly in the pricing of water services. As we noted earlier, tariffs fell significantly below the cost of operation of services within the city and the retail price of water had only been revised five times from 1970 to 1990. The failure to adjust prices in line with costs occurred despite the fact that it led to large operational deficits for the sector that had to be met out of the general budget. Even more telling, however, was the Federal Government's decision on charges for aquifer extraction in 1989. The newly formed national water authority, the *Comision Nacional de Agua* (CNA), undertook a study that attempted to estimate the cost of water in various zones across Mexico based on the relative scarcity of the water and the productivity of its various uses (CNA private communication). However, when the government set the price of the extraction rights it did not apply the cost estimate from the study to all users, instead it applied much lower rates to organizations which provided public services. As we will show, these rates are about 2% of what private parties pay to extract water from the same aquifer in the D.F.

Water's political importance is also reflected in legal restrictions on disconnections. Although Article 27 of the Mexican constitution of 1917 allows for the government to concession water rights to private persons, federal health legislation, passed in 1930s, bans the complete disconnection of residential users for nonpayment (field interviews). In compliance with this law the Federal District Financial Code states that service can be reduced to minimum "vital levels" but cannot be completely severed (Codigo Financiero del D.F., 1997). In practice, however, no residential consumers have ever had their service reduced for nonpayment in the D.F. This is in contrast to municipalities such as Monterrey in the north of Mexico, where residential consumers who fail to pay have their supply severely restricted as a matter of routine (field interviews).

DESIGN OF THE REFORM

Now we turn to the design of the reform. (In practice, as we will show in the following section, the reform that was actually implemented differed in important ways from the original plans.)

Organizational Structure for Regulation and Operation

As noted above, the national water sector reforms decentralized responsibility for regulating water and sanitation to the state level, with the creation of state water commissions. Responsibility for the administration and regulation of water resources at the national level in Mexico now rested with the National Water Commission (CNA), which the Salinas administration created in 1989 to oversee reforms in the sector across the country. It is the CNA, for example, that takes a primary role in planning and administering major, interstate, bulk supply works, and that licenses, proposes the tariffs, and collects charges for the abstraction of water from all water sources including wells in the D.F.

The *Comision de Agua del Distrito Federal* (CADF) was created in July 1992 as the agency responsible for the administration, operation and maintenance of the infrastructure necessary for the provision of water, sewerage and drainage services within the D.F. Its responsibilities overlapped with those of existing agencies, particularly the Treasury, DGCOH and the *delegaciones*. In the decree creating CADF these overlaps were recognized and the relevant organizations are instructed to work with CADF and "very slowly" transfer their responsibility to it (3rd Transitory Article of Decreto 14 July 1992). CADF was initially placed under the secretary of finance but in 1994 (i.e. while reforms were being implemented) it was transferred to the Secretary of Works and Services. Therefore the immediate effect of the reform was to add yet another organization to the administration of the sector as can be seen in Table 5.6.

Table 5.6

Administrative arrangements in the D.F. water sector post-reform (1997)

Agency	*Responsibility*	*Reports to*
DGCOH	Construction and maintenance of primary water and sewer networks	Secretary of Works and Services (D.F.)
Delegaciones	Construction and maintenance and operation of secondary water and sewerage networks.	Secretary of Government (D.F.)
CADF	Oversees contractors who are responsible for meter installation, meter reading and billing	Secretary of Finance (92–94) Secretary of Works and Services (D.F.) (94–present)
Treasury	Collections and transfer of budget to other entities	Secretary of Finance (D.F.)

Source: Authors' calculations.

CADF, upon its creation in 1992, took over the billing of the 20,000 largest users, who were mainly industrial or commercial entities. These customers represented only about 2% of the consumer base and consumed an estimated 20% of total volume, but they accounted for 53% of all collections.[16] They were targeted because they were easy to identify, measure and enforce payment upon. This policy largely ignored the residential users who were responsible for about 75%[17] of water use in the D.F., and therefore failed to encourage a more rational use of water among most customers.

Private Sector Contracts

Contract design

Despite the fact that the chosen reform was gradual and did not immediately involve any major challenge to existing agencies, the stated objectives were ambitious: billing all customers based on metered consumption; optimizing the use of resources; achieving cost recovery; reduction of aquifer extraction; improving the overall efficiency of the water sector; and achieving sustainable improvements in services quality. These goals were to be achieved through ten-year service contracts with private consortia. The service contracts were expected to improve the financial performance of the city's water system through more efficient billing and collection and by reducing physical losses (water produced but not delivered to customers) from an estimated 37% to 24% of total production over the life of the contracts.

Mexico City was divided into regions or 'zones', each of which would be let under a separate contract. In each zone, a private consortium would be awarded a ten-year

[16] Interview with Director General of the Comision de Agua del Distrito Federal (CADF, 1997).
[17] From DGCOH (1997b).

service contract.[18] The reasons for this were two-fold: to promote a spirit of competition among the contractors and to reduce the risk of contract failure should a single contractor renege on the requirements of its contract. The number of zones was not decided at the beginning of the process although the bidding documents did state that at least two zones would be awarded. The contractors were asked to bid on the assumption that they would service at least a quarter of the population in the D.F. and were instructed to include a table of discounts and surcharges in case they were assigned larger or smaller zones. The zones were to be roughly comparable in terms of population and have a minimum number of customers.

The contracts consisted of a 'menu' of tasks involving the operation and maintenance of the secondary water distribution and drainage systems, as well as several other associated activities, such as opening customer service centers. The tasks were grouped into three stages and the D.F. provided each of the bidders with a detailed work program complete with a projection of volumes of each operational task by *delegacion* and semester for each of the ten years of the contract period. The consortia were expected to perform all of the tasks on the menu by the time the contract period ended but prior to performing any task the CADF must issue a specific execution order. The private consortium for a given zone would have the right of first refusal to execute orders in its zone. This allowed the D.F. to sole-source a number of relatively small batches of work over time that otherwise would have been required by law to go to competitive tender. This was a conscious decision on the part of the designers for three reasons: (i) to give the contractors revenue security so they would undertake the other requirements that did not lend themselves to short term contracting (i.e. customer service); (ii) to avoid the high transactions costs associated with public bidding without sacrificing competition entirely; and, most importantly, (iii) to progressively move responsibility for all operations and management of the secondary network to the private sector (field interviews).

The tasks in the contracts were grouped into three stages:

Stage 1: Initial activities

- Mapping of the secondary water distribution network.
- Completion of a customer census.
- Installation of meters for all customers.

The objectives of Stage 1 were to obtain reliable information on users and the state of the distribution and drainage network and provide both operators and consumers with complete and reliable information on consumption levels.

Stage 2: Customer-oriented tasks

- Regularization of billing (meter reading, maintenance and the sending of bills).
- Shared role in collection of bills.
- Establishment of customer care centers and telephone care centers.
- Connect new customers.

[18] Federal law required each private consortium to be majority owned by domestic Mexican companies, which in the case of water paired up with a private international water company. According to the bidding documents, the Federal Law for the Promotion of Mexican Investment and Regulation of Foreign Investment requires that power to determine management and control rests with a domestic company. This requirement is backed by the Constitution.

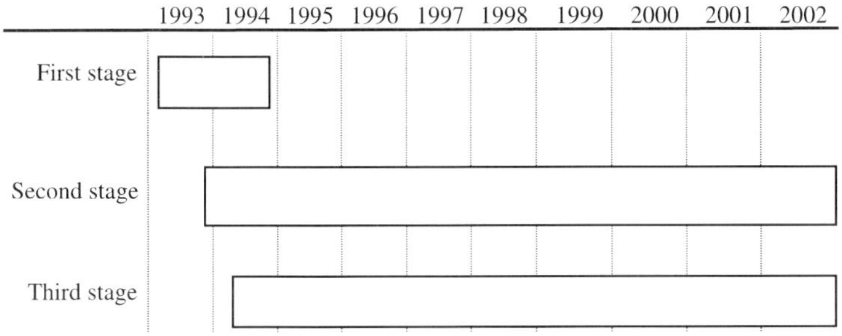

Source: CADF.

Figure 5.4. Schedule for implementation of contractual stages.

The objectives of Stage 2 were to increase revenues, raise consumer consciousness about the careful use of water and the punctual payment of bills, and ensure billing of all customers.

Stage 3: Network-oriented tasks

- Operation and maintenance of the secondary water and drainage networks.
- Detection and repair of visible and invisible leaks (water and drainage).
- Rehabilitation and extension of the secondary network (water and drainage).

The objectives of Stage 3 were to improve the efficiency and quality of water distribution and drainage service to consumers, recover water previously lost through leaks, and reduce operating costs.

In the first two stages of the contract the consortia are not directly exposed to commercial risk, since payments are on a fee-for-service basis. In the third stage some degree of performance-based remuneration was planned.

The contracts did not set out a firm timetable for progression through these stages. However, according to the calendar of tasks used for bidding, Stage 1 would begin at signing and run for two years, Stage 2 would begin approximately one year later and run to the end of the contract, and Stage 3 would begin approximately two years into the contract, and also run for the remainder of the contract. The planned schedule is illustrated in Fig. 5.4.

Influence of politics on reform design

The reforms were intentionally designed so that the more politically controversial objectives would only be addressed in the later stages. All of the key actors involved with the reform decision maintain that this gradual approach was the only politically viable option (field interviews).[19] Gradualism had many advantages from the government's

[19] There might have been nonpolitical explanations of this decision but they seem unlikely. It was not due to lack of technical capacity to design and implement a more complex form of contract since complex contracts already existed in Mexico City. DGCOH had contracted out the operation of several water treatment plants. Lack of information was a problem, but this alone would not necessarily constitute a barrier to a more immediate reform, as had been shown by the concession in Buenos Aires in 1992.

point of view. It allowed the new sectoral oversight body, CADF, time to build itself up while it was still non-threatening to more powerful interests who were expected to oppose later stages of the reform. Gradualism was also electorally convenient. The new contracts could be signed with large domestic firms, helping to assure their support in upcoming elections. At the same time the later stages of the reforms, where the contractors take over responsibilities for system operations and were likely to make a significant number of people employed by the *delegaciones* redundant, were not scheduled until after the 1994 presidential elections.

The decision to design a staged reform can also be explained by the PRI mind set at the time. The reformers never seriously considered the possibility that the PRI would not be in control of the D.F. government within the foreseeable future and therefore felt little pressure to ensure that reforms were completed quickly (Field Interviews). The decision to leave the implementation of the more difficult parts of the reform to future administrations, ultimately had significant costs, which we discuss in greater detail below.

The Transaction

The selection process

CADF issued a request for proposals in November 1992; bids were submitted in February 1993; a decision was made on the winning bidders in March 1993; and the contracts were signed in September 1993.[20]

Although the pre-qualification criteria and basic parameters of the contract were clearly specified, the number of zones was not decided upon until after the bids were received.[21] Bidders submitted price lists for each of the required actions in each of the 16 *delegaciones*. These detailed prices, as well as discounts and surcharges for zones larger or smaller than 1/4 of the D.F., were then fed into a computer model, developed by CADF, which computed the net present value of the cost of the actions for different numbers of zones and different allocation of zones across bidders. The model was then solved to minimize costs, which dictated the selection of a single private partner. The choice of four zones was not, therefore, a direct output of the model. Rather, the model was used to identify the costs associated with using different numbers of zones, to identify the best configuration of the zones, as matched with alternative private partners.[22] None of the bidders interviewed had a very clear idea of how the modeling process worked.

[20] CADF did not draw on external advisers to assist in the preparation of the contracts, nor did they consult widely with the major sectoral stakeholders, the DGCOH and *delegaciones*. They did hire Price Waterhouse to examine the financial portions of the bids submitted and make sure that all bidders complied with the rules for bidding so that all prices were strictly comparable (Ochoa, 1998).

[21] Criteria included a requirement for a majority domestic presence in the bidding consortia, a minimum capital of 100 billion pesos, and technical capacity and relevant experience in urban infrastructure provision in Mexico or overseas.

[22] Informed sources report that the models calculations indicated that including three zones would not raise costs dramatically above two but there was significant pressure to include a fourth company. While this was more expensive than three zones it was done (field interviews 1997–1998).

Table 5.7

Allocation of service contract zones in the D.F.

Zone	*Company*	*Consortium partners*	*Connections (Treasury list 1992)*	*Delegations*
A (north)	Servicios de Agua Potable (SAPSA)	Grupo ICA (Mexico); Compagnie Generale des Eaux (France)	373,350	Gustavo A. Madero Azcapotzalco Cuauhtemoc
B (north-central)	Industrias del Agua (IASA)	Socios Ambientales de Mexico (Mexico); Severn Trent (UK)	362,750	Venustiano Carranza Iztacalco Benito Juarez Coyoacan
C (southeast)	Tecnologia y Servicios de Agua (TECSA)	Grupo Bufete Industrial (Mexico); Lyonnaise des Eaux Dumez (France)	311,490	Iztapalapa Milpa Alta Tlahuac Xochimilco
D (west)	Agua de Mexico (AMSA)	Grupo Gutsa (Mexico); United Utilities (UK)	293,270	Alvaro Obregon Cuajimalpa Miguel Hidalgo Tlalpan Magdalena Contreras

Source: Compiled from CADF documents.

There were seven bids in total. The winning consortia and the zones in which they operate are set out in Table 5.7. The losing consortia were GMD (Mexico) and Biwater (UK), Tribasa (Mexico) and Thames (UK), and Geo (Mexico) and Aguas de Barcelona (Spain).

Each of the task prices was supported by a breakdown in the expected direct costs, indirect costs and profit. The bidders were asked to offer unit prices for each of the specified tasks, which would be adjusted for inflation over the course of the contract.

Bidding strategies and the pricing of risk

In order to determine their bids, the bidders had to price a basket of risks and responsibilities that were hard to calculate and to compare.[23] In each stage the contractors were obliged to carry out a number of actions for which they could not charge a fee directly and these costs had to be figured into the direct costs for which the contractor could charge a fee. Each consortium also had to evaluate the probability that they would, in fact, be allowed to carry out (and therefore charge for) each action in each stage. For example, in Stage 2, the contract states that contractors will be paid a separate fee for each of the following actions: (a) meter reading; (b) meter maintenance; (c) emission

[23] The bidders took quite different approaches to pricing their bids; some did detailed preparatory work to determine the costlier *delegaciones*, others took rough approximations (for example, bidding higher fees for *delegaciones* built on rocky ground), while others bid uniformly.

of bills; and (d) new connections. Regardless of whether these actions go forward, the contractors are obligated to (a) open a customer service center for each 50,000 customers in their zone, (b) establish dedicated phone service lines, and (c) receive and process payments. Thus the weighted probability of the basket of actions would then determine the amount of indirect costs that would be associated with each action. In essence this required an assessment of credibility for each of the promised actions in each of the stages of the contract.

Because the first stage involved uncontroversial actions — updating the user registry, mapping the system and beginning the installation of meters — there was little doubt that execution orders would be issued during this stage. In the second and third stages, however, this was not true, and the various consortia appear to have estimated and priced these risks differently. This variation in the pricing of risk, combined with different technology choices in the case of one contractor, led to wide variations in the prices bid for any given action. For example, one of the actions on the menu was fixing a household leak including replacement of up to 1.5 m of pipe, one contractor quoted a price of 673 pesos while another quoted 1866 pesos (CADF, 1998). Because the bids were evaluated on the net present value of the total estimated package over the ten years of the contract, these variations in the prices of individual works were not critically examined. As we will see below this would lead to criticism and political problems in the future.

IMPLEMENTATION EXPERIENCE

As we have seen the D.F. reforms, as initially designed, were modest in scope but the reforms as actually implemented were even more constrained. Initial start up delays slowed implementation of this gradual reform even more than expected, making unpopular changes more vulnerable to reneging. Weakness in the regulatory environment, particularly in sectoral coordination and the tariff setting process, has also affected outcomes. Finally, the original contracts were renegotiated in 1998, which has, in effect, changed their nature entirely from the original design. We first discuss the delays, then coordination and broader regulatory problems and finally the renegotiation.

Delays

The reform was subject to a number of delays. Although the winners of the bidding were chosen in March of 1993, the contracts were signed only in September of 1993 and the contractors did not begin operations until May of 1994. These delays were then compounded when the opposition party took power in 1997 and moved to renegotiate the contracts.

There may have been a delay at the outset because of a contract dispute, although accounts differ. According to Ochoa (1998) and some field interviews, the GMD/Biwater consortium disputed the contract award claiming that they should have been chosen over one of the other consortia. However, GMD representatives declined to discuss the matter and all high level officials interviewed, including the ex-mayor and the ex-secretary of finance, claimed to have no knowledge of the dispute (field interviews). A second factor in the delay was the need for the CADF to work out, internally and with the contractors

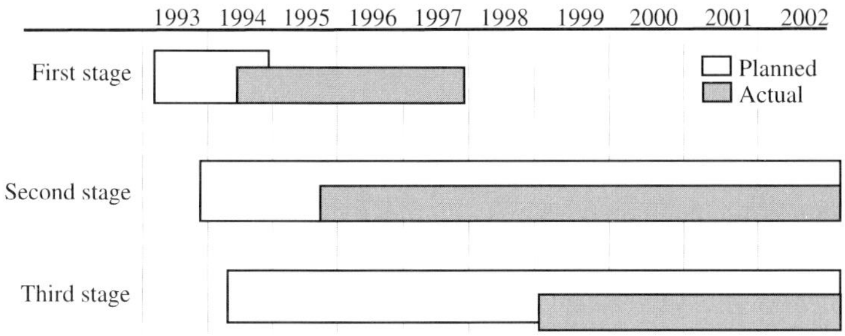

Figure 5.5. Revised schedule for implementation of contractual stages.

the operational details of how the individual execution orders would be issued and paid.

Political factors, however, were a more important source of delay. As noted above, Mayor Camacho was one of the leading contenders to become the PRI presidential candidate and fully expected to be chosen. He had little incentive to jeopardize his base of support within the D.F. by pushing ahead with reforms that might antagonize powerful allies and strong labor unions at the time of the selection of the PRI's candidate in 1993. As it turned out, on November 28, 1993, President Salinas announced that Donaldo Colosio was the official PRI candidate, and Mayor Camacho resigned the next day. This abrupt departure led to further delays until a new mayor could be appointed.

The devaluation of the peso in December 1994 led to further delays. Specifically, the companies dramatically slowed meter installation because the budget which had been allocated for the purchase of imported electronic meters by the D.F. government was insufficient at the new exchange rates. In total these early setbacks meant that by 1997 the program was at least two years behind schedule (see Fig. 5.5). As we will see below the renegotiation of the contracts in 1998 led to still further delays in implementation of stage three of the contracts.

Regulation: Monitoring, Coordination and Tariff Setting

Monitoring and coordination

The absence of a clear and effective set of monitoring and enforcement arrangements in the D.F. has had an important effect on the implementation of the contracts. CADF is the closest thing to a regulator, since it designed and monitors the service contracts, although it does not set tariffs. Although the decree that created CADF implied it would have wide oversight responsibility for the sector, it did not assign CADF the power to enforce its decisions. Rather, CADF was added to the existing organizations in the sector which has resulted in significant overlap and lack of coordination.[24] CADF has a more straightforward role vis-à-vis the private contractors. It assigns and monitors (or hires an experienced company to monitor) the execution orders and must verify the

[24] One example of this is the fact that DGCOH, the *delegaciones* and the contractors all maintain teams to repair reported leaks. Field interviews revealed that frequently teams from two or more organizations show up at the same site to repair a reported leak.

work before the private contractor is paid by the Treasury. It may assess penalties for unsatisfactory performance although this seems not to have been a significant factor to date.[25] Confusing relations further is the fact that CADF is in some ways in competition with the private operators. CADF reads the meters of and bills approximately 20,000 large users and operates its own customer care center. Approximately 65% of CADF staff are absorbed in this activity, which it does not plan to transfer to private contractors.

CADF's lack of budget autonomy (its budget is part of the budget of the D.F.) can hamper the efficient implementation of works by the contractors. CADF's budget is expected to cover both the operating costs of the CADF and the service fee payments to the private contractors and all payments from clients and to contractors go through the D.F. Treasury. CADF does not receive any direct benefit from increases in collections. As a result of this arrangement when CADF funds were short because it had not received its budget allocation from the D.F. Treasury, which was common in 1996–1997, it compensated by not issuing new execution orders (field interviews with contractors).[26]

All of the contractors agreed that they could have accomplished much more if they had been given longer term and more comprehensive execution orders. Typically an execution order is good for only about two months and when it expires the contractors should, in theory, wait for another to be issued before carrying out more work. They point out that many of the tasks they are performing (billing, collection, customer service) are processes and not discrete actions and it would be costly for them stop when the current execution order expires. It is therefore not uncommon for contractors to continue to carry out work and meet the expenses associated with that work without a contract, waiting several weeks or even months before a new execution order is issued and they are paid.

CADF's lack of authority in managing the sector has also meant that responsibility could not be shifted to the contractors as planned. Aside from expected bureaucratic turf battles, further sector management consolidation has not occurred due to political considerations. In 1996–1997, the D.F. government made a series of decisions that delayed the consolidation of sector management under CADF as foreseen in the original reform plans. The main reason for this was the fear that the resulting unemployment would be, politically, very costly. For example, the government chose not to implement portions of the second stage of the contract (e.g. passing new connections to contractors) and further delayed the passage of operation and maintenance to the contractors. The failure to pass these actions to the contractors means that the newly connected customers do not receive water bills (as the information about those new connections is not passed on time or not passed at all to the contractors) and that there are fewer revenues from execution orders to finance the costs they cannot directly bill, which has caused financial difficulties for at least one contractor.[27] However, these reforms were expected

---

[25] While we were unable to get reliable information from CADF on the extent to which they used penalties, contractors reported that being penalized was commonplace. Contractors maintained that many of the fines were of low value and for very minor infractions and they thought that a better approach would be to fine for significant failing in performance.

[26] By late 1998 most of the contractors said that delays had been minimized and that this was no longer a significant problem (field interviews 1997–1998).

to result in labor redundancies in the *delegaciones*. In an already tight political contest the government had no desire to antagonize public labor unions and add to already increased unemployment. post-reform.

Tariffs

Tariffs are proposed by CADF but decided on by the D.F. Assembly and, although all tariffs were increased in the early years of reform, they have since fallen for most residential users. As Fig. 5.6 shows, between 1996 and 1998 the metered tariff fell even in nominal terms for users consuming between 30 and 100 m^3 per bimester, these are predominately residential consumers in the middle and upper classes. In real terms Fig. 5.7, an even larger group of consumers, all those using between 30 and 220 m^3 per bimester or about 52.2% of all residential consumers, have seen 1998 tariffs substantially lower than in 1996. It is difficult, however, to know exactly how many customers actually noted these changes because so few customers received bills based on actual meter readings in 1996 (most 'metered' bills were actually based on historical readings, not actual readings).

This tariff reduction for metered residential customers was combined with an increase in the cross subsidy from nonresidential to residential consumers. Numerous documents, including water bills and public reports produced by the CADF and the

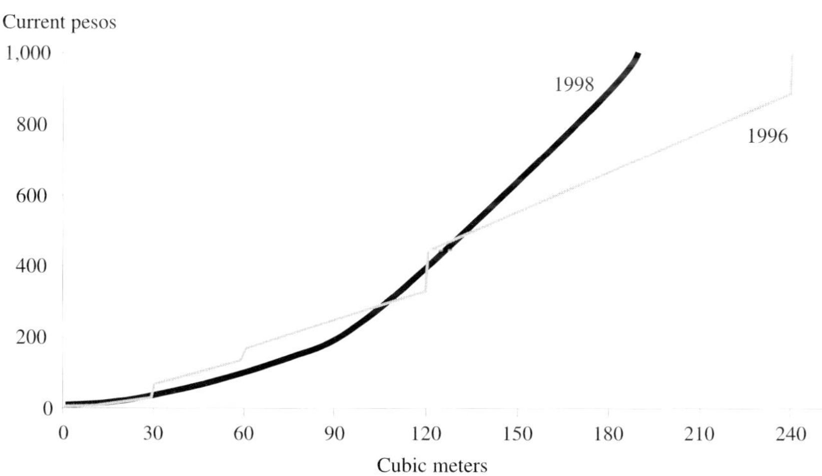

Source: Codigo Financiero del Distrito Federal 1996 and 1998.

Figure 5.6. Nominal residential metered tariffs in D.F., 1996–1998.

[27] The contractor in question was by far the most aggressive in pricing its bid. While this made them number one in the bid assessment it has meant that they are working on a substantially narrower margin than their competitors and therefore are much more susceptible to financial difficulties due to the lack of implementation of any given action in the contract. They have reportedly lost millions of dollars due to this problem during the second stage.

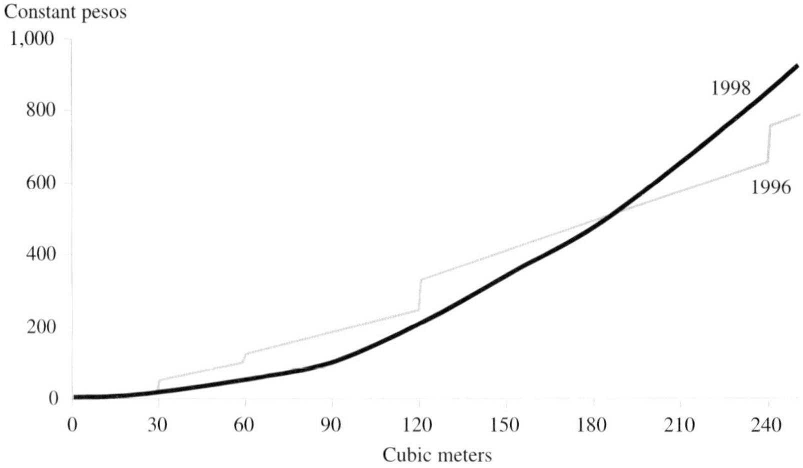

Source: Codigo Financiero del Distrito Federal 1996 and 1998.

Figure 5.7. Real residential metered tariff in D.F., 1996 and 1998.

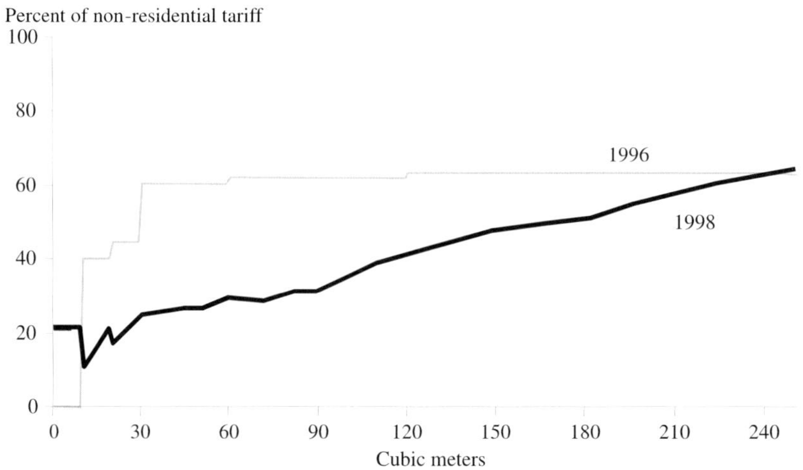

Source: Codigo Financiero del Distrito Federal 1996 and 1998.

Figure 5.8. Residential metered tariffs as a percentage of nonresidential metered tariffs, 1996 versus 1998.

DGCOH refer to this lowering of rates as a boon to residential consumers.[28] At the same time nonresidential consumers have faced steadily rising prices in real terms in each revision, leading to the growing price differential shown in Fig. 5.8. Since non-residential consumption accounts for roughly one-quarter of total consumption, average real metered prices have dropped.

Two factors explain the fall in real metered water tariffs in this period. First, in

[28] DGCOH (1997a) and CADF water bills 1997.

December of 1994 the peso was devalued, eventually falling to about 50% of its former value against the dollar. Because the 1995 water tariffs had been set earlier in 1994 they did not keep pace with the high inflation that occurred after devaluation. In an effort to limit the fall in real prices the D.F. assembly (controlled by the PRI) approved a large rise in the water tariff for 1996. By the time the new rates came into effect in 1996 the economy had entered a steep recession and any price rises were extremely unpopular. As we pointed out above, in 1996 most residential consumer's meters were not regularly read. Thus the combination of the price changes and receiving bills based on actual usage (which was often much higher than 'historical' readings) meant that many consumers perceived a large price shock. This in turn contributed to the second factor affecting prices, the increasing politicization of tariffs.

The D.F. Representative Assembly was granted the power to approve (and modify) the D.F. budget, including proposed water tariffs, in 1994. Although the PRI had a majority in the assembly the public outcry over the tariff change following the devaluation increased the assembly's sensitivity to raising tariffs. Politically, it was difficult for representatives to the assembly to justify a rise in water tariffs during a period of rising inflation, falling wages and diminishing purchasing power.[29] The assembly, and the PRI more generally, became more sensitive to this in the run-up to elections of the mayor and the assembly of July of 1997. In 1996 the D.F. Assembly voted for a large cut in water tariffs for the middle and upper middle class. The tariff cut seems particularly driven by the new political dynamics in the D.F. By early 1996 it was clear that the major threat to the PRI in the D.F. was the left-wing PRD which is especially strong among lower-income groups. If the water tariff decision in the D.F. is any indication, it appears that the PRI consciously decided to abandon the fight for the lower-income voters and concentrate instead on voters in the middle to upper income brackets who they feared might also defect to the PRD (Fig. 5.9). In fact metered rates for the poorest consumers (less than 20 m^3) actually rose, while those consuming 20–25 m^3 had no change. The rise in price for the poorest metered consumers was mirrored by price hikes of about 19% in the same year for non-metered residential customers, the majority of whom were lower income. In contrast wealthier metered customers, those consuming between 35 and 75 cubic meters per bimester, received large price reductions, even more than the price rises of the year earlier.

Renegotiation

Implementation of the contracts was severely disrupted by the entry of the first democratically elected, and opposition led, government of the D.F. When the new PRD administration took power in December 1997 they were very skeptical of the contracts. Initially, they believed that the contracts had not been properly designed. There was a serious risk that the new government might try to cancel the contracts altogether. After a careful and critical review, the new administration decided to maintain the contracts

[29] For example, in real terms the minimum wage fell by 19% and the average manufacturing wage fell by 24% between 1994 and 1996 (World Bank, 1998).

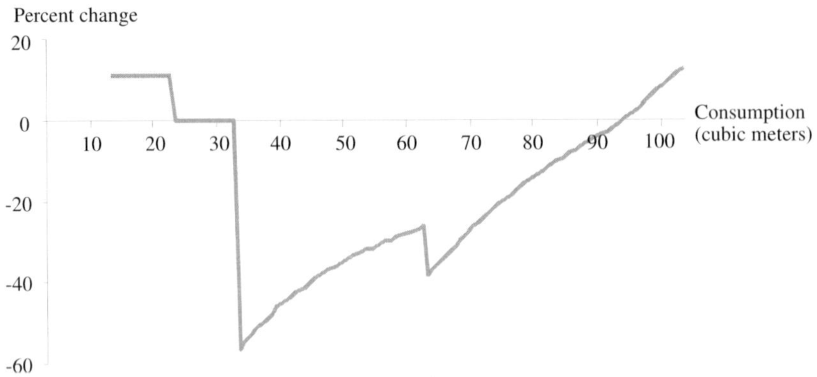

Source: Codigo Financiero del D.F., various years.

Figure 5.9. Residential metered tariff change, 1996–1997.

but amended the parameters of stage three (see details below) and, more significantly, revised the prices applicable to the execution orders in the third stage.

As noted above, the bidding process had resulted in significant variations in prices between contractors for the same works. This was considered unacceptable by the new government, which initially attempted to renegotiate all of the prices. A majority of the contractors (although not all) refused to renegotiate prices for ongoing works such as metering and billing, but did agree to review prices for stage three, which had yet to start. The new administration obliged all bidders to charge ‘market rates’ which coincided, in most cases, with the lowest bidder’s rates on nearly all actions already in the contract for stage three. They also set prices for new actions that they wanted to add.[30] Reportedly, the administration took the position that the general contract simply gave the contractors the right of first refusal for new actions and that they could either accept these new prices or see the work contracted out to third parties.[31]

As a result of the renegotiations, Stage 3 progressed but it is highly unlikely that the contractors will be given full control of operations and maintenance of the secondary network as originally envisioned. Instead the new administration chose to use the contractors to supplement existing organizations to improve the system by making them more active in leak repair and in upgrading or repairing the secondary network on a fee-per-action basis. In other words, the spirit of the original stage three — more responsibility of the private sector for improving operations — has now been lost. New actions have been added, but since these are on a fee-per-action basis they do not redistribute risk and reward between the public and private actors.

[30] This was true for most, but not all, actions. In at least one case the government agreed to prices above those of the lowest bidder. The implication being that the bidder severely underestimated the actual cost of the work in their original bid. However, the government did not allow this low (below cost) bidder to collect the newly agreed price for the action, but instead insisted that they be reimbursed at the price they bid originally, despite the fact that the government agreed to a higher price with the three other consortia.

[31] This does appear to be the stipulation in the bidding documents (‘base de licitacion’).

REFORM OUTCOMES

In this section we analyze the impact of the reforms on information, metering, billing and collection, service and efficiency of operations. Since the reform is still underway our conclusions about the outcomes are necessarily provisional.

Information

One important accomplishment of the contracts was to improve information about the water distribution system and the client base in the D.F. Thanks to the reform the D.F. now has an electronic map of the entire water system for overall planning, a customer census that has allowed regularization of previously unregistered connections and meters for most customers.

It seems reasonable to attribute the information improvements to the contracts since earlier efforts to use the existing government agencies to improve information, register customers, expand metering and read the meters once installed had failed. The dispersion of authority across numerous public agencies and the entrenched nature of their workforce had made reform difficult. The split in responsibilities between the *delegaciones*, which connected customers, and the D.F. Treasury, which billed them, allowed many customers to be added to the system but not to the registry. The participatory programs described earlier added to the problem by extending semi-formal connections in many poor neighborhoods. Moreover, although prior to the contracts up to 53% of registered consumers had meters, because of technical problems and corruption most users paid a flat fee or were billed on 'historical' meter readings (field interviews).

The water census, completed in 1996, yielded an average increase in existing, although unregistered, connections of about 22% (Table 5.8). The increase in registered connections was far from uniform across the city and the large variations show the extent of the information deficit that existed previously, particularly in the southeast (zone C) where one out of every three connections was not registered.

Metering, Billing and Collection

Nearly 1.2 million meters were installed as a result of the reforms (Table 5.9). As of June 1998, 64% of customers were billed on a metered basis while another 16% were billed the average of the metered customers in their zone.[32] This is a major improvement on the pre-reform state of affairs where, because of the large number of unregistered connections, at least 22% of customers did not receive a bill at all and those that did were being billed a fixed rate, regardless of usage. Again the southeastern zone, which is the poorest and has the lowest water and service quality, has experienced serious difficulties in metering and billing, as of August, 1998 only 37% of their customers receive a

[32] The D.F. financial code states that once metering has reached 70% in a given zone then the remaining un-metered customers can be billed for the average metered use in the zone.

Table 5.8

Changes in registered connections and metered billing in the D.F.

Company	*Zone*	*Registered connections*[a] *1992*	*Registered connections*[b] *(August 1998)*	*Percentage change in total number of registered users*	*Connections billed on metered basis (August 1998)*	*Percentage of total with metered billing*
SAPSA	A (north)	373,350	410,681	10	316,020	77
IASA	B (north-central)	362,750	411,269	13	314,838	77
TECSA	C (southeast)	311,490	446,249	43	167,159	37
AMSA	D (west)	293,270	346,605	18	258,785	75
CADF	Large users	n.a.	19,689	n.a.	13,576	69
		1,340,860	1,634,493	22	1,070,378	65

[a] According to Treasury list.
[b] According to CADF.
Source: CADF.

Table 5.9
Installation of meters in the D.F.

Year	*Number of meters installed*		*Total cumulative number of meters installed*
	Type A	*Type B*	
1993	–	–	1,699
1994	168,007	45,903	215,609
1995	189,365	20,108	425,082
1996	177,779	134,665	737,526
1997	196,438	117,036	1,051,000
Mid-1998	38,909	66,548	1,156,457

Source: Authors' estimates based on data provided by CADF and Secretaria de Obras y Servicios.

metered bill (Table 5.8). To address previous technical problems and corruption, the new meters can be read electronically and are less subject to failure.

Billing has improved substantially since meter reading and billing was taken over by the contractors under CADF (Fig. 5.10), with the total number of bills growing by about 26% over the period. The amount billed has also risen, by 205% in nominal terms.

Despite these improvements, information on payment of water bills is still poor, a result of the sector's fragmented administration. Customers can pay their water bill at one of the 24 customer centers run by the contractors, one of the Treasury's offices or one of hundreds of designated bank branches. Only the Treasury is in a position to aggregate payment data for the entire system. For about a year after the contractors began taking over billing there were no information sharing arrangements

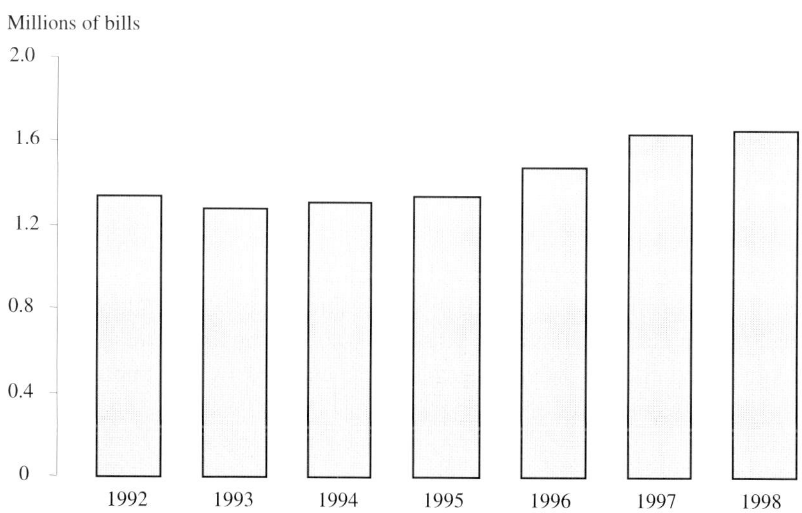

Source: Patron de Tesoreria and CADF .

Figure 5.10. Bills issued, 1992–1998.

Table 5.10

Collection ratio 1994–1998 (thousands of pesos)

	1994	*1995*	*1996*	*1997*	*Est. 1998*
Total billed	914,844	1,115,375	1,667,476	2,376,804	2,787,472
Total collected	766,000	807,521	1,140,000	1,589,701	1,930,240
Collection rate (%)	84	72	68	67	69

Source: CADF.

and contractors sent out bills but had no idea which customers actually paid at the Treasury's offices or at the banks. Only in late 1998 did the Treasury begin sending information on receipts to the contractors. Unfortunately, this information is still received late and continues to suffer from problems originating from mistakes in the manual capture processes of the Treasury so that many payments cannot be matched with the corresponding receivables.

Although the reforms did much to improve billing, they did little to increase the incentive for bills to be collected. The private companies are paid only for producing and sending out the bills, not for collecting money; although they have put effort into reminding and encouraging customers to pay. While their efforts have helped somewhat, the responsibility for collection still rests with the D.F. Treasury. Not surprisingly, the Treasury's incentive is to maximize the district's resources. Since the Treasury collects all taxes and fees levied by the District Government, it focuses its attention on the single largest source of revenue, general taxes. General taxes, particularly property and payroll taxes, accounted for 25% of total revenues in 1996. In contrast, revenues from water collection were only 5% of revenues that year. Also not surprising is the fact that the Treasury has always focused its efforts in water collections on the very large customers who provide high returns to their collection efforts (Ley de Ingresos, 1995). This lack of significant improvement has led the CADF to institute its own program to help collect past arrears, sending letters to people who owe more than 500 pesos, but it is too early to know how effective this will be.[33]

This separation of responsibilities and misalignment of incentives has meant that total collections have barely improved. Although in nominal terms collections increased by over 150% between 1994 and 1998 (Fig. 5.11), they have increased by only 7% in real terms over the same period (Fig. 5.12). The collection rate (total collected/total billed) has actually fallen during the reforms (Table 5.10) reflecting the increased billing activity of the private contractors which has not been matched by the public sector agencies in charge of collection.

[33] The response to over 300,000 letters sent in July 1998 suggests that only about 3% of outstanding arrears were paid, in part because 40 to 50% of the customers turned out not to be in arrears.

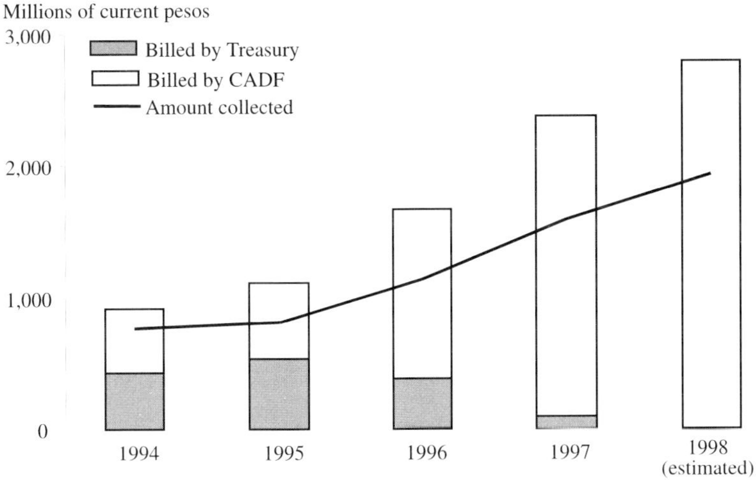

Source: Computed with data from CADF and Treasury. Contractors included in CADF.

Figure 5.11. Nominal amount billed and collected.

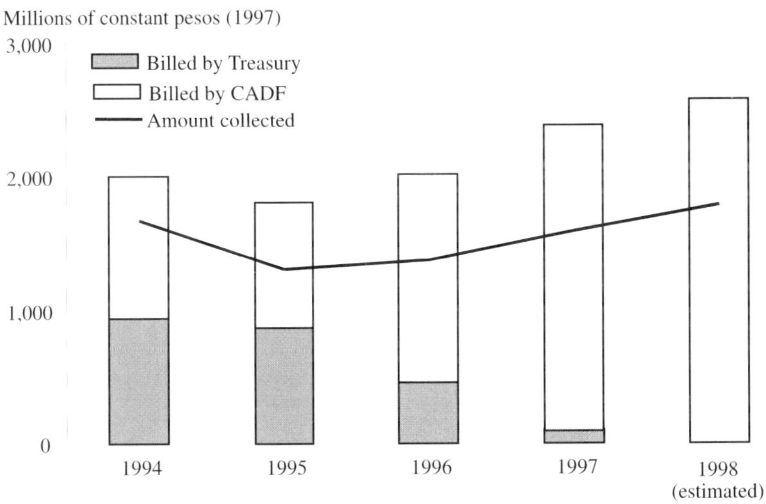

Source: Computed with data from CADF and Treasury. Contractors included in CADF.

Figure 5.12. Real amounts billed and collected.

Service

The effect of the contracts on service has been mixed. The contractors reportedly respond more rapidly to questions on billing and meter readings or to reports of leaks or lack of water. Furthermore, the operators have opened a total of 24 customer service centers and each has telephone hotlines in their area (Ochoa, 1998).

Service problems remain, however, including poor quality of water and interrupted service, especially in the southern and southeastern portions of the city. As we saw above the reliability of water services is still the responsibility of the *delegaciones*; hence the contractors have had little effect on this aspect of service, which has not improved markedly. Even by 1998, neighborhoods in nine of the sixteen *delegaciones* continued to suffer routine cuts in service, although the severity of the problem and the number of people affected varied widely. We can illustrate the problems by examining two of the worst affected *delegaciones*, Tlalpan and Iztapalapa. In Tlalpan 43 out of 192 neighborhoods received water only two to three days a week usually for eight to ten hours a day; an estimated 30% of the population were affected (DGCOH, 1998, and field interviews). In Iztapalapa about 40% of the population were affected by the cuts; some received water as infrequently as one day a week for as little as six hours (field interviews).

In addition to the poor service in these sections of the city the quality of the water is also extremely poor. In part this is due to the groundwater in the area which has high concentrations of manganese but in part it is also because the southeastern section of the city is the last to receive water brought in from the Cutzamala system. The combination of frequent interruptions in supply and poor quality local water means that the water delivered is often of very poor quality and in some sections of the southeast residents have to run their tap for several minutes before using the water as it initially runs black due to the high sediment content. To combat these problems the DGCOH was implementing a plan to treat water at the well-head in these locales and there is a longstanding plan to extend the southern branch of the Cutzamala aqueduct to the southeast of the city.

A related effect of such poor service is to reduce revenues. Because of interruptions in service and the poor quality of the water many of these customers are charged on a fixed rate, regardless of the fact that many also have meters, in line with CADF's policy of not introducing metered billing where service quality is poor. The flat rates are set according to the income tax classification of the area and for most of the affected areas is between 11.75 and 17.65 pesos per bimester.

Operating Efficiency

Sector management

The reforms did little to change the organization of the sector, and fragmentation of administration of water services remains a serious problem. As shown in Table 5.6, CADF and the private operators were simply added to the existing organizations operating in the sector. Although the initial decree that created CADF gave it wide sectoral oversight powers, it did not give it any powers of enforcement over the existing organizations, nor did it change the organization or responsibilities of these organizations (DGCOH, Treasury and the *delegacion* water departments). As a result there continues to be significant overlap, duplication and lack of coordination, which, in turn, lead to inefficiencies in the use of resources (see footnote 24).

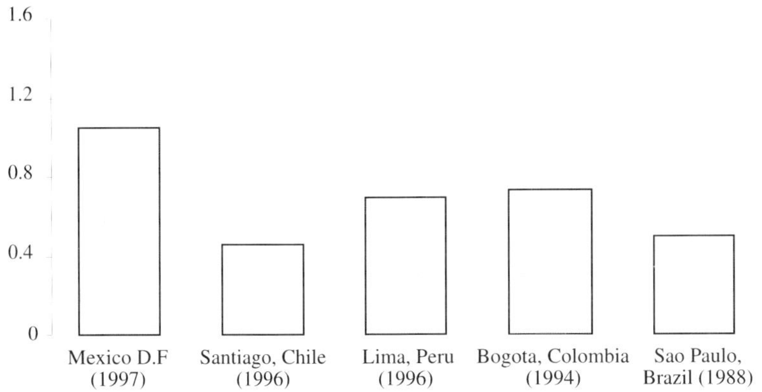

Source: For Mexico, authors' calculations; for Santiago, EMOS; for Lima, SEDAPAL; and for Bogota and Sao Paulo, Yepes and Dianderas 1996.

Figure 5.13. Comparison of working ratios for various countries.

The inability or unwillingness of successive D.F. governments to unify and streamline the many public agencies in the sector or to link budgetary resources to clear efficiency indicators has blunted their ability to use private sector participation to improve efficiency. Since the reforms effectively added the private contractors to the existing agencies, they have not resulted in cost savings. As a proxy for administrative efficiency we examined the working ratio (operating costs/operating billed revenues) of the system. As can be seen from Fig. 5.13, the D.F. compares unfavorably with a range of water systems in Latin America.

Ongoing political reforms in the D.F. are unlikely to change this situation. The plan is to move to local election for municipal mayors (*Delegado's*) in stages between 2000 and 2003. As a result of the transitional arrangements, the first mayoral term in office was shortened to just three years from the usual six and the mayor faced reelection in 2000. Beginning in 2001 the municipal mayors are to be elected in each of the 16 *delegaciones*. This means that the current government has little incentive to push reforms, since they will incur significant up-front political costs, in terms of lost jobs in the public sector, but deliver gains only in the long term.

It was in this context that the first PRD administration renegotiated the third stage of the contract, profoundly altering the role of the private sector in the reform plan, supplementing the public agencies (i.e. detect and repair leaks, carry out system maintenance, etc.) not supplanting them. In this way the PRD could protect itself from accusations of causing unemployment while still seeking to improve the service in the water sector. It also left the *delegacion* water departments as a useful political tool in terms of employment and politically motivated service delivery.

Barring an unforeseen attempt to create a centralized, autonomous sector administration, the political changes over the next few years will almost certainly further fragment administration of the water system. Since some districts routinely vote for one of the three main parties (PAN, PRD, PRI) it is likely that different *delegacion* water systems

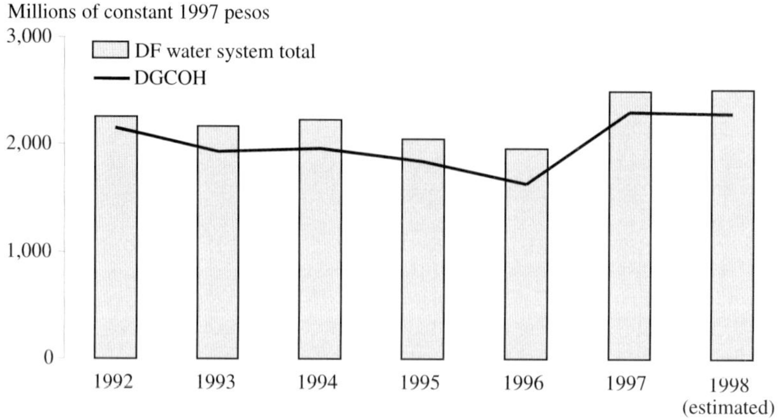

Note: Operational costs includes labor services, materials and s upplies, general services, aid/subsidies and transfers and consumer/capital goods. It does not include the money expended in the account "Public Works" as this was considered an investment rather than an expense.
Source: DGCOH, CADF and delegations.

Figure 5.14. Real operating costs of the D.F. water system.

will be run by rival political parties. This may politicize water issues even further and make it more difficult to delegate responsibilities to the private sector.

Operating costs

Not surprisingly, operating costs have not decreased as the reformers intended; rather they have increased in real terms over the reform period. The budgets of the DGCOH and the *delegaciones* were not significantly affected by the reforms, and the DGCOH accounted for an average of 90% of the total operating budget in water from 1992 to 1998 (Fig. 5.14). Hence, it would be unrealistic to expect costs to fall when the largest entity in the sector has not been required to reduce its operating budget over the reform period.[34] Since the contractors have not yet been given the operational responsibility envisioned for Stage 3 in the original design of the contracts, there is little time left to correct the situation and achieve any significant reduction in operating costs in the rest of the project time frame.

Cost recovery

Since costs did not fall and tariffs did not increase, it is not surprising to find that the system still does not cover its operating costs, although the level of cost recovery did improve modestly from 64% in 1993 to 71% in 1998 (Table 5.11). Because the water system is financed as part of the government budget rather than on a commercial basis and all loans and financial charges are handled directly by the D.F. Treasury, we do

[34] There were some reductions in the water budget in 1995–1996, but these were in response to problems caused by the devaluation and were not sustained.

Table 5.11

Operational cost recovery in the D.F. water system, 1993–1998 (thousands of current pesos)

	1992	*1993*	*1994*	*1995*
Revenue collected	471,400	596,000	766,000	807,521
Operating expenses	878,252	926,544	1,017,073	1,262,310
Operational cost recovery ratio (%)	54	64	75	64

	1996	*1997*	*1998*
Revenue collected	1,140,000	1,589,701	1,930,240
Operating expenses	1,624,505	2,494,149	2,709,792
Operational cost recovery ratio (%)	71	64	71

Source: Cuenta Publica, CADF, DGCOH.
Operational cost recovery ratio = operating revenue collected/operating expenses.

not have full information on the total outstanding debt or interest charges. We cannot, therefore, examine how far the system is from generating enough revenues to recover all costs including debt service, and provide a reasonable return on investment.

Accomplishments and Failings of the Contracts

Although the outcomes were far less than intended, the D.F. was still better off with the private contracts than it had been before. The private contractors produce much better information on the customer base and they install and read meters with greater accuracy and more consistency, they bill more customers more accurately and they are more responsive to consumer complaints. It seems unlikely that even these modest changes could have been achieved without private participation. Prior attempts to reform the public agencies in the water sector had had no success, and there was little political will for a reform that would have required restructuring and unification of the government agencies in the sector.

The contracts failed to address some of the most serious problems of the sector, however, and the prospects that these will be tackled in the future are few. The reform failed to improve efficiency, reduce costs, raise revenues or curb waste. And, as we discuss in the next section, serious water resource issues and equity problems remain.

BEYOND THE REFORM: UNRESOLVED KEY POLICY ISSUES

The reform planned — and, even more so, the reform implemented — in Mexico City was very partial in scope. It was not expressly designed to tackle the most serious water resource problems confronting the city though, as originally planned, it would have had an important effect in curbing excess consumption and waste. Similarly while the reform, at least as planned, was intended to increase efficiency in service provision, which might have been expected to improve both service quality and cost options for all segments of society, it was not specifically intended to improve the lot of the poorest of the city's citizens, who also happened to be those with the poorest services at the time of the reform. With these provisos in mind, this section considers some of the incremental effects of the reforms as implemented on these broader policy issues. We first consider its consequences for the negative externalities from water consumption and water pollution; second, we consider its effects on the poor.

Social Costs of Water Consumption

The reform did little to change the incentives of residential consumers to conserve water, despite its high opportunity cost in the D.F. The increasing rate of over-exploitation of the aquifer was reduced, but this appears to have been partly accomplished by simply failing to meet expanding demand.

Over-exploitation of the aquifer has been recognized as a problem for over half a century and it has generated a host of negative externalities, including:

(1) the cost to building owners of damage to their property from subsidence due to over-extraction;
(2) the cost to government of damage to public infrastructure (roads, pipes, etc.) from subsidence due to over-extraction;
(3) the cost to future generations of current unsustainable use manifest through higher costs in the future or diminished quality of life;
(4) the cost of foregone use of water by the surrounding communities whose water is being taken to the D.F.;
(5) raising the cost of water extraction for nearby communities due to higher pumping costs caused by the lower water tables (hence deeper wells) which are the result of over-exploitation by the D.F.

These costs have never been calculated for Mexico City, but as we will show, they are probably very high, much higher than the prices charged by the municipal system.

The period of the reforms did see a slowing in the growth in production of bulk water from the aquifer, but not as a result of any change in residential demand. Between 1988 and 1998, total water produced by the D.F. water system remained almost unchanged at about 35 m^3/s, despite a rise in population of an estimated 650,000 people over the decade. This, combined with a policy of substituting water from Cutzamala for groundwater, reduced the rate at which water was pumped out of the aquifer from 23.5 m^3/s to 19.1 m^3/s. This reduction in pumping, however, is still not enough to stop the subsidence of the city. The poor data available on consumption patterns make it difficult to pinpoint the exact source of this reduction, but it appears to be the result of a reduction in *nonresidential* demand combined with the failure to meet rising residential demand.

As we have seen the tariff rate for large metered users (who are mainly nonresidential) increased by nearly 65% in real terms between 1991 and 1998, at a time when real prices for residential users, who are responsible for three quarters of consumption, were falling. This nonresidential price rise, combined with the effects of a severe recession in 1995–1996, appears to have reduced nonresidential demand. Thus, estimated average

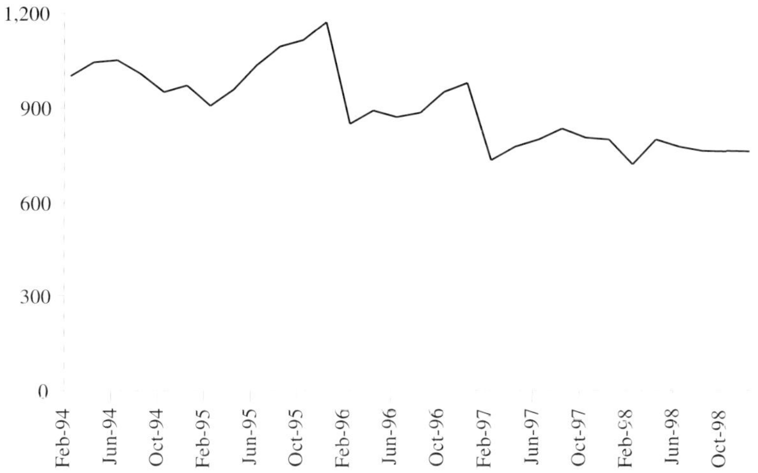

Source: Author's calculations using CADF data.

Figure 5.15. Average consumption of large users, 1994–1998.

consumption for the 20,000 largest users dropped by 20% from 1994 to 1998 (Fig. 5.15).[35] This translates into over 23 million m^3 per year or about 0.74 m^3 per second.

Although the state of the data makes measurement difficult there is little reason to expect that residential water demand declined during this period, given that real prices were not rising. Rather, a growth in rationing and an expansion in the proportion of the population with lower propensity to consume probably explain much of the flat level of production not explained by the drop in nonresidential consumption. Much of the D.F.'s population growth has been in the southern districts, which have lower per capita consumption in part because they have the highest rate of rationing of water and in part because they are poorer and hence are less likely to have gardens, washing machines, multiple bathrooms, etc. From 1990 to 1995 (the last year of available statistics) the population of the northern and central *delegaciones*, which have much higher per capita consumption of water, shrunk by nearly 1% per year, while the southern *delegaciones* grew by an average 2.7% per year (Instituto Nacional de Estadistica Geografia e Infromatica, 1997).

The most important reason for the failure to make further progress in curbing consumption is that water tariffs were far less than the opportunity cost of water. We have already shown that water tariffs did not cover operating costs, although it can be argued that operating costs were much higher than those of an efficiently operated utility. We can illustrate how far tariffs were from the economic cost of water by using the charges CNA applies for extraction rights from private wells, which are based on a 1989 study by the CNA of the relative scarcity of water and its alternative uses. In 1997, for example, the CNA charged private well operators 7.96 pesos per m^3 that they pumped out of the

[35] The implied price elasticity of demand is well within the estimates found in numerous demand studies. For a good summary of these studies see Abdala (1996).

Table 5.12

Financial and economic costs of raw water in the D.F., 1997

Source	*Production (millions m^3)*	*Public-sector price (pesos)*	*Economic price*[a] *(pesos)*	*Public-sector cost (millions pesos)*	*Economic cost (millions pesos)*	*Difference*
Wells	603.6	0.14	7.96	84.5	4,804.6	4,720.1
Cutzamala	313.5	1.42	7.96	445.1	2,495.2	2,050.1
Lerma	154.2	0.85	7.96	131.1	1,227.5	1,096.4
Springs	25.5	0.14	7.96	0.0	25.5	25.5
Total	1,096.8			664.3	8,730.7	8,066.4
Weighted average		0.61	7.96			

[a] as estimated by CNA.
Source: CNA, authors' calculations.

aquifer, while the DGCOH paid only 0.14 pesos per m^3, or less than 2% of the private cost. The discrepancy is not especially surprising in that the charges, while proposed by the CNA, must be approved by congress. Since most of the water system operators are still owned and run by publicly owned municipal companies or departments, a rise in the charge to public operators would mean a rise in the operating budgets of these public entities and a consequent rise in consumer prices or government subsidies.

If we assume that the price CNA charges the private users is a reasonable proxy for the real economic cost of water from the aquifer in the D.F. (and according to CNA it is probably an underestimate), then we can compare the economic cost of water from the aquifer with the price paid by the public sector. That calculation yields an effective subsidy of over 4.7 billion pesos in 1997 (Table 5.12), more than the entire budget of all the D.F. water agencies that year. If we add to this the CNA's estimate of the economic cost of water from other sources (Cutzamala or Lerma), then the total subsidy rises to over 8 billion pesos. If we further assume that operating costs and physical losses (which are about 37%) remain constant, then the average tariff would have to rise to 15.3 pesos per m^3, or about US \$1.93 at 1997 exchange rates, to cover costs. In contrast the actual average tariff (calculated as revenues billed divided by m^3 water produced minus physical losses) in 1997 was 3.4 pesos m^3 or about US \$0.43.

To understand the importance (and the economic cost) of the current high rate of physical losses (37%) we recalculated the prices that would have to be charged if losses fell to 20%. The price that would need to be charged to cover the economic cost would fall from 15.3 pesos to 12 pesos. To look at it another way, if physical losses could be reduced to 20% this would lead to a saving of over 186 million cubic meters per year or 5.9 m^3 per second. This is more than the current Cutzamala extension project is bringing to the city at a cost of over 500 million dollars. However, as long as the D.F. administration does not itself face the full cost of developing and transporting new external water supplies to the city — or the full costs associated with subsidence in the city from overuse of the aquifer — the incentives to pursue such savings are likely to remain weak.

Costs of Wastewater and Sewage Disposal

The disposal of wastewater is a second issue that was not directly addressed by the reforms, and remains a major problem for the city. The sewerage system contributes about 26–28 m^3 of wastewater per second, which is equivalent to 75–80% of the total water produced (Joint Academies Committee on the Mexico City Water Supply, 1995).[36] There are 21 treatment plants around the D.F. with a nominal capacity of 5.9 m^3/s. In 1997 these functioned at an average of 50.8% of capacity or about 3 m^3/s, which suggests that just over 10% of all wastewater was treated. The treated wastewater is used mainly for the watering of public parks (83%), as well as some industrial and commercial uses (12%) and a small amount for irrigation (5%). The remaining 90% of the wastewater is pumped north, out of the D.F. and into the state of Hidalgo where it is used to irrigate 80,000 hectares of farmland.[37] This state of affairs leads to two problems, the negative health effects of irrigation with untreated sewage and pollution of an abundant water source.

Health effects

In 1991 the federal government, recognizing the health risk of irrigating with untreated sewage, banned the practice for surface crops or crops that are eaten raw. This legislation reduced the health threat, but it violated the principle that the polluters should pay for the cost of dealing with their pollution to give them an incentive to pollute less. It passes some of the cost on to consumers in the form of higher prices for these foods. Although some of these consumers may be in Mexico City, the source of the pollution, they are unlikely to be aware of the link between untreated wastewater and food prices, so it would not affect their willingness to pay for sewage treatment. The rest of the cost is borne by farmers who either have to invest in new crops or find clean water to irrigate the same crops. These farmers, who are not even in the same state, are unlikely to have much influence over the D.F. policy towards sewage treatment. Finally the failure to treat the D.F. sewage means that the rivers and streams that receive the runoff are still polluted, which is a threat to the health of people who live near these waters and use them for drinking, fishing, bathing and washing clothes.

Pollution of water sources

The failure to treat the wastewater has consequences for the aquifer as well. Because the D.F. drainage system combines the runoff from rainfall with sewage, storm runoff cannot be used to recharge the aquifer. Since Mexico City is located at the bottom of a valley, much of the city is naturally predisposed to flooding. As a result the thousands of cubic meters of rainwater which fall on the city during the rainy season (up to 120 m^3/s) is pumped out of the city at great cost.

---

[36] Because the drainage system in the D.F. is combined (i.e. sewage and storm water) the amount of wastewater produced fluctuates dramatically depending on the season.

[37] Joint Academies Committee on the Mexico City Water Supply (1995) and DGCOH (1997a).

The DGCOH has carried out studies of ways to improve the system's capacity to treat and reuse water as well as to recharge the aquifer through injection wells or to recharge reservoirs and has developed medium- to long-term plans to increase treatment and recharge. Without a proper accounting of the economic costs of water usage and a rise in consumer tariffs to cover these costs, however, it is unlikely that such efforts will have much effect.

Effects on the Poor

While we cannot measure the income effects of the reforms directly because of the poor state of the data, we can extrapolate their impact on people in different socio-economic strata by examining how their water bills changed due to the reforms. We can also draw some inferences about the extent to which changes in bills were accompanied by changes in service quality. Briefly, the primary effect of the reform in this context was to improve the efficiency of billing (which, for the prevailing tariff structure, meant an increase in bills for some poor households), without offering any significant compensatory improvement in service quality, or the promise of such improvements.

Effects by socio-economic group

One of the effects of the reforms was to increase metering, which changed the size of consumer's bills as shown in Fig. 5.16.[38] Prior to the increase in metering most bills bore little relation to consumption and instead were based on flat rates for different zones, those in the lowest tax zone (presumably the poorest) were not billed at all. As a result of the flat rates many consumers received similar bills despite highly varied consumption.

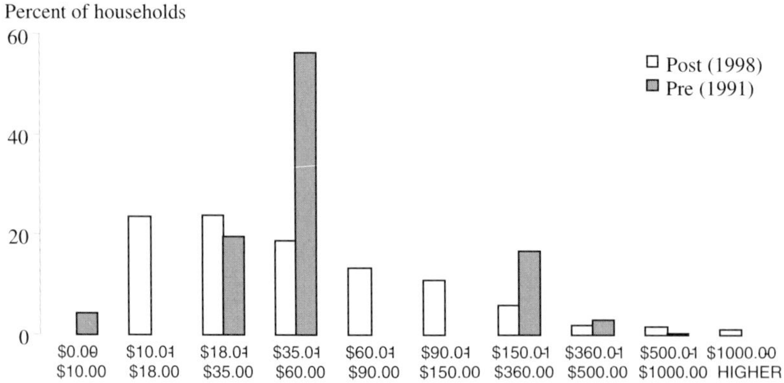

Source: CADF and authors' calculations using INEGI's data.

Figure 5.16. Distribution of residential water bills in the D.F., pre- and post-reform.

[38] The 1991 data are based on the distribution of tax codes which determined the flat rate applicable to water consumption, while the 1998 data are from actual billings.

For example, in 1991, 55% of consumers received bimonthly bills of between 35 and 60 pesos. In 1998 most bills were based on actual, metered consumption. As Fig. 5.16 shows, this raised water bills for the lowest consumption levels, which are often (though not always) the lowest income levels, and left bills unchanged or lower for many of the middle and upper consumption levels.

Service quality and consumption are both highly correlated with income in the D.F., since many of the wealthy who have high water consumption due to modern conveniences (multiple toilets, washing machines, swimming pools, etc.) live in the central and western portions of the city which have good quality service. We can therefore illustrate how the reforms affected different socio-economic strata of society by examining three fairly homogeneous *delegaciones* that broadly represent the lower income/consumption (Milpa Alta), middle income/consumption (Azcapotzalco) and upper income/consumption (Benito Juarez) areas. Naturally, with areas this large it is impossible to categorize all people in the *delegacion* but the findings can be taken as broadly representative. Table 5.13 compares the costs of water for households in constant 1998 pesos for the pre- and post-reform periods. The pre-reform prices are calculated based on fixed charges by tax classification while post-reform prices are from 1998 billing data.

Low-income consumers

Consumers in Milpa Alta were billed about 12 pesos a bimester (two months) in 1998. This was a big change from 1991, because Milpa Alta is in the lowest rated tax zone, their tariff was zero. So for residents of Milpa Alta the change was to actually receive a water bill, although it is still not related to their consumption. Milpa Alta, which in 1998 remained completely unmetered, is an extreme example of the effect on most low-income consumers. The majority of low-income consumers were paying a flat rate of ten pesos per bimester prior to reform. For the most part these areas continue to receive non-metered fixed bills, hence it is unlikely that their consumption patterns have been affected by the reforms. The only noticeable effect of reforms in most poor areas is a rise in the number of people who actually receive a water bill.

Middle-income consumers

If Azcapotzalco is any guide, the water bills for over 40% of middle-income consumers were lower with the introduction of metering, while 34% faced modestly higher bills. Our data do not tell us the extent to which bills went down because they were being overcharged before meters were introduced or because real prices declined.

High-income consumers

Similarly, many consumers in high-income areas such as Benito Juarez found that their water bill stayed the same or decreased in real terms. In Benito Juarez some 37% of consumers were paying 50–75% less in 1998 than they did in 1995.

Table 5.13

Changes in real household bills in the D.F., 1991–1998 (%)

	Amount billed ('98 pesos)				
	0	*10.00–12.00*	*12.01–18.00*	*18.01–35.00*	*Over 35.0*
Low income [a]					
Pre-reform (1991)	100	–	–	–	–
Post-reform (1998)	–	100	–	–	–
	Amount billed ('98 pesos)				
	10.00–18.00	*18.01–35.00*	*35.01–60.00*	*60.01–150.00*	*Over 150.0*
Middle income [b]					
Pre-reform (1991)	–	–	98	2	–
Post-reform (1998)	17	26	21	28	8
	Amount billed ('98 pesos)				
	10.00–18.00	*18.01–35.00*	*35.01–60.00*	*60.01–150.00*	*Over 150.0*
High income [c]					
Pre-reform (1991)	–	–	11	70	19
Post-reform (1998)	12	25	27	26	11

Source: CADF and authors' calculation based on INEGI data.
[a] Delegation: Milpa Alta (100% unmetered connections in January 1998).
[b] Delegation: Azcapotzalco (81% metered connections and 19% average metered in January 1998).
[c] Delegation: Benito Juarez (81% metered connections and 19% average metered in January 1998).

Affordability

At current tariffs, piped water is likely affordable but not always available for the poor in the D.F. leaving many consumers to rely on more expensive sources. Using the 1998 tariffs, combined with census data, we estimate that non-metered water bills for even the poorest households with any reported income[39] would equal about 2% of their monthly salary, which is well within the bounds of any traditional definitions of affordability.[40] Our calculations for a metered household that consumed 15 m^3 per month also equals just 2% of the monthly salary.[41]

[39] This includes only households that report income, obviously for the 3.9% who report no income any positive price is unaffordable.
[40] Using 1998 minimum salary and income (reported as number of minimum salaries) from the 1995 census and the lowest fixed tariff. We have assumed, conservatively, that a household had no more than one income earner.
[41] 15 m^3 was used because it is double the estimated minimum safe level of 50 liter per day (Joint Academies Committee on the Mexico City Water Supply, 1995) assuming five member households which is the average in the D.F. There is also evidence that the poorest in the D.F. currently consume only 20 liter per day (Joint Academies Committee on the Mexico City Water Supply, 1995).

However, the poorest sections of the D.F. are also the least well served by the current system in terms of the quantity of water, the quality of water and the continuity of service. For example, the *delegacion* water department in Iztapalapa estimates that they have a deficit of over 1 m^3 per second, equivalent to 86,400 m^3 per day of unmet demand (field interviews). We have already seen that interruptions in service are commonplace in poorer areas. To compensate for these deficiencies people in these areas substitute expensive bottled water for drinking or buy water from private water trucks. The cost of water from these vendors in September 1998 ranged between 33 and 66 pesos per cubic meter.[42] This is in contrast to the 1 peso per cubic meter price for all customers who use less than 10 cubic meters per bimester and a maximum of 20.3 pesos per cubic meter for residential consumers. To put it another way: the marginal cost per cubic meter for people in some of the poorest neighborhoods of the D.F. is between 64% and 228% higher than the marginal cost for the largest residential consumers.

In sum, then, there were a number of key policy areas in the sector which the reforms did not explicitly address, including resource scarcity, treatment of wastewater and improvement of access and quality of service for the poor. It is difficult to imagine any future reform being effective without dealing with these issues.

CONCLUSIONS

Although the reforms generated a significant amount of information about the system which has begun to replace largely fictitious 'estimates' with real numbers, the overall impact on the sector has been very limited by comparison with the magnitude of the problems it faced. The main problems that existed prior to reform were not addressed: (i) fragmented, costly and inefficient administration of the water system resulting in high costs, poor service and low returns; and (ii) economically irrational pricing of water and pollution resulting in high levels of consumption and waste and low levels of treatment of waste water. Certainly these problems were institutionally and politically difficult to address, involving challenges to powerful interest groups (the existing agencies, labor, and consumers who had long enjoyed significantly underpriced water), and requiring fundamental and politically difficult decisions about water pricing. In addition, the D.F. itself had neither clear jurisdiction nor clear incentives for resolving the major problems of water resource and pollution management. The difficulty of reform has risen with the political changes of the mid-nineties, and is likely to rise further with the implementation of a fully democratic local government that could further fragment administration along party lines.

The service contract model used in Mexico City offers only low-powered incentives to the private partners to improve efficiency or service quality. The contractors assumed little to no commercial risk and receive no significant rewards or penalties based on their performance. The public agencies operating in the system had no incentive to work together, nor were their budgets based on clearly defined efficiency indicators.

[42] The cost of a 6 m^3 tanker in Itztapalapa in 1998 was between 200 and 400 pesos implying 33–66 pesos per m^3 (field interviews).

Moreover, the progressive shift of responsibility to the private operators envisioned in the original contract did not happen, and is now highly unlikely. Accordingly, relatively few of the expected benefits will actually be captured.

The reforms also failed to exploit fully the opportunities for competition presented by the use of competitive bidding for the contracts, and the division of the city into four zones. Although the D.F. had the necessary ingredients for competitive bidding, the structure of the auction did not really use competition to reduce costs to consumers or motivate efficiency. The decision to choose four contractors on the basis of bids for actions which might never, and in the event will never, take place, meant that the bids were not strictly comparable. As a result the expected benefits of competition, particularly lower costs and more accurate information about prices, were not achieved. Since the zones were also not structured to be comparable, it is relatively difficult to use yardstick competition. In particular, the network in zone C in the southeast is in worse repair than the others, and both water quality and service quality are much lower. Nevertheless, it might be possible to benchmark *delegaciones* from zones that share similar socio-economic and water system characteristics, such as Milpa Alta and Cuajimalpa, which are both rural, low density areas with low metering, and Benito Juarez and Cuahtemoc, which are relatively wealthy, with high per capita consumption and a high degree of metering. If these comparisons could be tied to a system of rewards and penalties it would help improve performance. In practice, little attempt was made to take advantage of the potential for yardstick competition, or to identify ways of making legitimate comparisons between zones.

The experience of the D.F. suggests a number of lessons for policy makers elsewhere.

(1) It is possible for a water system to be in a state of 'crisis' for very long periods of time without triggering genuine reform, especially if water and wastewater services are not priced in accordance with their opportunity costs. This is largely because the resource is not visible to the general public and therefore its increasing scarcity is not widely appreciated. This lack of knowledge makes politically costly reforms, such as price rises, extremely difficult.

(2) Gradualist reforms that defer politically difficult decisions or leave them to future governments involve a high risk that the reform will not be implemented as originally intended.

(3) Service contracts, which are often portrayed as a stepping stone to management contracts, leases or concessions, do not necessarily create circumstances that make further reform more likely. In fact, the experience of firms to date, particularly with politically motivated pricing and unpredictable renegotiation, may actually make further private sector involvement less likely.

(4) The use of bidding for the market — or the use of multiple zones — are not in themselves sufficient to sustain a competitive environment. For competitive pressures to be created and maintained, attention must be paid, first, to making the basis of the bidding highly transparent, and, second, to the development of clear benchmarks, and agreed processes for benchmark comparisons, and associated allocation of penalties and rewards.

(5) Fragmentation of water system management among many, poorly coordinated,

politicized public agencies makes reform much harder to implement and the realization of potential benefits less likely.

(6) In the absence of prices that correctly signal the scarcity and cost of a resource, including the relative prices of treated to new water, attempts to change behavior will not succeed. The price signals must apply to all actors in the system, including governments, operators and consumers; thus, large central government subsidies to the local government will be a barrier to greater efficiency.

Thorough-going reforms aimed at achieving a more equitable and sustainable water delivery system in Mexico City would need to focus not only on major efficiency improvements in the operation of, and investment in, services within the city, but also, more broadly, on substantial reform to the way in which water resources are priced and allocated in the city and the surrounding region, as well as in the design, management and pricing of wastewater services. This would need to include reductions in federal subsidies for new production, higher prices for system operators as well as consumers, and large investments in the treatment and storage of wastewater. None of these changes is likely without strong political leadership, both within the D.F., and in broader national water policies.

APPENDIX A

Table 5A.1

Electoral reforms in Mexico

Year	*Reform*	*Result*
1977	Reformed the federal law of political organizations and electoral process (LOPPE) by: • Increasing the number of majority districts for federal deputies from about 200 to 300 seats. • Setting aside an additional 100 seats to be allocated to opposition parties on a proportional basis.	This reform raised the level of opposition participation in the lower house from about 17% (1964–1976) to about 26% (1977–1985).
1982	President de la Madrid established new policy of accepting municipal-level right-wing (PAN) opposition victories but state-level victories still not allowed.	The acceptance of opposition control of municipalities was a large change but the de la Madrid administration would not accept electoral gains of left-wing parties in Oaxaca.
1986	President de la Madrid issues new electoral code. Basic changes include: • Winning party cannot obtain over 70% of total lower chamber seats. • Proportionately allotted seats increased from 100 to 200 (total lower house increased from 400 to 500). • Opposition can gain 40% of seats (200) without winning any majority districts. • Party that gains largest number of majority seats must be given a simple majority of the lower house. • 50% of senate to be elected every 3 years (instead of whole senate each 6 years)	More increases in representation of the opposition but they are given by the government not won at the ballot box.

Source: Adopted from Camp (1996), with additions from Gomez Tagle (1993).

REFERENCES

Abdala, Manuel Angel, 1996. "Welfare Effects of Buenos Aires' Water and Sewerage Services Privatization." Processed.

Alonso, Jorge, Alberto Aziz and Jaime Tamayo, eds., 1992. *El Nuevo Estado Mexicano: III Estado, Actores y Movimientos Sociales.* Mexico, DF: Universidad de Guadalajara, Nueva Imagen y Ciesas.

Baer, M. Delal, 1993. "Mexico's second revolution: pathways to liberalization". In: Roett Riordan, ed., *Political and Economic Liberalization in Mexico: at a Critical Juncture*? Boulder, CO: Lynne Rienner Publishers.

Bailey, John and Jennifer Boone, 1994. "National solidarity: a summary of program elements". In: Wayne Cornelius, Ann Craig and Jonathan Fox, eds., *Transforming State–Society Relations in Mexico.* San Diego, CA: Center for U.S.–Mexican Studies, University of California at San Diego.

Basáñez, Miguel, 1993. "Is Mexico headed toward its fifth crisis?" In: Roett Riordan, ed., *Political and Economic Liberalization in Mexico: at a Critical Juncture*? Boulder, CO: Lynne Rienner Publishers.

Brook Cowen, Penelope, 1997. "Getting the private sector involved in water: what to do in the poorest of countries?" *Viewpoint* Note 102, Washington, DC: World Bank.

CADF (Comision de Agua del Distrito Federal), 1997. "Progress report". Mexico, DF: CADF. Processed.

CADF (Comision de Agua del Distrito Federal), 1998. "Progress report". Mexico, DF: CADF. Processed.

Calderon Alzati, Enrique and Daniel Cazes, 1991. *Prontuario de Resultados: Elecciones Federales de 1988.* Mexico, D.F.

Camp, Roderic Ai, 1996. *Politics in Mexico.* New York: Oxford University Press.

Centeno, Miguel Angel, 1994. *Democracy within Reason.* University Park, PA: The Pennsylvania State University Press.

Cornelius, Wayne, Judith Gentleman and Peter Smith, eds., 1989. *Mexico's Alternative Political Futures.* San Diego, CA: University of California at San Diego.

Cornelius, Wayne, Ann Craig and Jonathan Fox, eds., 1994. *Transforming State–Society Relations in Mexico.* San Diego, CA: Center for U.S.–Mexican Studies, University of California at San Diego.

Coulomb, Rene, 1989. "La participacion popular en la provision de los servicios urbanos: Estrategias de sobrevivencia o practicas autogestionarias?" In: Martha Schteingart and Luciano d'Andrea, eds., *Servicios Urbanos, Gestion Local y Medio Ambiente.* Mexico, DF: El Colegio de Mexico.

Crespo, Jose Antonio, 1994. "Dominación y hegemonia en los sistemas partidistas de México y Japón." Foro Internacional (México) 34: 437–456, July-September, 1994.

DGCOH (Direction General de Construccion y Operacion Hidraulica), 1994. *Agua 2000, Estrategia para la ciudad de Mexico.* Mexico, DF: Departamento del Distrito Federal.

DGCOH (Direction General de Construccion y Operacion Hidraulica), 1997a. *Compendio, 1997.* Mexico, DF: Departamento del Distrito Federal. Processed.

DGCOH (Direction General de Construccion y Operacion Hidraulica), 1997b. *Plan Maestro de Agua Potable del Distrito Federal 1997–2000.* Mexico, DF: Departamento del Distrito Federal.

DGCOH (Direction General de Construccion y Operacion Hidraulica), 1998. *Programa de Tandeos en las Delegaciones del Distrito Federal.* Mexico, DF: Departamento del Distrito Federal. Processed.

Dresser, Denise, 1994. "Bringing the poor back in national solidarity as a strategy of regime legitimation". In: Wayne Cornelius, Ann Craig and Jonathan Fox, eds., *Transforming State–Society Relations in Mexico.* San Diego, CA: Center for U.S.–Mexican Studies, University of California at San Diego.

Gamboa de Buen, Jorge, 1994. *Ciudad de Mexico, una vision.* Mexico, DF: Fondo de Cultura Economica, S.A.

Gobierno del Distrito Federal, 1997. *Codigo financiero del Distrito Federal.* México, D.F. Gazeta Oficial del Gobierno del Distrito Federal.

Gobierno de México, 1992. Decreto 14, Julio 1994. México, D.F. Gazeta Oficial del Gobierno de México.

Gomez Tagle, Silvia, 1993. "Electoral reform and the party system 1977–90". In: Neil Harvey, ed., *Mexico Dilemmas of Transition.* London: Institute of Latin American Studies, University of London and British Academic Press.

Haber, Paul, 1993. "Cárdenas, Salinas and the urban popular movement". In: Neil Harvey, ed., *Mexico Dilemmas of Transition.* London: Institute of Latin American Studies, University of London and British Academic Press.

Haber, Paul, 1996. "El arte de la reestructuración y sus implicaciones políticas: el caso de los movimientos urbanos populares". In: Maria Lorena Cook, Kevin Middlebrook and Juan Molinar Horcasitas, eds., *Las Dimensiones Políticas de la Reestructuración Económica*. Mexico, DF: Cal y Arena.

Harvey, Neil, ed., 1993. *Mexico Dilemmas of Transition*. London: Institute of Latin American Studies, University of London and British Academic Press.

Instituto Federal Electoral (IFE), 1991.*Estadistica de las Elecciones, Elecciones Federales de 1991*. Mexico, DF: IFE.

Instituto Nacional de Estadistica Geografia e Infromatica, 1997. *Distrito Federal Conteo de Poblacion y Vivienda*. Mexico, DF: INEGI (CD ROM).

Joint Academies Committee on the Mexico City Water Supply, 1995. *Mexico City's Water Supply: Improving the Outlook for Sustainability*. Washington, DC: National Academy Press.

Klesner, Joseph, 1996. "Broadening toward Democracy?" In: Laura Randall, ed., *Changing Structure of Mexico: Political, Social and Economic Prospects*. Armonk, NY: M.E. Sharpe.

Lustig, Nora, 1994. "Solidarity as a strategy of poverty alleviation". In: Wayne Cornelius, Ann Craig and Jonathan Fox, eds., *Transforming State–Society Relations in Mexico*. San Diego, CA: Center for U.S.–Mexican Studies, University of California at San Diego.

Molinar Horcasitas, Juan and Jeffrey Weldon, 1994. "Electoral determinants and consequences of national solidarity". In: Wayne Cornelius, Ann Craig and Jonathan Fox, eds., *Transforming State–Society Relations in Mexico*. San Diego, CA: Center for U.S.–Mexican Studies, University of California at San Diego.

Ochoa, Ontoniel, 1998. "El cambio estructural del sector agua del Distrito Federal 1992–1997". Processed.

Purcell, Susan Kaufman, 1995. "The interdependence of political and economic reforms in Mexico". *SAIS Review*, 15: 1–13 Winter–Spring, 1995.

Ramirez Saiz, Juan Manuel, 1992. "Entre el corporativismo social y la logica electoral. El estado y el movimiento urbano popular (MUP)". In: Jorge Alonso, Alberto Aziz and Jaime Tamayo, eds., *El Nuevo Estado Mexicano: III Estado, Actores y Movimientos Sociales*. Guadalajara: Universidad de Guadalajara, Mexico.

Reding, Andrew, 1989. "Mexico under Salinas: a façade of reform". *World Policy Journal* 6: 685–729.

Roett, Riordan, ed., 1993. *Political and Economic Liberalization in Mexico: at a Critical Juncture*? Boulder, CO: Lynne Rienner Publishers.

Salinas de Gortari, Carlos, 1990. *Segundo informe de gobierno*. México, D.F. Dirección de Publicaciones.

Schteingart, Martha, ed., 1997. *Pobreza, Condiciones de Vida y Salud en la Ciudad de Mexico*. Mexico, DF: El Colegio de Mexico.

Schteingart, Martha and María Torres, 1997. "Politicas de agua y drenaje en la ciudad de mexico, y su aplicacion en las colonias estudiadas". In: Martha Schteingart, ed., *Pobreza, Condiciones de Vida y Salud en la Ciudad de Mexico*. Mexico, DF: El Colegio de Mexico.

Teichman, Judith, 1995. *Privatization and Political Change in Mexico*. Pittsburgh, PA: University of Pittsburgh Press.

Ward, Peter, 1989. "Government without Democracy in Mexico City. Defending the High Ground". In: Wayne Cornelius, Judith Gentleman and Peter Smith, eds., *Mexico's Alternative Political Futures*. San Diego, CA: University of California at San Diego.

Wiarda, Howard, 1988. "Mexico: the unravelling of a corporatist regime?" *Journal of Interamerican Studies and World Affairs* 30: 1–28.

World Bank, 1997. *World Development Indicators*. Washington, DC: World Bank (CD ROM).

World Bank, 1998. "Mexico: enhancing factor productivity growth". *Country Economic Memorandum*, Report Number 17392-ME. Washington, DC: World Bank. Processed.

Yepes, G. and A. Dianderas, 1996. *Water and Wastewater Utilities: Indicators*. Washington, DC: Water and Sanitation Division, World Bank (2nd edition). Processed.

Thirsting for Efficiency: The Economics and
Politics of Urban Water System Reform
Mary M. Shirley (Ed.)

CHAPTER 6

REFORMING URBAN WATER SUPPLY: THE CASE OF CHILE

MARY M. SHIRLEY, LIXIN COLIN XU AND ANA MARIA ZULUAGA

INTRODUCTION

In the late 1980s Chile intended to privatize the Santiago Metropolitan Sanitary Works Enterprise (Empresa Metropolitana de Obras Sanitarias, or EMOS). Instead EMOS was ultimately reformed under public ownership, by means of a regulatory framework that mimicked the design of a concession with a private utility. (Table 6.1 lists the reform components.) In particular, the new regulation set tariffs so as to assure at least a 7% return on assets, created a neutral regulator independent of ministry intervention, and gave EMOS the right to appeal the regulator's tariff decisions. As we shall see, the changes allowed significant price increases, which helped fund improved maintenance and enabled EMOS to expand as Santiago's population grew. The reform produced net benefits in economic welfare that are surprisingly large given the relatively good initial conditions. These gains largely accrued to government, although consumers benefited from almost 100% coverage of expanding demand, better water pressure and fewer interruptions of service. Consumers also had to pay higher prices, but the effects were ameliorated by a direct subsidy. Employees gained from wages closer to market wages. By reducing information asymmetries, increasing managerial incentives and signaling commitment, the reform fostered important efficiency gains in EMOS.

Many observers regard Santiago's water system reform as an important example of successful reform under public management. This chapter addresses the central question of why this reform improved performance, when similar attempts to reform under public ownership have met with failure in other countries. It also addresses the question of why EMOS was not privatized, but was reformed, and what were the costs, if any, of not privatizing. To answer these questions we analyze how institutions, in the broad sense of rules and norms delimitating and guiding human behavior and social interactions,

Table 6.1

Principal components of the reform in Santiago's water regulation

	Old	*New*
Tariffs	Ad hoc increases set by the Ministry of Economy (issued as Presidential decrees) within a ceiling (no more than 10% return on fixed assets).	Tariffs indexed to inflation and adjusted every five years according to a formula based on marginal costs. New tariff formula introduced gradually from 1990 to 1995 starting with a 30% real increase in 1990.
Subsidies	Cross subsidies for smaller amounts of consumption (first ten cubic meters of consumption had lower tariffs).	Direct subsidy paid to the company by the municipality from federal budget for all customers within a low-income category since 1990. Covered an average of 60% of first 15 (later 20) m^3 of consumption per month.
Regulator	National Service of Sanitary Works (SENDOS), an autonomous public entity attached to the Ministry of Public Works, directly managed 11 of the country's 13 regional water offices. EMOS was autonomous.	Superintendent of Sanitary Services, a regulatory body attached to the Ministry of Public Works, responsible for setting tariffs, setting and enforcing technical standards, and granting concessions for water supply and sewerage services since 1990.
Company status	Autonomous entity under public law.	State-owned corporation under private commercial law since 1988.

See Republic of Chile, Ministry of Economy, 1990a,b; Ministry of Finance, 1981, 1989, 1991; Ministry of Public Works, 1988a,b; 1990 and 1991.
Source: authors' calculations.

influence choice and economic performance (North, 1990). We examine the gains from reform comprehensively, including consumer surplus and effects on labor, buyers and government compared to a counterfactual.

As we shall show, political circumstances led to a regulatory framework designed to attract private investors to the water system, and motivate them to operate efficiently and expand the system. Chile's unique electoral institutions sustained this framework under state operation after democracy was restored. Chile's bureaucratic institutions and norms, combined with Santiago's relatively low-cost water system, allowed it to implement prices that in effect increased quasi-rents for investment in the system, while minimizing the risk of inefficiency or monopoly rents. The regulation was enshrined in detailed legislation backed by constitutional safeguards. Since political institutions make it hard to change laws in Chile, this legal foundation had strong credibility. The lack of regulatory discretion also protected the regulator from politically motivated intervention, while a direct consumer subsidy reduced political incentives to press for lower tariffs.

The next section describes the circumstances that made reform politically desirable and feasible despite relatively good initial conditions in the water system, and explains why EMOS was not privatized but was reformed under public ownership. We then analyze the characteristics of the reform in terms of an implicit 'regulatory contract' supported by fundamental bureaucratic and judicial institutions. The chapter then assesses how the contract addressed agency problems by improving information, raising incentives and signaling commitment. Following that, we measure the effect of the reform on economic welfare and other performance measures and show that the gains were relatively large despite good initial conditions. Finally, we conclude with a discussion of the implications of our findings.

THE DECISION TO REFORM

We might assume that governments would seek water reform where there is a pressing need to expand coverage of unmet demand, improve the operation of a poorly run company, or reduce a fiscal drain. The slow pace of reform in cities where the water system has been in crisis for years, however, suggests that serious sector problems alone are not what motivates water reforms.[1] Rather, we expect that reform occurs when a macro-economic crisis shifts political circumstances so that reform becomes politically desirable, in the sense that the political benefits outweigh the political costs, and politically feasible, in the sense that those favoring reform can win control of the veto gates and overcome opposition.

Under many circumstances the political costs of water system reform are likely to be high. One reason for this is that reform often reverses a long history of underpricing of water services. Since capital investments in water are very durable, a water company may be able to operate for years without recovering its fixed costs (Chapter 2). As a result many governments, including Chile's, have chosen to keep water tariffs low for long periods, sometimes for so long that the company begins to under-invest in maintenance as well as expansion. Political costs of water reform may also be large in cases where water has high political saliency. Some countries, such as Mexico, have laws or constitutional prohibitions against cutting off consumers who fail to pay their water bills. Another reason why the political costs of water reform may outweigh the benefits is that the beneficiaries of water reforms tend to be poor, recent migrants and the opponents tend to be wealthier, better organized workers and consumers who are already connected. For all these reasons we expect that water reform follows a change in the net political benefits from reform.

The rest of this section describes the relatively good conditions in the water system in Santiago, explains how the political equation changed in Santiago to make reform politically desirable and feasible, and discusses the political circumstances that led to the unusual reform that resulted.

[1] For example, planned reforms were not implemented in Lima in the 1990s, even though at least a quarter of the population was not connected and those with connections had no water for 15 hours a day on average. In Mexico City, depletion of the aquifer has been causing the city to sink yet low tariffs and deficient bill collection encouraged high levels of water consumption until recently. See Chapters 4 and 5.

Circumstances in the Water Sector Leading to Reform

Absent politics, the circumstances in the water sector that would be likely to provoke reform are (i) water shortages, (ii) large unmet demand, (ii) deficient management, and (iv) financial problems. As we show below, none of these problems except finances were especially serious in EMOS' case. Initial conditions in Santiago's water sector were relatively good: raw water was cheap and abundant, the demand for water and sewerage was largely met, and the company was reasonably well managed. EMOS' main problems were due to a lack of funds for maintenance and expansion.

Water shortages

At the time of the 1990 reform there were no unpredictable shortfalls that might motivate consumer dissatisfaction, nor was there a need for costly new investments to expand supply in the near term. EMOS' main source of water was relatively abundant, good quality and cheap surface water from the Maipo River, supplemented by a lake and 90 deep wells. The system was sustainable, in the sense that EMOS' own water sources were sufficient to meet Santiago's projected demand for the next 10 to 20 years (World Bank project documents).[2] Supply was also sufficient even during low periods, although reliance on snow melt during summer led to some variability (field interviews with engineering faculty members, University of Chile). Furthermore, alternative sources were available, although expensive.[3]

Demand

Unmet demand for water or sewerage was not a motivating factor for reform. Since 1980, about 98% of EMOS' potential clients had water connections, and 90% had sewer connections (see Table 6.2).[4] These data overstate coverage somewhat since clients in EMOS' definition did not include households in informal settlements that had been built without networks nor did it incorporate several poor communities just outside its concession boundary.[5] The population without connections to the piped water

[2] Supply was 1997 was considered sufficient to cover current average consumption until 2005 (field interviews with the Superintendencia staff). Note that all field interviews took place in 1997.

[3] Water rights are tradeable but costly. Water rights are almost entirely allocated to private parties. The water market functions poorly, with few transactions and high transaction costs, in part because of the high cost of physically changing canal flows and in part because much of the water rights are not legally inscribed but are exercised on the basis of traditional holdings. (Chile has allowed private water rights since the 1920s.) EMOS purchased 33 shares (a share is equivalent to at least 22 liters per second 85% of the time) between 1990 and 1993 at an average cost of US $10,000 per share. This is a small purchase, as evidenced by the fact that EMOS owns 1369 shares of the Río Maipo water, while total shares are 8133. There have been too few transactions to know how typical this purchase price would be (Hearne and Easter, 1995; Rios and Quiroz, 1995).

[4] These numbers overstate coverage somewhat because the indigent population without legal housing (about 5% of the city's population) are excluded.

[5] The legal responsibility for building housing connections and networks ready for hookup to EMOS' infrastructure lay with the developer, or in the case of poor, informal settlements, with the local government. Santiago's 40 odd municipalities lacked the funds to build networks or housing connections.

Table 6.2

Coverage of water and sewerage in Santiago

Year	*Client population (000)*	*Percent with water connection*	*Percent with sewerage connection*
1980	3,597	98.6	85.0
1981	3,675	97.8	83.9
1982	3,750	98.5	84.6
1983	3,860	98.6	84.6
1984	3,926	98.6	87.0
1985	4,049	99.0	87.3
1986	4,287	99.0	88.1
1987	4,250	99.2	90.0
1988	4,422	99.2	90.3
1989	4,593	99.2	90.5
1990	4,816	98.0	90.3

Source: EMOS, Statistical Bulletin, various.

system in Santiago was overwhelmingly poor. Most lived in informal settlements built without a secondary network and would probably not have been able to finance the cost of connecting, which could be more than US $1000 (field interviews). The Pinochet administration's program to eliminate illegal settlements and rental tenements in the city center added to the number of unconnected consumers by leading many poor people to relocate to the far south of Santiago in areas with little existing infrastructure (Gilbert, 1993). These unconnected citizens had little political influence, as we explain below.

Demand for sewage treatment was low at the time of the reform, even though only about 1% of wastewater was being treated (raised to 3% in 1995). Santiago had higher rates of waterborne disease than the rest of the country because food crops were irrigated with contaminated water, but consumers had little information about the health risks.[6] More recently, with the outbreak of cholera in Peru in 1993 and the rising economic importance of Chile's agricultural exports, concerns over the health and economic effects of irrigating food crops with untreated wastewater have increased.[7]

Management

Mismanagement is often a motivation for reform, but this was not a significant problem in Santiago. In 1990, at the time of the reform, EMOS served a water and sewerage

[6] Polluted wastewater was used to irrigate 130,000 ha, including 7000 ha used for growing raw vegetable crops for immediate consumption in the metropolitan region. Santiago averaged 150 cases of typhoid per 100,000 inhabitants a year from 1985 to 1991 (compared to a rate of 50 cases per 100,000 for all of Chile) and epidemiological analysis suggests that the higher rate of endemic typhoid in Santiago was causally linked to irrigating vegetables with polluted water (World Bank, 1994).

[7] The Santiago metropolitan region produces about 40% of Chile's exportable fruit (fruit represents about 85% of agricultural exports for human consumption) (World Bank, 1994).

concession area that covered most of the Santiago metropolitan region, with water connections to some 4.7 million people.[8] EMOS was regarded as efficiently managed and performing well in comparison with other water companies in the region (World Bank, 1992, p. 34).[9] Connections were almost entirely metered, and bill collection rates were over 80%, high by regional standards. Unaccounted-for water or UFW was high at 31% of total water produced in 1989, but this rate was down from 38% in 1987.[10] Furthermore, according to World Bank, 1992, high UFW was not the result of mismanagement but of failure by the Ministry of Finance to permit EMOS sufficient investment funds to maintain its system, something we discuss in the next section.

Overstaffing was not a problem thanks to a 1977 effort by the Pinochet government to reduce the size of the public sector and expand competitive procurement from private providers through outsourcing. By contracting out such activities as meter distribution, reading and repair; network maintenance; and transport, EMOS reduced its staff from 3200 workers in 1977 to 1700 by December 1989, or from an average of 2.4 staff per 1000 water connections in the early eighties to 2.1 in 1990.[11]

Financial problems

Finance was the one area where EMOS had serious problems. The company ran a loss before taxes from the early 1980s until 1987, largely because its tariffs were not allowed to increase in real terms. Tariffs during this period were substantially below World Bank estimates of EMOS' long-run marginal costs (World Bank, 1992, p. 33).[12] EMOS' rate of return on total assets in 1989 was only 2.7%, which was still better than the average return of all Chile's public water companies at that time: negative 1.6% (Table 6.3).

EMOS' low cash flow, combined with borrowing constraints imposed by the Ministry of Finance to keep the debt of state-owned enterprises low, explain why the company under-invested in maintenance and expansion. Some of EMOS' facilities were old and much in need of repair and replacement by 1990. For example, it had collection and treatment works dating to 1917 and 20% of its network of pipes exceeded their usable life of 30 years. This contributed to 52 pipe breaks per 100 km in 1989, compared to a US average of 17. The World Bank estimated in 1986 that EMOS would need to invest

[8] The population in EMOS' concession area at that time represented 87% of the population of Santiago; the rest of the city was served by a small municipal and a private water company. The private company, Lo Castillo (later Aguas Cordillera), operated in a largely high-income neighborhood with about 5% of the population of Santiago. The municipal company, Maipu, is in a middle- to low-income area of Santiago and served about 7% of the metropolitan population.

[9] EMOS was created in 1977 as an autonomous decentralized government company.

[10] Defined as the difference between the volume of water produced and the volume billed (i.e., the losses due to leaks, metering inefficiencies, theft, and the like). In 1984 unaccounted-for water averaged 12% in the US and 15% in Canada.

[11] Some observers suggested that EMOS might have been understaffed in certain skills as a result of these reductions (World Bank, 1986a, and field interviews).

[12] The World Bank estimated the long-run marginal costs of EMOS' 1982 to 1985 water and sewerage investment program (at a discount rate of 11%) at 25.7 US \$ cents/m^3. In comparison, its actual tariff rate for water and sewerage was 10.5 US \$ cents/m^3 in 1987 and 11.5 in 1988 (World Bank, 1992, p. 53).

Table 6.3
Rate of return on assets of water companies in Chile (percent after tax profits/total assets)

	EMOS	*All state enterprises*	*Private and municipal*	*All water companies*
1988	3.6	−1.4	−0.1	−1.4
1989	2.7	−1.6	13.4	−0.8
1990	4.8	−1.1	3.9	−0.8
1991	6.6	−0.2	1.6	−0.1
1992	8.4	0.9	4.0	−0.9
1993	9.7	3.4	6.5	3.6
1994	11.5	5.1	7.2	5.2
1995	11.0	5.9	10.0	6.3

Source: Republic of Chile and Superintendency of Sanitary Services (1995).

US $118.6 million from 1987 to 1989 to meet projected demand and maintenance needs; but its actual investment for that period was only $24 million (all figures in constant 1989 dollars).[13] Although service was still reasonably good, without additional funds the company would not be able to keep water pressure up, avoid even more frequent service interruptions and expand the system to keep up with population growth (field interviews).

Political Circumstances Leading to Reform

Crisis

We expect that reform is more likely to occur after a crisis alters the political costs and benefits to the decision-makers. The crisis which played this role in Chile was hyperinflation followed by the 1973 military coup, which overthrew the elected Socialist President Salvador Allende and brought to power the dictatorship of Army General Agusto Pinochet. The Pinochet government's ideology was based on free market principles and advocated small and efficient government; early in its tenure the Pinochet administration decided to sell EMOS.[14]

[13] World Bank, 1986b. These projected investment requirements are probably an overestimate since they were based on demand estimates that assumed a lower price than actually was charged (a 9% real increase in 1987, 6% in 1988, and 9% in 1989, compared to actual real increases of 17% in 1987 and 10% in 1988 and 1989). Nevertheless, most observers agree that EMOS was seriously under-investing in maintenance and expansion during this period.

[14] In keeping with these principles the administration removed all quantitative restrictions on trade and reduced tariffs to 10% across the board, privatized virtually all state-owned finance and manufacturing enterprises and much public infrastructure, and replaced a public pay-as-you-go social security system with one based on private pension funds. It also eliminated most subsides and transfers and required profitable SOEs to pay dividends, while, as we have seen, pushed public agencies to reduce their in-house staff and procure services competitively.

Political benefits

Although Pinochet was committed to privatize EMOS, it does not seem to have been a particularly urgent or important component of the government's program. As we have seen, most of the usual motivations for change were absent or weak in Santiago's water system. Moreover, many of the main beneficiaries of reform were not important constituencies of the Pinochet government.

The main potential beneficiaries of a water reform are: (i) users who would otherwise not be connected, (ii) those already connected consumers who value the benefits from improvements in service quality more than the cost from higher prices, and (iii) private investors who benefit from the expansion in construction of water infrastructure or the opportunity to operate the firm. In the case of Santiago the first two groups of beneficiaries were not a powerful force for reform. Their interests did not coincide: EMOS' existing customers could expect to see their water bills increase if EMOS expanded more rapidly. This reduced their motivation to organize and jointly pressure for reform. More importantly, even if they had organized, they would not have had a sympathetic hearing, since neither new nor existing EMOS customers were important constituents of the Pinochet administration. The military government's strongest support in general was rural and upper income (Méndez, 1990). Santiago in particular was not an important source of support. Evidence for this can be found in the city's voting record. The metropolitan area had the second highest vote of any region in the country against Pinochet's new constitution in the 1980 plebiscite — 36% voted no compared to 30% nationwide (*El Mercurio*, September 12, 1980). The city also had the second highest vote against the continuation of the Pinochet regime in the 1988 plebiscite — 58% of Santiago voted against Pinochet versus 53% in the rest of the country (*El Mercurio*, October 7, 1988).

Although Santiago was not a constituency, the Pinochet government had an interest in national water reform and in the sale of EMOS. Other Chilean water companies were performing much worse than EMOS and operated in cities where the government had more supporters. Hence there was an interest in improving their regulation, and since water is regulated nationally any change would also affect EMOS.

Furthermore, the administration had one constituency group in Santiago that wanted reform. If EMOS had more capital to invest, contractors and developers would benefit from more contracts with EMOS as well as from expanded real estate development. The municipal area had been reduced from 100,000 ha to 60,000 in the mid-1980s to control sprawl, which meant that most new developments in Santiago took the form of greater density within the EMOS concession area, rather than movement into areas where rival water companies might operate. This restriction combined with advances in anti-seismic construction fostered a surge of high rise development in Santiago. Because of its deteriorating financial situation, these contractors were not confident that EMOS could expand its infrastructure fast enough to meet demand or maintain enough pressure to accommodate new development (field interviews). The developers and construction contractors were part of the constituent base of the Pinochet government: according to surveys before the election of 1989, the Pinochet administration had strong support from upper-income groups (Méndez, 1990) and business associations (Weyland, 1997).[15] Developers and contractors help draft the water legislation and strongly supported EMOS' privatization (interviews with members of the Chilean Chamber of Construction).

Political costs

The major opposition to reform came from EMOS' workers. Although layoffs due to outsourcing had occurred under public ownership, EMOS' workers believed they still had much to fear from privatization. Firing state enterprise workers was difficult, so previous reductions had been voluntary, through early retirement and agreements with private contractors to hire EMOS staff whose job was outsourced.[16] EMOS' workers feared outright dismissals under a private owner (interviews with union officials). Moreover, even if they could retain their jobs, EMOS' workers expected that privatization would reduce their job security.[17]

Worker opposition to the reform was not a deterrent, however. EMOS' workers were not part of the Pinochet government's constituency. Trade unions in general had been vocal opponents of the administration, and most of EMOS workers had opposed the regime according to field interviews with union leaders.[18] Moreover, in other privatizations the Pinochet government had overcome opposition from workers who might otherwise have engaged in disruptive strikes or demonstrations, such as those in ports, electricity or telecommunications, through a combination of compulsion and compensation (World Bank, 1995).[19]

Why was EMOS not Privatized but Reformed?

In this section we first explain the political circumstances that delayed and ultimately stopped the intended sale of EMOS in 1989. We then discuss how political changes combined with Chile's electoral and political institutions led the government to privatize the water regulation without privatizing ownership of the assets of Santiago's state-owned water company.

[15] Weyland (1997) cites the strong support of the leading confederation of businessmen (CPC, the Confederation of Manufacturing and Trade) for the military government's economic program.

[16] EMOS' ordinary retirement package was one month for each year of service; workers asked to retire were given two months (field interviews). Activities which might still be outsourced included bill collection, engineering and the company kitchen.

[17] Field interviews. Line workers represent approximately 60% of EMOS' staff and have their own unions. (EMOS has three unions: one for operators and other line workers, one for clerical staff and one for professional and technical staff. Almost 100% of eligible staff are unionized.) EMOS has had low turnover except for the outsourcing.

[18] Weyland (1997) argues that workers in general were strong opponents of the Pinochet government because it had limited trade union rights.

[19] The usual compensation was the chance to buy 10% of the shares at the offer price by borrowing on workers' severance pay.

Why was EMOS not privatized?

The sale of water assets had low political saliency for the Pinochet administration and for that reason it moved more slowly to privatize EMOS than other utilities. One reason for this low saliency was the relatively minor water problems in Santiago combined with the relatively small proportion of beneficiaries who were constituents of the regime. A second cause was financial: privatization in other sectors would bring much higher revenues than in water. EMOS was one of the few water companies that would command a good price and in 1991 the sales value of EMOS was estimated at about US $150 million. In contrast the 4 telecommunications transactions raised over $540 million and 3 of the 14 electricity transactions generated US $220 million.

In addition, the leadership believed that they had a longer window of opportunity to privatize water than they ultimately had. Although a plebiscite on the continuation of the regime was scheduled for 1998, Pinochet was convinced that he would win this vote and stay in power until 1997 (Constable and Valenzuela, 1989).[20] For all these reasons, new water legislation was only enacted in 1988–1990, after much of telecommunications and electricity had been privatized.

The defeat of Pinochet in the October 1988 plebiscite left the regime with little time to pass all the planned water legislation and sell EMOS before national elections were due to be held on December 14, 1989.[21] The government was already in negotiations with a Spanish buyer for EMOS and could conceivably had gone ahead and sold the company before the elections. The privatization of EMOS should have attracted private investors willing to pay a good price. Private investors viewed the country as a relatively good risk according to Chile's scores in investor risk ratings in the second half of the 1980s.[22] In addition, institutions to protect private property in water were in place, as evidenced by Chile's historical tradition of private water rights, the fact that a private water company was already operating in Santiago, the successful privatization of other infrastructure, and the constitutional protections against expropriation. Moreover, Chile's constitution and electoral rules described below would have made it very difficult for a new regime to reverse privatization, or repeal or rewrite the water legislation.

Against these protections, however, potential buyers would weigh pre-electoral uncertainty and that might have led them to deeply discount their bids. National polls as early as March 1989 suggested that the military backed candidate, Hernan Büchi, was not a sure winner of the plebiscite (Table 6.4). Private investors voiced fears that if the opposition won, they would drastically increase expenditures and accelerate inflation (Weyland, 1997). There were also concerns that the military might intervene if their candidate lost (Constable and Valenzuela, 1989). To privatize EMOS at a deep discount right before the elections was also politically risky since the firm was widely regarded as a company that

---

[20] Constable and Valenzuela (1989) further argue that the General regarded the polls suggesting his defeat as biased and ignored them.

[21] Ninety percent of eligible voters participated in the plebiscite and 55% voted against the continuation of the Pinochet regime (*El Mercurio*, October 7, 1988).

[22] Chile's ICRG score, which gives a value to private sector assessments of firm risk, averaged 16.8 from 1985 to 1989, compared to 15 for Mexico and 16.8 for Korea during the same period (World Bank, 1995, p. 208).

Table 6.4

Polling results before the December 1989 presidential elections (percent of Chileans surveyed)

	March	*October*	*December*
Attractive personality:			
Büchi	37.6	30.2	20.5
Aylwin	31.9	36.6	44.5
Inspires confidence:			
Büchi	35.8	30.7	23.1
Aylwin	36.0	44.1	52.2

Source: Méndez, 1990.

should command a high price (field interviews). The administration feared that it could harm public support for the military backed candidate (field interviews).

Another explanation for why the Pinochet administration did not sell EMOS before the elections was worker opposition. As we discussed, worker opposition would not normally have been a barrier to privatization. In this case, however, the unions' threat of an illegal national strike of water workers raised the political risks of selling EMOS so close to the election. Water union leaders were able to convince two members of the Chilean military junta that privatization of water in 1989 would be a mistake.[23]

In the event, the Pinochet government decided not to sell EMOS until after the election. This proved to be too late. The opposition candidate, Patricio Aylwin, a Christian Democrat who headed a 17-party coalition (Partidos Concertados por la Democracia, or Concertación) won a decisive victory over the candidate supported by the military, Hernan Büchi (Unión por el Progreso, or UPP), and the new government took office on March 11, 1990.[24]

The Aylwin government faced a different political equation: the political cost from selling EMOS did not outweigh the political benefits. The first cost was ideological: President Aylwin's stated view before the elections was that state enterprises which provide basic services should "stay in public hands" (Gonzalez Parra, 1990, p. 36). His party's position had been that any privatizations after the 1988 plebiscite should be reversed (EIU, December 1, 1989) and sales of all SOEs came to a virtual halt after he took power (EIU, January 1, 1993). A second cost was that the opponents of privatization of water were important constituents of the Aylwin administration. Besides the workers in EMOS and other state-owned water companies, the sale was opposed by members of the winning coalition who now headed regional and local governments and wanted to appoint their supporters to posts in the water companies.[25] Third, the main

[23] According to field interviews, the representatives of the Navy and police force opposed water privatization after meeting with EMOS' workers and swayed the rest of the junta.

[24] Aylwin got 53.8% of the popular vote; Büchi got 28.7; and a centrist independent candidate (Errazuriz) got 15%.

[25] The sale of one water company was widely regarded as the precursor to the sale of the remaining companies according to field interviews.

supporters of a sale, the building and construction industry, were not constituents of the new regime.

Why was EMOS reformed?

A central question in this case is why Chile introduced a reform designed for a privatized company despite the election of a government that opposed EMOS' privatization. The answer is that the new regulation would have been very difficult to reverse because of Chile's unusual political institutions, as cogently described by Baldez and Carey (2000). The Chilean constitution, which was written by the Pinochet government and ratified in a plebiscite in 1987, is an example of deck stacking as defined in McCubbins et al. (1987, 1989). It ensured that the preferences of Pinochet's constituents were likely to prevail even after the military regime had lost power. The constitution required an absolute majority in Congress to change laws, and at the same time made it almost impossible for a new opposition regime to control such a majority (see Baldez and Carey, 2000). Reversal was difficult first because of the 'designated non-elected Senators', of which four were appointed by the Military Junta, three by the Supreme Court, and two by the President, while the rest were ex-Presidents who had served six years. Since in 1990 most of these senators were appointees of the Pinochet government, they would have opposed changes to the water legislation.[26] A second factor making reversal hard was an electoral system with two member districts that gave preference to the top candidates on whatever two of the different party lists received the most votes. This system encouraged parties to form coalitions and split most districts' votes between the two coalitions that represented the left and the right (ibid).[27] By "systematically over-representing the parties of the right", the electoral system made it unlikely that a left-leaning government such as Aylwin's could win enough seats to change existing legislation (ibid, p. 188).

Thus, even though Aylwin's coalition, Concertación, won a majority in the lower chamber of Congress in the 1989 elections, the distribution of seats meant that it could not overturn the water legislation without the — highly unlikely — agreement of some of the opposition or appointed senators (Table 6.5).

Even if the new government could not reverse the reform, it could have dragged its feet in implementing it. Instead Aylwin chose to implement the changes expeditiously and vigorously. The reform was politically desirable in part because the political costs

[26] Pinochet replaced many of the Supreme Court justices before leaving office with younger men. Supreme Court justices serve until they are 75 (except the president of the court, who serves for life), although the President can remove them 'for bad conduct' with the Court's agreement (Republic of Chile, 1980).

[27] Briefly, each district elects two members of congress from lists of two candidates presented by the parties or coalitions. The total votes for both candidates on each party's list are totaled first and the first seat is awarded to the more preferred candidate from the list with the most votes. Votes for that list are then divided by two and if the quotient is higher than any other list's total votes, the second candidate gets the second seat; otherwise the seat goes to the first candidate on the list with the next highest total votes. As the authors note, this system assures that the candidates of the top two coalitions will each get a seat unless the top coalition's list gets more than double the vote of the second place coalition's list. Parties which are not part of the two largest coalitions are virtually disenfranchised.

Table 6.5
Distribution of congressional seats in Chile, 1989 (percent)

Coalition or group	*Chamber of deputies*	*Senate*
Concertación	58	47
UPP	40	33
Independents	2	
Appointed senators		19

Source: Baldez and Carey (1997).

were low. Workers would receive compensation for going along with the reform, since they could keep their protections as public workers yet collectively bargain.[28] Moreover, the workers were not veto players because they could not credibly threaten to leave the coalition over something as mild as the water reform.

Another reason why net political benefits were positive was that the main beneficiaries, the urban poor, were an important part of the coalition's constituency. The Concertacìon's platform, characterized by the slogan 'growth with equity', called for retaining many of the prior administration's market-oriented/fiscally responsible policies, but combining these with greater social spending and poverty reduction, (Weyland, 1997). One objective was to reduce the housing shortage of the urban poor, which would require EMOS to increase further its rate of expansion. Santiago's poor could credibly threaten to withdraw support for the coalition if the Aylwin government did not deliver on its social promises.[29] The reform would benefit them directly through increased coverage and indirectly by generating funds for government spending.

Institutional factors were important in creating another motivation for the reform. Constitutional constraints on increasing government expenditures and shifting funds among ministries made it hard for the new administration to reduce its large debt overhang and keep its commitment to expand social spending. The administration was mindful that voters in general and private businesses in particular were concerned that the new govern-

[28] Some bureaucrats responsible for supervising EMOS and the other water companies were also opposed to the creation of a new regulatory structure that would cost them jobs and power, but they were probably not an important constituency of the new government. Their fears were well founded. EMOS and the other water companies were supervised by the National Sanitary Works Service (SENDOS, Servicio Nacional de Obras Sanitarias) of the Ministry of Public Works. With reform, supervision was transferred to the state enterprise holding company (CORFO), while the new Superintendence for Sanitary Services took over the regulatory function. Most of the 200 bureaucrats in SENDOS were no longer needed and were offered early retirement or other, less desirable, jobs (field interviews).

[29] For example, in the 1989 congressional elections the two poorest voting districts in the Metropolitan Region cast 19 and 24% of their ballots for the socialist party or other candidates outside the two main coalitions, compared to 11% in the metropolitan region as a whole (*El Mercurio*, Dec. 16, 1989).

[30] In polls just before the election voters rated inflation and price increases as their number 4 fear of an Aylwin presidency (27.5% rated it number one) and economic recession ranked 6th (11.9% rated it first). In contrast inflation and price increases were the 6th ranked fear for a Büchi presidency (15% rated it first), and economic recession was 8th (9.7% rated it first) (Méndez, 1990). Polls also suggested strong popular support against substantial changes in the free-market economic model of the Pinochet administration (*Economist Intelligence Unit*, 1990 and various years). Sáez Lozano (1996) found that any increase in voter pessimism about the economy favored the center-right parties and hurt the Christian Democrats in the 1993 elections.

ment would fuel inflation through deficit spending.[30] By moving ahead with the reform they could raise tariffs enough to allow EMOS to pay dividends that the Aylwin government could use to expand spending in areas important to its constituent base without spurring inflation. At the same time the subsidy to poorer consumers could effectively defuse the adverse political consequences of steep rises in water rates by reducing the cost to a broad range of poor and middle-income constituents (see Appendix A).

The amount of additional revenues the administration got from EMOS' dividends totaled US $75 million from 1990 to 1994. This amount would not have been large enough to motivate the Pinochet administration to prefer reform to privatization, since the sale of EMOS would generate both the $150 million purchase price and tax revenues. But these additional discretionary funds were important to the Aylwin government since they helped it overcome the inflexibility created by the constitutional rules while avoiding the politically distasteful step of privatization.[31]

CHARACTERISTICS OF THE REFORM

One characteristic of particular interest in the EMOS case is that the reform was designed for a private firm but implemented in a state-owned enterprise. Evidence from similar efforts in other developing countries suggests that contracts between government and an SOE designed to mimic private agreements typically fail (Shirley and Xu, 1998, 2001). Regardless of ownership, a contract can be expected to improve operating efficiency and investment only if it: (i) reduces government's information asymmetry vis-a-vis management, (ii) provides management with greater incentives to comply with the contract, and (iii) provides credible signals that government is committed to enforce the contract and adhere to any promises it has made.[32] These necessary conditions are usually not met in contracts with SOEs. First, the failure to auction the contract or otherwise simulate competition robs government of a tool to reduce information asymmetries and motivate performance. Second, government incentives tend to make for weak monitoring. Monitors are typically low paid civil servants who cannot effectively demand information from more powerful SOE managers and who may not be rewarded for finding politically embarrassing failures. Third, incentives to SOE managers are usually low-powered. It may be politically difficult to provide substantial bonuses to government employees or government may prefer low-powered rewards out of a fear that reneging by the SOE will go undetected. Threats of takeover, bankruptcy or firing for poor performance are seldom credible in SOEs. Fourth, commitment is a problem in any contract with a sovereign authority. The problem is exacerbated in contracts with SOEs because

[31] Many of the state enterprises which had been sold since 1983 were heavily indebted and had been privatized without their debts. As a result, by 1990 the state enterprise holding company (Corporación Nacional de Foment, CORFO) held US $500 million in debt. In contrast, EMOS had a very small debt; in 1990 its debt : equity ratio was only 10%. See Baldez and Carey (2000), for details on the constitution-imposed constraints on raising spending or shifting funds.

[32] Although this hypothesis was originally formulated for cases of PSP and privatization, the same conditions have been shown to apply to contracts with SOEs (see World Bank, 1995). See also Williamson (1976, 1985), and Laffont and Tirole (1986, 1993).

state enterprise is an involuntary partner to the agreement and cannot demand safeguards against government reneging as a condition for signing. Moreover, state enterprises are not usually granted access to neutral enforcement mechanisms, such as courts. Finally, institutions to support an effective contract, such as strong budget constraints on the SOE or norms of efficient conduct in bureaucracy, are usually absent in developing countries.

This section examines how Santiago's 'regulatory contract', in the sense of the implicit and explicit agreements between government and enterprise, overcame the agency problems and improved information, incentives and commitment under state ownership.

Information

The new contract with EMOS had the potential to increase government's information through more and better accounting, greater competition, and better supervision. Although the opportunities for increased competition were not fully exploited, accounting and supervision were improved. All the water companies were now required to produce audited annual accounts according to generally accepted standards. Although EMOS had produced extensive statistics on its operations before incorporation, it did not have a balance sheet or income statement until its first annual report after its incorporation in 1988.[33]

The reform greatly strengthened monitoring, by replacing a hands-on government agency with an independent, arms-length regulator. Before reform, EMOS was under the supervision of SENDOS (Servicio Nacional de Obras Sanitarias) an arm of the Ministry of Public Works. Since SENDOS was preoccupied with managing the regional water companies, EMOS was largely autonomous. The reform created an independent regulatory agency, the Superintendency of Sanitary Services (SSS), designed to signal government's commitment to potential private investors and assure that privatization of a monopoly would not lead to consumer exploitation. The organizational design of the SSS was detailed in a law that mandated a small, professional staff with above average civil service salaries.[34] Unlike SENDOS, the SSS had no management responsibilities. Its focus was on tariff setting and monitoring compliance with investment plans and standards. The SSS' maximum professional staff cannot exceed 45 according to the law (although through short-term contracts it had grown to 61 professionals in 1997); in contrast SENDOS had over 300 staff when it was dissolved in 1990.[35] Although

[33] The World Bank had produced estimated accounts for EMOS prior to 1988 for use in its project documents and supervision reports, but these could not serve the same monitoring purpose as an annual report. They were not produced on a regular basis, did not track clearly with EMOS' statistics, involved judgments by outsiders, and were not always accepted by EMOS management or the government. In the preparation of this case we were unable to match the World Bank's version of EMOS accounts with those produced by the company for the same years.

[34] Its total responsibilities are to (i) set tariffs, (ii) set and enforce technical standards, (iii) monitor, award and revoke water and sewerage concessions, and (iv) control liquid industrial effluent. (Republic of Chile and Ministry of Economy, 1990a.)

[35] The SSS is organized by function, with most staff assigned to setting and regulating standards (22, of which 15 are engineers) and setting tariffs (15, of which 9 are engineers). The term engineers includes commercial engineers who have the equivalent of an economics/business degree. (Republic of Chile, SSS, 1995.)

the staff of both agencies were civil servants, the salaries for SSS professionals were similar to those of bank or stock market regulators (*fiscalizadoras*), and hence higher than those which SENDOS could pay. All SSS professional staff had at least a B.A., which was not true in SENDOS, and most were engineers. There is still room for improvement, since few of the SSS staff are civil engineers or economists, and most do not have a background in water and sanitation (field interviews, former and current Superintendency staff).

The SSS monitors EMOS to assure compliance with quality standards and investment plans, and to measure costs and efficiency in the context of tariff adjustments.[36] Information on quality and investment was always good; the main improvement was in accounting information. Companies must submit regular information on costs and service to the SSS and can be fined if they fail to do so. As we explain in the next section on incentives, tariff policy was designed to minimize the effects of information asymmetries by reducing the regulator's dependence on company information and the company's ability to manipulate the computations. Nevertheless, the information demands for tariff revision are still large, but less frequent since tariffs are revised only every five years.

The reform created the potential to use competition to extract more information, but this opportunity has not been exploited. The new legislation transformed all water and sewerage services into concessions. Since existing service operators were granted indefinite rights to their current concession areas at the time the legislation was approved, much potential for competition was lost. However, there can still be competition for new concessions, as well as for older concessions that had been revoked because the company failed to meet regulatory standards. Thus far, the regulator has been loath to revoke concessions that failed to meet standards, and has seldom used competition to award new concessions either.[37] None of the concessions in Santiago have been bid.[38]

Since the reform introduced standardized accounting, the regulator could also compare among firms to determine best practice. Yardstick competition among the Chilean water companies would be least effective for EMOS since it is larger and better managed than the other water companies in Chile. EMOS could be compared with international standards, but as we shall show, the international standards used for tariff setting were secret and hence less useful for yardstick competition. This secrecy makes it hard for company management to perceive how they would be rewarded for, e.g., bringing water losses down.

[36] Quality is also monitored by the Ministry of Health.

[37] The law allows new concessions to be awarded without competition at the regulator's discretion unless a competitor mounts a challenge. An interested provider produces a concession proposal, which is then published in the official paper, and if no competitor chooses to challenge it within 60 days, it can be awarded without bids (although a call for bids can still be issued at the discretion of the government). Ley General de Servicos Sanitarios No. 382. (Republic of Chile, Ministry of Public Works, 1988b.)

[38] Although the private water company in Santiago was sold (to a private electricity company), the Superintendency did not consider that grounds for rebidding the concession. Concessions have been bid in other parts of Chile.

A third way to reduce government's information asymmetry was to give interest groups with an incentive to inform government about abuses of EMOS' monopoly position, such as water consumers, positions on the board or standing in regulatory decision-making. However, as is typical in SOEs everywhere, directors were chosen so as to reward political factions rather than to increase government's information. Regulatory decisions are not open, and no interests are represented besides government and enterprise.

Incentives

The most important change in incentives was the result of the new tariff policy, which we discuss first. Several supportive institutions were key to the incentive effects of tariffs, as we also explain.

Tariff policy

Before the reform, tariff decisions were internal to the bureaucracy, ad hoc, and vulnerable to political manipulation. As we have already seen, tariffs were increased only infrequently, when the political liability of unmet demand and deteriorating service exceeded the political cost of raising tariffs. The reform not only led to higher tariffs, it also increased transparency, consistency and public accountability.

All of the formulas and definitions of the variables to be used for tariff setting are spelled out in great detail in the legislation, down to how to calculate peak and off-peak monthly tariffs. As we shall see, this legalism and lack of regulatory discretion has proved a good fit with Chile's institutions. Tariff increases became automatic and the effect was a sharp rise in water and sewerage prices phased in over five years (see Figs. 6.1 and 6.2).

Tariff policy had been designed both to signal to a private investor that government was committed to not expropriating their return on capital through under-pricing and to curtail the chance of monopoly rents. Tariffs were calculated every five years to cover the long-run marginal cost of a 'model' or benchmark company, and then readjusted to permit a 'reasonable' return on assets, i.e., to cover long-run average costs and allow at least a 7% return on capital. To reduce the risk of monopoly rents the construction of the model company was a black box, which was expected to make it harder for the company to manipulate information.

The result was a complex process (see Fig. 6.3). First, a benchmark efficient firm is estimated for each water company, based on the company's actual geographic, demographic and technological conditions, but setting certain parameters (such as unaccounted-for water, collection rates and capital costs) at 'efficient' levels.[39] This 'model' company

[39] These parameters were set through a survey of five companies in Chile plus international standards. For example, the billing collection rate is set at 100%, unaccounted-for water at 20%. Capital costs are set using a CAPM model where the risk free rate is set equal to the state bank deposit rate and the risk premium is based on international rates.

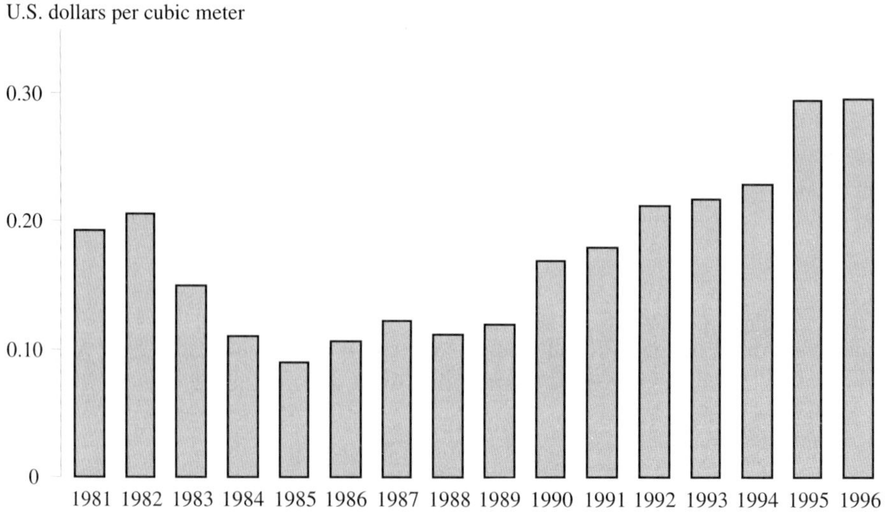

Source: Superintendency of Sanitary Services.

Figure 6.1. Average price of water in Santiago.

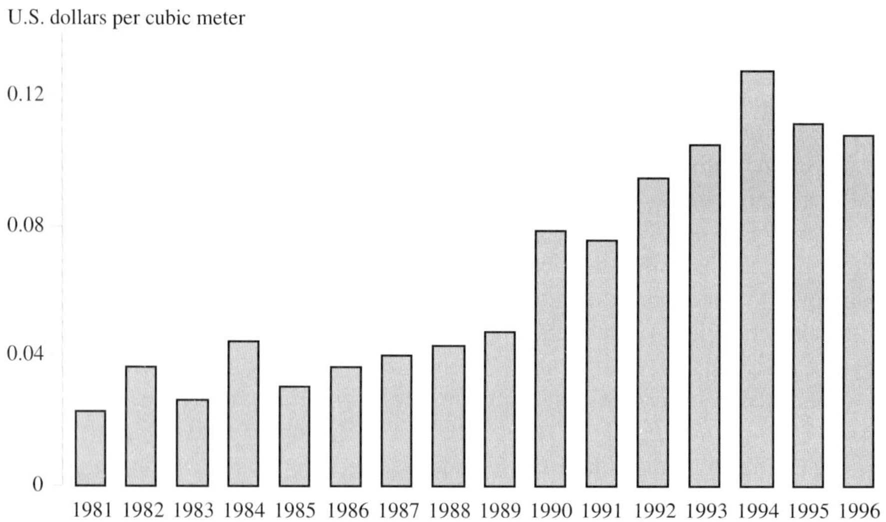

Source: Superintendency of Sanitary Services.

Figure 6.2. Average price of sewerage in Santiago.

is derived from a computer program that uses a mix of international and local standards adapted to the circumstances of the individual company. In the second step, the tariff is set to cover new investment plus operation and maintenance costs associated with meeting forecast demand growth, and to allow the benchmark firm to earn a minimum annual return on assets of at least 7%. Third, the tariff calculated in the first step

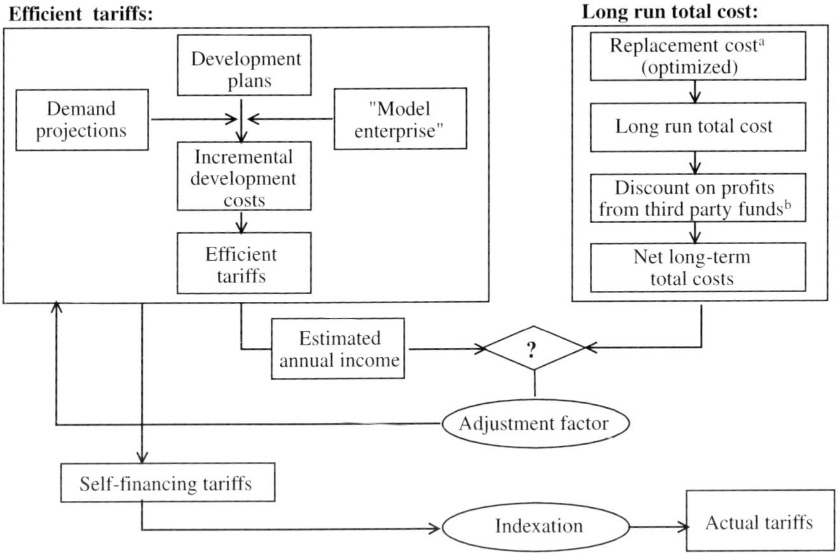

[b]Replacement cost of a plant large enough to meet demand for the five year p eriod of the tariff.
[a]Discounts on profits from funds from outside the enterprise.
Source: Superintendency of Sanitary Services, 1995.

Figure 6.3. Tariff setting process.

is adjusted by a factor derived by comparing projected average long-run costs with projected revenues from step two for the next five years to assure that the company can break even over the five years. Finally, the water tariff is indexed to a price index.

The tariff has incentive properties similar to a price cap since it is set on the basis of an efficient model company for five years and indexed for inflation. If EMOS can be more efficient than the model, it earns additional profits, giving the company an incentive to maximize its efficiency. At the end of the period tariffs may be adjusted downward to force the company to share its gains with consumers.

The incentive properties of the price cap was reduced somewhat because government demanded most of the returns be paid as dividends. For the first three years after tariff reform the government required EMOS to pay 100% of its net earnings for the previous year as dividends. Subsequently EMOS had to remit 65% of the prior years profits as dividends; the remainder could be retained for investment.[40] Because the dividend payment is estimated at the beginning of the year based on the previous year's profits and profits were increasing rapidly, there was sufficient surplus for investment and other uses. EMOS management and staff had an incentive to increase returns, since they got profit sharing bonuses equivalent to 10% of profits for the first five years of the reform, and higher real wages throughout (Fig. 6.4). Although these wage increases might seem to suggest a soft budget, they apparently just brought EMOS salaries in line with the

[40] In 1998 the government announced a plan to return to 100% dividends.

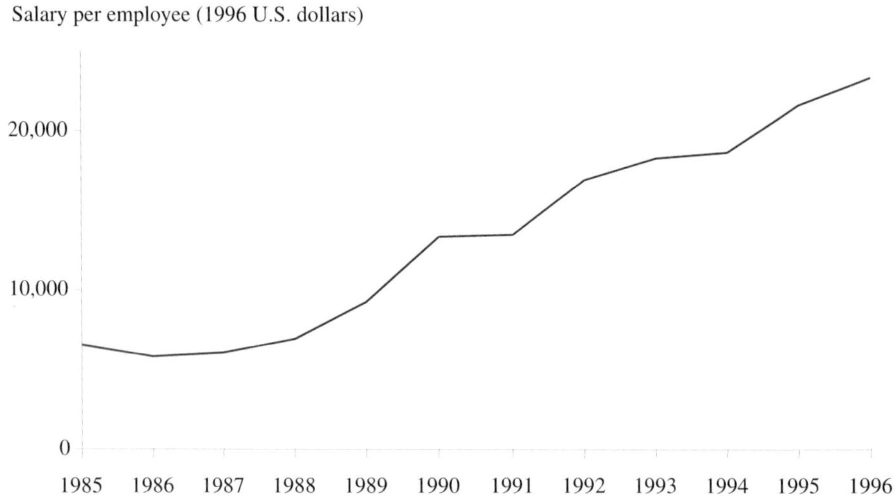

Source: EMOS annual reports.

Figure 6.4. Average salary in EMOS.

market; a recent consultant study suggests that EMOS wages in 1996 were close to market standards.[41]

Supportive institutions

The incentive effects of higher revenues were supported by several strong institutions that help explain why the increased funds generated from the tariff led to performance improvements, despite agency problems and continued information asymmetry. Among these were Chile's civil service norms and traditions of professionalism and honesty in public service. These norms were strong in EMOS. Field interviews suggest that its reputation of being one of the best performing water companies in the region motivated EMOS' management to strive to improve performance before and after reform. This tradition of professionalism also explains why the Aylwin government selected an experienced, well qualified and competent manager to run EMOS and allowed her to select an equally competent team.[42]

[41] The study (by Langton Clarke Consultants) has the following comparison of 1996 annual salaries:

	EMOS	*Market*
Professionals, managers	US $30,854	US $30,883
Technicians	US $13,954	US $15,098
Administrative	US $11,701	US $14,743
Service (guards, drivers, etc.)	US $11,003	US $9,576

[42] Most EMOS staff were supporters of the new government and democracy, and reported this was a further incentive to increase effort to try to make reform under state ownership succeed (field interviews).

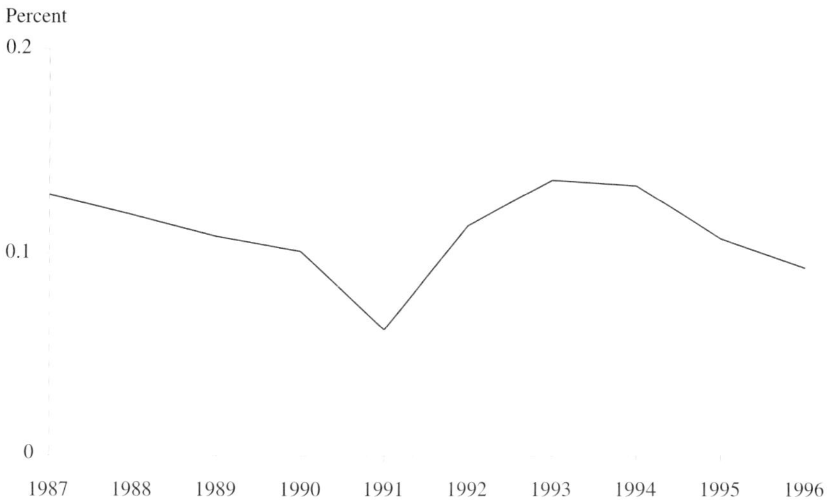

Source: EMOS annual reports.

Figure 6.5. Debt to equity ratio: EMOS.

Also important was the regulation introduced by the Pinochet government that had created hard budget constraints in the government and state enterprises. Borrowing was strictly curtailed, most transfers and subsidies were eliminated, and state enterprises were required to pay dividends. As part of this effort, the staff of EMOS had been reduced and most services were procured through outsourcing. SOEs were not allowed to exceed a debt–equity ratio of 15% and EMOS maintained a very low debt–equity ratio throughout the period (Fig. 6.5). Investment increased during this period, as we discuss below, and was financed almost entirely by internal funds (Fig. 6.6).

Finally, EMOS continued to face the prospect of privatization. The debate over privatizing EMOS continued after Aylwin's election. (EMOS was sold in 1999.) Studies have found that SOE performance improves with the prospect of privatization, especially when staff expect to receive shares in the company.

Credible Commitment

A weak point in most contracts between government and SOEs is the lack of credible commitment, especially on the part of government actors (Shirley and Xu, 1998). If a contract specifies no enforcement mechanism or other recourse against government reneging, management will not expect to receive the promised incentives or to be penalized for poor performance. Mangers will not increase effort, instead they will try to exploit their information advantage to bargain down their targets ex post. Once again the design of a contract intended for a private operator led to a different outcome from the usual in Chile.

The regulatory contract for EMOS signaled government commitment by specifying neutral and automatic enforcement. Any water company could appeal a dispute over tariffs to the SSS within 30 days. If the SSS and the company fail to reach agreement,

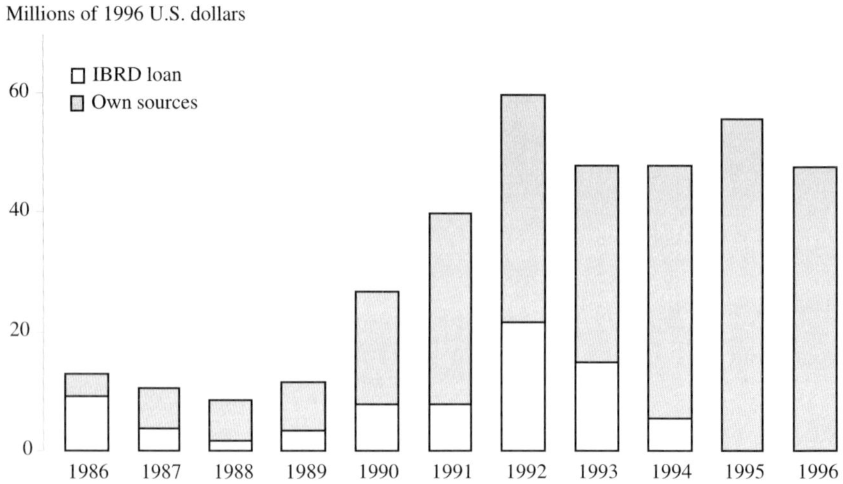

Source: Authors' calculations based on EMOS annual reports.

Figure 6.6. Sources of investment funds in EMOS.

a panel of three arbitrators — one appointed by the company, one by the government and one jointly agreed — would have 37 days to reach a decision that both sides must accept. Decisions could also be appealed to the courts, which are regarded as honest and independent, although slow, but an SOE would be less likely to be able to use this route.

The fact that state-owned companies can appeal the decision of a government regulator increases the credibility of the contract. Moreover, this ability to arbitrate a tariff decision has not been a theoretical privilege. So far water tariffs have been set twice under the reformed procedures. Once in 1990 when the system was first implemented and again in 1995. The first tariffs were gradually phased in over five years and did not generate appeals since they led to an average real increase in tariffs in the first year alone of almost 25% for all water companies, 20% for EMOS. The second round of tariff setting led to much lower increases: on average the real annual increase awarded companies was 5.9%, while EMOS' tariffs were reduced by 1.2% a year in real terms for 93% of its customer base and raised by 6.8% for the remainder (Republic of Chile and Superintendency of Sanitary Services, 1995). Five companies took disputes over tariffs to arbitration, including EMOS, two other SOEs and the two largest of the six private companies (one of which operates in Santiago). EMOS' dispute was on the capital costs, which SSS had set at 7.5% and EMOS argued should be 10%. The arbitrators decided on 9.16%.

A second source of credibility was the lack of regulatory discretion. The tariff calculations were detailed in laws that, as we have seen, were hard to change. The SSS relied on a computer model designed by external consultants to comply with this law, which could not be easily manipulated for political purposes.

Credibility was also enhanced by a third tool, a means tested subsidy, which reduced the political pressure against tariff increases. On average 60% of the bill for the first 20 m^3 of consumption of qualified low-income households was paid by taxpayers;

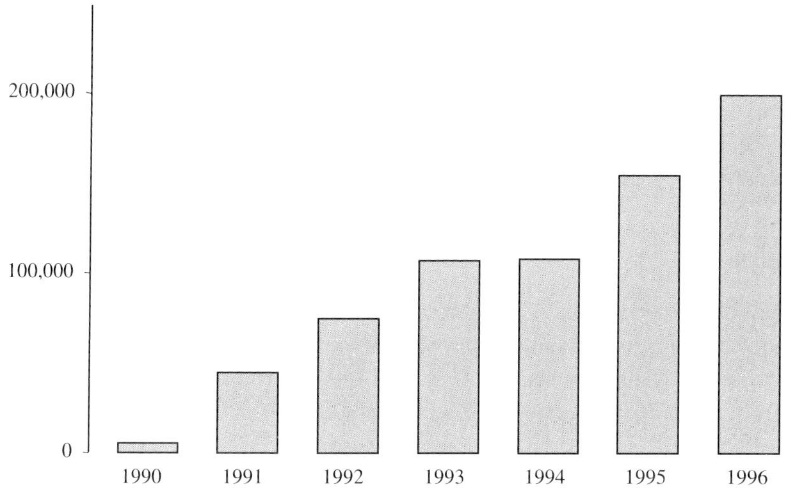

Source: MIDEPLAN, 1997.

Figure 6.7. Number of water subsidy recipients in Chile.

government transferred the funds directly to the water companies (see Appendix A for details). The number of recipients has grown with tariff increases (Fig. 6.7), an easing of the subsidy's limits and requirements, and a public information campaign by the water companies. In 1996 about a third of recipient households were in the lowest-income quartile and another third in the second-lowest. This spillover of benefits to middle-income groups added to political constituency affected by the subsidy.

A number of the institutional safeguards described above supported the government's credible commitment to the regulatory contract. These included the set of laws and regulations that underlay the contract plus the political institutions that made it difficult to overturn this legislation, and norms of behavior and legal strictures supporting efficiency and competitiveness in state-owned enterprises. The new government signaled that it would support these norms by, for example, moving quickly to implement the tariff system and appointing a well-qualified insider to run the company. Although there were violations of the implicit contract, such as political appointments to the board and ministerial interference in some internal decisions, these did not threaten the tariff, which was the most important change from the previous regulation.

Finally, factor endowments offered safeguards. Cost of usage was low since Santiago's principal water sources were sustainable at current and projected rates of extraction, and the system is gravity-fed. As a result, even with the 25% increase in tariffs between 1990 and 1996, water tariffs in Santiago were still relatively low compared to the other water companies in this book.[43]

---

[43] Santiago's average revenues collected per m^3 distributed (minus UFW) was US $0.29 in 1996, higher than Buenos Aires or Mexico City (at US $0.22/$m^3$), but lower than Lima (US $0.32), Abidjan (US $0.51) or Conakry (US $0.74). Unlike Santiago and Buenos Aires, Mexico City's tariffs do not cover marginal cost and return on investments.

EFFECTS OF THE REFORM ON PERFORMANCE AND WELFARE

We present evidence in this section that, despite relatively good initial conditions, the reform led to improvements in performance and welfare that were important in magnitude. This section first presents partial performance indicators and then describes net economic benefits.

Performance Effects

The new tariff policy allowed an increase in real investment (see Fig. 6.8) that enabled EMOS to extend water and sewer connections while maintaining an aging system.

Market coverage for water reached 100% soon after the reform and stayed there, while sewerage coverage increased from 88% at the start of the reform to 97% from 1994 onwards (Fig. 6.9). Since demand for connections accelerated as the Aylwin government increased the pace of public housing construction while private construction rose with rising incomes during this boom period, EMOS had to increase the number of new connections to keep coverage constant (Fig. 6.10). EMOS also expanded its concession area to include poor municipalities on the periphery of Santiago, subsidizing customers who could not afford to pay their requisite share of the cost of connecting (Alfaro, 1997).[44]

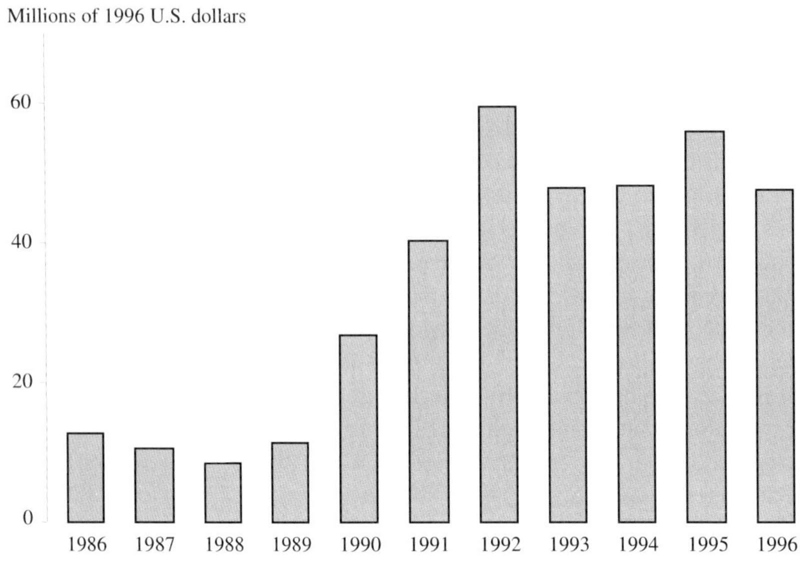

Source: EMOS annual reports

Figure 6.8. Investment: EMOS.

[44] EMOS also allowed middle-income consumers to pay the cost of connection in 12 to 20 installments; low-income families (those certified for subsidies, see Appendix A) to pay in 60 installments; and the very poor (again with certification) to pay a fraction of the cost in ten installments. Alfaro (1996).

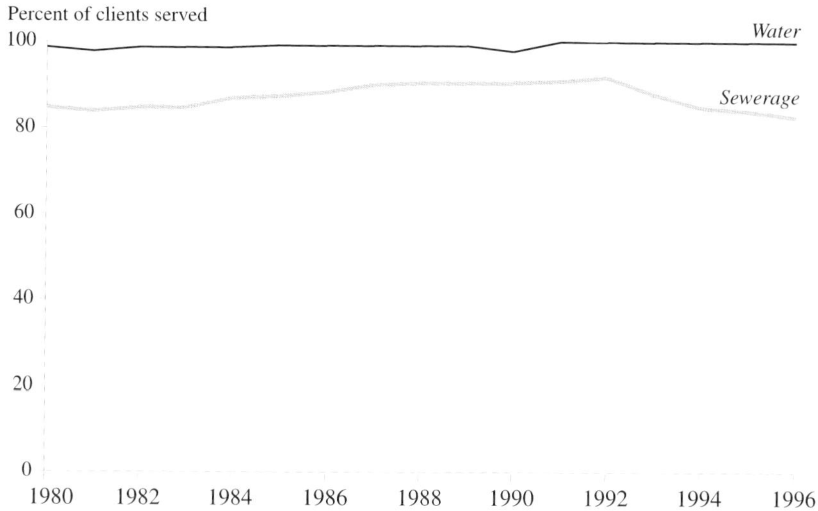

Source: EMOS annual reports.

Figure 6.9. Market coverage: EMOS.

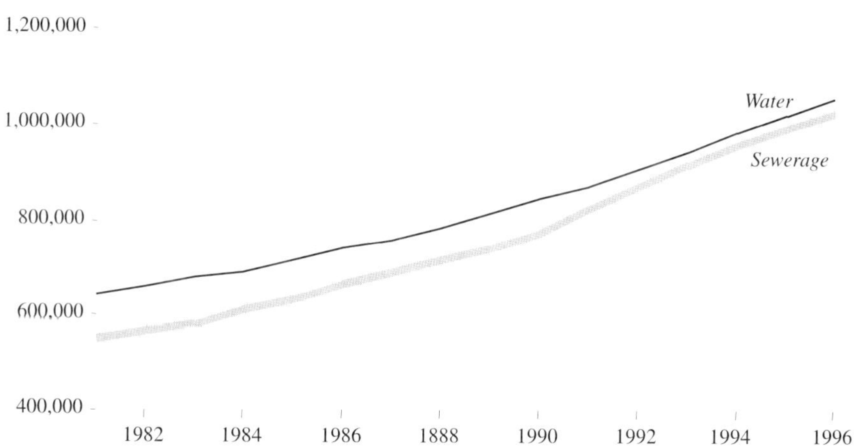

Source: EMOS annual reports.

Figure 6.10. Number of connections: Santiago.

Another benefit came from EMOS' ability to finance investments in maintenance, which allowed it to upgrade pipes and thus reduce water losses and improve pressure. The number of pipe breaks per kilometer dropped from over 0.52 in 1988 to 0.31 by 1994. As a result, the trend in unaccounted-for water (UFW) which had been largely caused by physical losses, was reversed and UFW was brought close to international standards of 20% of production (Fig. 6.11).

The main cost to consumers from the reform was higher prices. Since all connections were metered and billing was efficient, price increases had a direct effect on consumer

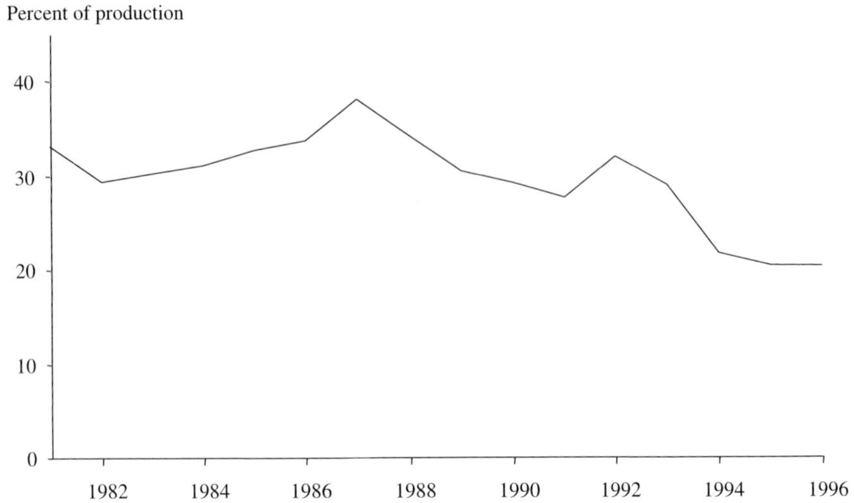

Source: Authors' calculations based on EMOS annual reports and s tatistical bulletins.

Figure 6.11. Unaccounted-for water: EMOS.

welfare, although this was partly offset by the subsidies described earlier. Nevertheless, Fig. 6.12 points to a continuously declining trend in average monthly water consumption per connection, which suggests that consumers responded to rising prices by cubing consumption. EMOS launched a campaign during this period to educate consumers on reducing water wastage. It could also be that the rise in housing construction reduced the extent of shared housing.[45]

Partial indicators suggest that the reform increased EMOS' already high productivity even though total factor productivity did not change much. Although the number of employees increased slightly, the workers per thousand connections declined to 1.75 in 1996 (Fig. 6.13).

The growth in intermediate inputs also suggests that productivity was improving. Real growth in the value of intermediate inputs per cubic meter of water production declined from an average of 9.1% a year to less than 4% a year after the reforms (Fig. 6.14).

Despite these improvements in partial productivity indicators, EMOS' total factor productivity (TFP) showed little change, with inputs hovering at around 60% of outputs from 1988 to 1996 (Fig. 6.15). These numbers are misleading, however, since they count the cost of EMOS' increased investment in maintenance, but fail to take into account the value of the improved quality of service, including better pressure and fewer interruptions.

[45] An estimated 16% of housing was shared in some way in 1982. This probably increased with the destruction of much rental housing in the 1985 earthquake and a faster rate of removal of squatter settlements than of construction of new subsidized housing during the Pinochet administration (see Gilbert, 1993).

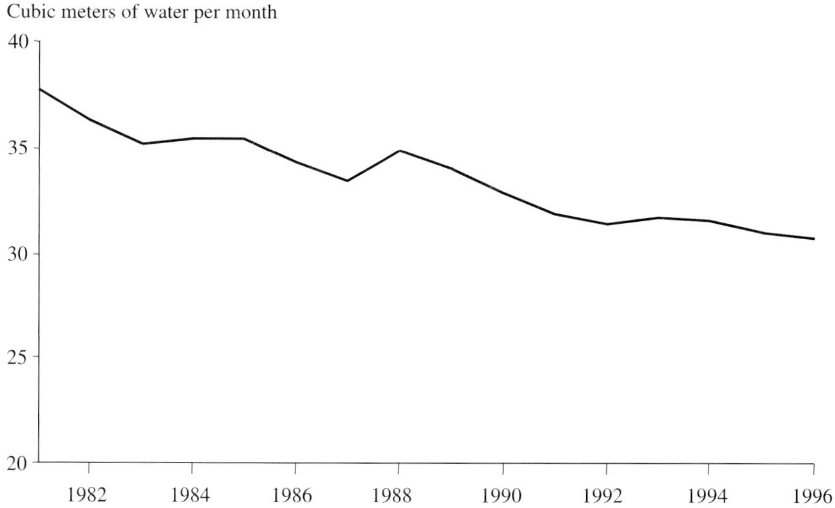

Source: Authors' calculations based on EMOS annual reports.

Figure 6.12. Average monthly water consumption per connection: Santiago.

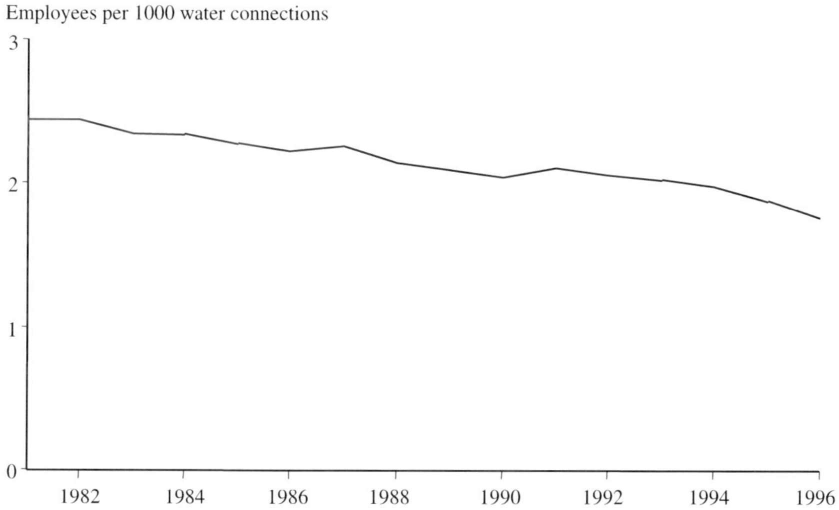

Source: Authors' calculations based on EMOS annual reports.

Figure 6.13. Employees per connection ratio: Santiago.

Welfare Consequences of the Reform

We calculated the welfare effects of the reform by comparing the net benefits from the actual reform with a counterfactual assuming no reform.[46] As we shall show the net

[46] We used a partial equilibrium cost–benefit methodology developed by Jones et al. (1990).

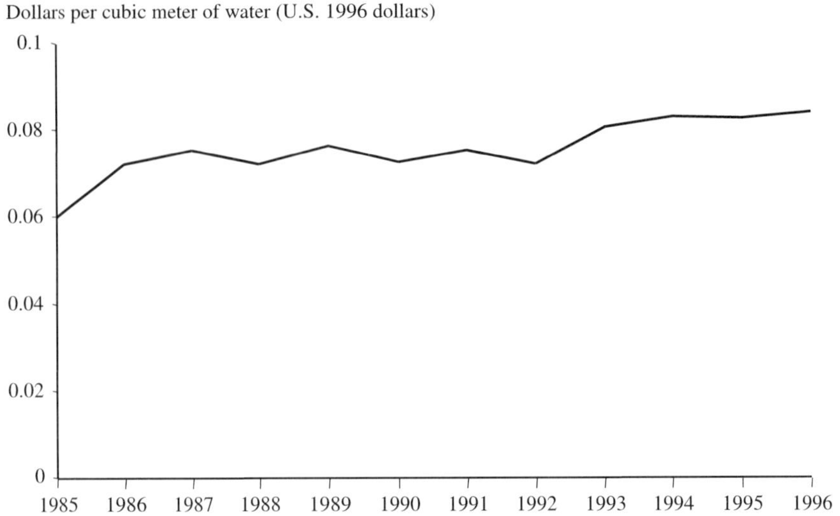

Source: Authors' calculations based on EMOS annual reports.

Figure 6.14. Ratio of intermediate inputs to output.

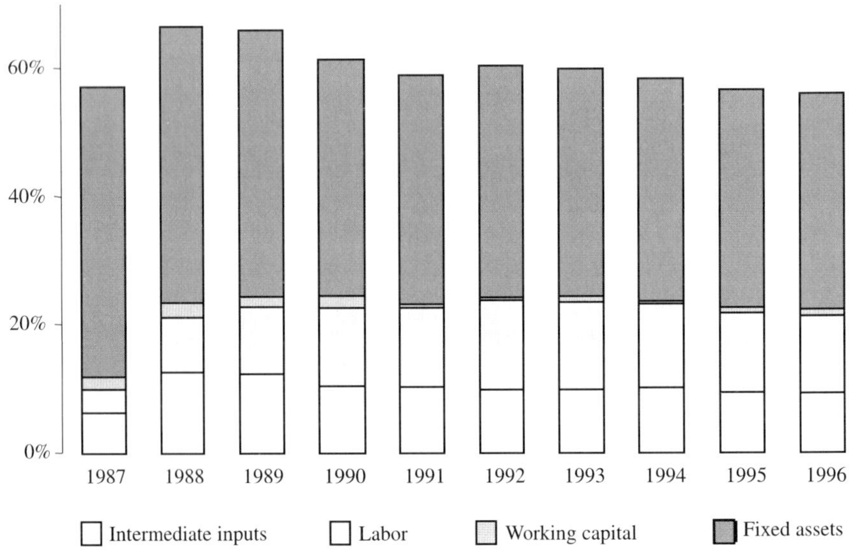

Source: Authors' calculations.

Figure 6.15. Total factor productivity.

gains are important, even when compared with Buenos Aires, a city with much less favorable initial conditions. This section begins with a brief summary of the methodology; it then describes the welfare gains; it concludes with some sensitivity analysis.

Methodology

To construct the counterfactual for 1989 to 1998, we projected the key parameters — including price and quantity of water and sewerage, labor productivity, unit costs of labor and intermediate inputs — on the basis of their linear trends from 1981 to 1988. We projected the counterfactual EMOS investment based on two assumptions: (1) that the ratio of output to fixed capital would stay at 1988 levels; and (2) that any additional investment needed to maintain that ratio would be financed by retained earnings.

In order to have a 'factual' time period of at least ten years we also projected the factual scenario for two years beyond the end of our observations: 1997 and 1998.[47] We projected investment using the same two assumptions as for the counterfactual, and fixed the other key ratios equal to those of the last year of the observed period, 1996.

We calculated consumer surplus using an elasticity of −0.26 and assuming that the shape of the demand curve was the same shape for the factual and the counterfactual.[48] We also assumed that at the start of the reforms in 1988 total demand was equal to all potential demand in the EMOS service area, so that unconnected customers were treated as excess demand, which seems reasonable given the low price for water that year (US $0.14). We further assumed that new connections have been rationed by willingness to pay, in other words that most wealthy people who would be willing to pay more than the price of connecting were already connected at the time of the reform. This seems reasonable given that the government was building subsidized housing and EMOS was subsidizing the costs of connection for poorer people in new sites and service areas. For more on methodology see Appendix B.

This cost–benefit methodology has its limitations. Although we tried to take into account as many relevant details as possible in constructing our factual and counterfactual scenarios, we necessarily had to leave out many unquantifiable factors. Thus, the analysis does not take into account benefits from improvements in quality mentioned above or social costs and benefits, which we discuss in the section below on privatization. In addition, our assumption that without reform EMOS would have followed its pre-reform trends may bias the welfare gains upward. We therefore consider some alternative assumptions in our sensitivity analysis below.

[47] Although another study using this methodology (Galal et al., 1994) used a much longer projection period, we concluded that further projection would entail too much prediction error, and the large discounting factors beyond the year 1998 would make this omission largely unimportant to our results.

[48] This elasticity comes from an unpublished study of water demand in Santiago, which finds a price elasticity of −0.516 for households which consume 1–20 m^3 of water, −0.2 for 20–30 m^3, and −0.25 for more than 30 m^3. We know the consumption patterns among households in Santiago: 5% of households use 1–10 m^3 per month, 40% use 10–30 m^3 per month, and 55% use more than 30 m^3 per month. This suggests that the weighted average elasticity is: $0.10 \times (-0.516) + 0.35 \times (-0.20) + 0.55 \times (-0.25) = -0.259$. We did not include the demand elasticity of industrial and commercial users because this category accounts for only 5.8% of total use according to EMOS, Annual Report 1990. This elasticity is similar to that assumed for Buenos Aires, Argentina (−0.32).

Welfare gains, winners and losers

The reform led to cumulative domestic gains in the ten post-reform years of about US $214 million net present value (NPV) in 1988 prices (Table 6.6).[49] Since these gains are really a flow of funds that goes on in perpetuity, we can get some sense of how significant they were by expressing them as a percent of another flow, namely EMOS' annual sales. The gains were large, equivalent to 52% of EMOS' 1988 sales every year in perpetuity. By way of comparison, the gains were approximately US $64 per capita (NPV in 1988 dollars), while the estimated NPV of gains (in 1996 dollars) from ten years of concession contract in Buenos Aires, using the same methodology, were about US $150 per capita (Chapter 3). This despite the fact that Buenos Aires' concession reduced prices by almost 27% and expanded water coverage from 70 to 81% and improved efficiency in very poor initial conditions.

In contrast to Buenos Aires, however, where consumers were the big winners, the biggest winner from EMOS' reform was the government. Thanks to increased taxes and dividends, government gained a NPV of roughly US $181.5 million in 1988 prices, or the equivalent of 44% of EMOS' 1988 annual sales in perpetuity.[50]

Consumers did less well, reaping a gain of roughly US $3 million (NPV in 1988 prices) or about US $0.40 per connection. Although the consumer surplus was reduced by price increases, this was offset by the gain from the pace of increased connections. Consumers' gains are understated somewhat because we do not include the benefits of higher pressure and reliability. This gain is very sensitive to our assumption that connections were rationed by willingness to pay, as we explain below.

Employees also benefited, thanks to higher wages, by about US $29 million (NPV in 1988 prices).[51] The net present value of a typical worker's gains was about US $1710 a year in real terms. Finally, EMOS' few private shareholders also gained modestly, by US $0.7 million.

Table 6.6

Winners and losers from reform (net present value)

	Government	*Domestic investors*	*Workers*	*Consumers*	*Total domestic*
Cumulative welfare gains (1988 US dollars millions)	181.5	0.7	29.0	3.07	214.3
Annual welfare gains/ 1988 sales (percent)	44	0.2	7.1	0.8	52.3
Annual welfare gains for each group (1988 US dollars)			1710 (per employee)	0.39 (per connection)	64.80 (per capita)

Source: authors' calculations.

[49] Domestic gains are almost the same as total gains since foreign participation was very small.

[50] The average exchange rate for 1988 was 245.01 pesos/US $.

[51] The gains for workers are computed as the net discounted present value of the wage differences between the factual and the counterfactual. The wage of the counterfactual is projected on the basis of the historical trend.

Sensitivity analysis. Although we tried to make plausible assumptions about the counterfactual scenario, we also conducted a sensitivity test to examine whether the main qualitative results remain intact when we change one key parameter from the base (Table 6.7). The qualitative conclusions, that there were large gains and that government is a major winner, remained robust with respect to alternative assumptions.

Consumer surplus is the most sensitive to changes in parameters. If we assume that excess demand was rationed randomly instead of by willingness to pay, then the consumer surplus increases from a NPV of US $3 million to US $155 million (1988 prices). This scenario is unlikely, however, since, as we have seen, most of those who were unconnected were poor. If we assume that the price elasticity is −0.2 instead of −0.259, then consumer gains are roughly US $30.31 million (NPV in 1988 prices). The former elasticity is closer to that of industrial and commercial consumers and other heavy water users. In contrast, if the price elasticity was higher, −0.3, which is probably the upper bound for Santiago, then the consumer surplus becomes negative, −US $9.49 million.

Other changes in assumptions have less effect on the results. Thus, we tested what would have happened if the counterfactual EMOS had used intermediate inputs as efficiently as the post-reform EMOS did, instead of projecting their use on the basis of the company's historical trend. Under this assumption, the efficiency gains from the reform decline only slightly compared to the base case (Table 7). We also tested what would have happened had unit labor costs under the counterfactual grown as much as in the factual instead of following historical trends. In this scenario, government's revenues in the counterfactual are less because EMOS' costs are higher, so its gains from the reform are greater, while workers', gains from the reform disappear. The net total welfare gains drop slightly compared to the base.

What if EMOS Had Been Privatized?

In this section we speculate on how a private EMOS might have differed from the actual. Such speculation is not idle since EMOS was privatized in 1999. Imagining what would have happened had EMOS been sold in 1990 helps illustrate some ways in which incentives are likely to differ under private operation.

First, most observers consulted in Chile expected that a private EMOS would have built a sewage treatment plant five to ten years after sale. Instead, EMOS' current plan is to have 100% sewage treatment in 2024.[52] The argument favoring faster action by a private EMOS is based on two observations. One is that EMOS had strong incentives to build treatment facilities since tariff regulations assure at least a 7% return on its investment, plus the company could sell treated wastewater to farmers. The second is that the project was delayed because of government imposed constraints on EMOS'

[52] The wastewater collection system of Greater Santiago is made up of a network of 6500 km of sewer mains that evacuate a current average flow of 13 m^3/s. The flow is expected to increase to 25 m^3/s by the year 2024. The wastewater collected is discharged without treatment in more than 40 points along three major natural channels that drain the metropolitan area. Some areas are highly polluted, especially areas irrigated by the Zanjon de la Aguada and the central and lower Mapocho River. In 1990, the total agricultural area irrigated with the polluted water from the canals was on the order of 130,000 ha.

Table 6.7

Welfare gains under alternative assumptions

	Base	*Random rationing*	*Elasticity 0.20*	*Elasticity 0.30*	*Counterfactual intermediate inputs equals the actual*	*Counterfactual labor equals actual*
Government						
NPV 1988 US dollars millions	181.5	181.5	181.5	181.5	164.9	203.0
Percentage of 1988 EMOS sales	44.3	44.3	44.3	44.3	40.3	49.5
Domestic investors						
NPV 1988 US dollars millions	0.8	0.8	0.8	0.8	0.8	0.8
Percentage of 1988 EMOS sales	0.2	0.2	0.2	0.2	0.2	0.2
Workers						
NPV 1988 US dollars millions	29.0	29.0	29.0	29.0	29.0	0.0
Percentage of 1988 EMOS sales	7.1	7.1	7.1	7.1	7.1	0.0
Consumers						
NPV 1988 US dollars millions	3.1	155.6	30.3	−9.5	3.1	3.1
Percentage of 1988 EMOS sales	0.8	37.8	7.4	−2.3	0.8	0.8
Total domestic						
NPV 1988 US dollars millions	214.3	366.9	241.6	201.7	197.8	206.9
Percentage of 1988 EMOS sales	52.3	89.5	58.9	49.2	48.3	50.5

Note: Total (domestic and foreign) welfare gain is not presented because it is generally very close to total domestic welfare gain.
Source: authors' calculations.

borrowing and slow, bureaucratic decision-making in the public sector which a private firm would not have faced. There are two main counter-arguments. The first is that a private owner would not have acted out of fear of a strong political reaction against raising tariffs to cover sewage treatment, which is expected to double charges for water service.[53] The commitment mechanisms we have described and the presence of a water subsidy, however, make it more likely that the government would have increased tariffs as mandated by the law. The second counter argument is that private operators would have been put off by farmers' apparently low willingness to pay for treated water. However, estimates of willingness to pay vary.[54] Moreover, it seems plausible that the government would outlaw irrigation of all foodstuffs with untreated sewage once a supply of treated wastewater becomes available. In any case sales to farmers are not necessary to make sewage treatment profitable as long as the tariff policy is sustained.

Drawing on World Bank (1994) we estimated that full sewage treatment 25 to 30 years earlier would have led to the range of net benefits shown in Table 6.8.[55] In 1991 the government put into effect a number of emergency measures in the face of the threat of cholera spreading from Peru that reduced the number of cases of cholera and typhoid attributable to irrigation with untreated sewage.[56] The benefits in Table 6.8 are based on these reduced rates of mortality and morbidity

By far the largest potential benefit comes from the reduced risk to earnings from agricultural exports should these emergency measures fail. Based on the consequences of trade restrictions on Chile's fruit exports because of a health scare in 1989, World Bank (1994) assumed that an eventual cholera outbreak would cause huge export losses, which are calculated to be equivalent to annual losses of at least 5% of fruit exports or US $50 million.

Another benefit comes from the elimination of cases of typhoid and cholera attributable to irrigation of food crops with untreated sewage, which is calculated as the average cost of medical treatment plus earnings lost due to lost work time per case.[57] The table also calculates the net effects on farmers productivity. The treatment of sewage will reduce the fertilizer value in wastewater, and eliminate it in areas deprived of wastewater by interception, which has a cost to farmers. However, farmers benefit

[53] EMOS expects the combined water and sewerage tariff, which is currently US $0.40 per m^3 of water, to increase by US $0.10 m^3 because of wastewater treatment the first year of the project, by US $0.16 m^3 the fifth year and by US $0.40 m^3 by the tenth year.

[54] According to surveys, farmers were willing to pay about one to three Chilean pesos/m^3 for untreated irrigation water from the Puclaro reservoir in 1991/92 (Aninat et al., 1993). Farmers might be willing to pay more now since prohibitions on irrigation of low-growing foodstuffs were put in place after this survey. A few recent purchases of raw water in the SMR for municipal and industrial uses suggest much higher prices, equivalent to 64 pesos/m^3 (or US $0.16/m^3).

[55] This analysis draws heavily on World Bank (1994), Chapter 2, "Water Pollution in Santiago: Health Impacts and Policy Alternatives."

[56] Specifically, the government prohibited irrigation of low growing vegetables with untreated sewage and restricted transporting irrigated vegetables and fruits outside of the SMR, intensified water quality monitoring, chlorinated irrigation water in canals, temporarily prohibited serving raw vegetables in restaurants, and intensified communication campaigns about the risks of vegetables and the need to wash and cook them.

[57] Between 1985 and 1990 the average annual direct cost of, for example, typhoid cases attributable to sewage irrigation of vegetables was $1.4 million per year.

Table 6.8

Annual cost and benefits of sewage treatment in Santiago (in millions of 1994 US $)

Total benefits [a]	24.1–79.0
Avoided cases of cholera	2.9–14.7
Avoided cases of typhoid	0.3–1.4
Net impact on farm profitability [b]	5.8
Avoided losses of farm exports	10.0–50.0
Reduced consumer costs [c]	4.1
Sale of treated wastewater	1.0–3.0
Total costs [d]	61.6
Net total benefits	(37.5)–17.4

[a] Low end of range assumes emergency measures fail every one year in five, high end assumes annually.

[b] Brzovic (1993) calculated the annual loss from reduced yields at US $0.6 million and from the reduction in irrigated areas at US $4.3 million, against the revenue gain to farmers from producing vegetables on prohibited acreage estimated at US $6.4 million a year.

[c] Assumes a rollback in price increases for crops which currently cannot be irrigated with wastewater.

[d] Assumes an average annual sewage flow of 18 m^3 and a cost per m^3 treated of US $0.109 including investment, amortization, operation and maintenance.

Source: World Bank (1994), Brzovic (1993), Fernandez (1993), Poblete (1993), and authors' calculations.

because vegetables could be grown in areas where they are currently banned because the only water available for irrigation is contaminated, which raises their profits somewhat, although it also reduces consumer prices. The calculations do not consider the effects of price increases to cover the cost of sewage treatment since the reduction in consumer surplus is offset by the increase in EMOS' revenue. Finally, the table assumes that treated water can be sold, producing additional revenues for EMOS.

Against these benefits Table 6.8 deducts the cost of sewage treatment, resulting in net benefits between −US $37.5 and +US $17.4 million in 1994 prices.[58] Although the lower range is negative, it should be kept in mind that the benefits do not include a reduction in hepatitis and other gastroenteric diseases, benefits to tourism and fishing in coastal waters and the effects to image and trade negotiations.

Besides perhaps building the sewage treatment plant, there were other ways we might expect a private EMOS to have behaved differently.[59] In particular there were regulatory weaknesses that the public EMOS had little motivation or capacity to exploit. One such weakness resulted from SSS vulnerability to capture (Sánchez and Sanhueza, 1997).

[58] EMOS' recent studies suggest that the total lifetime cost for treating 100% of the SMR wastewater using conventional technology will be US $574 million or US $61.6 a year.

[59] It could be that a private EMOS would have been motivated to increase operating efficiency faster since it could have captured more of the rents than the state-owned EMOS, which had to make large dividend payments to government. But since EMOS' performance was already good, the welfare gains from this would have been small.

SSS officials could have been influenced by prospective employment opportunities in the regulated sector if EMOS had been private. Although the SSS was better paid than its predecessor, professors of hydraulic and environmental engineering at the University of Chile reported that most of their graduates were not attracted to the SSS because salaries were low and career prospects poor compared to the private sector. The law does not place any restrictions on who can hold top executive positions at the SSS or what they do when they leave. The risk of capture was low as long as most of the regulated companies were state-owned, and hence could offer few inducements.

The SSS was also vulnerable to outright misbehavior by a private EMOS since its ability to enforce its decisions was weak. If a regulated company failed to comply with SSS' technical standards, requests for information, or requirements under its concession rules, the SSS could fine the company and in extreme cases revoke the concession. However, the fines are set by law and were widely regarded as too low to motivate compliance, while in most circumstances revoking the concession would have been seen as draconian and politically unacceptable.[60] This lack of credible sanctions was less of a problem with state enterprises like EMOS because the SSS could appeal to other government agencies such as the Ministry of Public Works for support in forcing compliance.[61] A private EMOS would not have been subject to orders from other government agencies.

In sum, there were some costs to not privatizing EMOS if we accept that a private firm would have built the sewage treatment plant. However, these costs are reduced by the likelihood of opportunistic behavior on the part of a private operator.

CONCLUSION

The reform of EMOS illustrates how a regulatory contract that addresses problems of information, incentives and commitment, combined with strong supportive institutions, can produce net benefits even in a relatively well performing water system. It also shows how institutional factors enabled Chile to overcome the agency problems that usually lead to contractual failure in SOE performance contracts.

In a situation such as Santiago's, where service was relatively good, such that the political costs of doing nothing were low, reform seemed unlikely. Yet a crisis brought in a government ideologically committed to the sale of EMOS. After the Pinochet government passed legislation designed to attract a private investor, motivate efficiency gains and prevent capture, Chile's political and constitutional constraints made reversal virtually impossible. Since there were net political gains to the Aylwin government, the reform was not only tolerated but vigorously pursued. On the political benefits side,

[60] The proposed reform of the SSS would increase fines sharply. For example, the present fine for providing false information is the equivalent of US $80,000 and would increase to US $800,000; the fine for poor quality water would increase from US $15,000 to US $80,000 (field interviews).

[61] Indeed, in a randomly selected week from the SSS' weekly report of fines, EMOS was cited for three infractions, apparently minor judging from the amounts of the fine. The private firm operating in Santiago, Aguas Cordillera, was cited for ten infractions, three of them serious, again judging from the amounts of the fines (Republic of Chile and Superintendency of Sanitary Services, 1997).

poor constituents could be connected, service could improve, and needed revenues could be generated. The existence of a subsidy that was means targeted but leaked over to middle-income consumers reduced opposition to tariff increases across a broad range of voters. Although workers were initially opposed, their opposition was weak to begin with, muted by their support for the new government and lack of alternatives, and offset by the compensation of collective bargaining, bonuses and other benefits.

A regulatory framework grounded in law that gave limited discretion to the regulator on pricing was a good fit with Chile's electoral, legislative and legal institutions. A price cap reinforced the existing incentives motivating efficiency in EMOS. Supportive norms and institutions mandating a hard budget reinforced the high powered incentives in the regulatory contract.

Would Chile have done better if it had privatized EMOS? If everything else remained the same, except that a private investor built the sewage treatment plant, then the health of Santiago's population would have improved, a serious risk to its food exports would have been mitigated and a source of contamination of its beaches and oceans would have been eliminated. The benefits of these externalities are hard to estimate, but our best guess suggests that the gains would have outweighed the costs. However, the assumption that all else would have been equal with private ownership of the country's largest water company seems dubious. A private investor would have had much stronger incentives to exploit the vulnerabilities of the SSS. The net effect of these actions is hard to predict. Recent legislation will correct some of the weaknesses of the regulator, through higher salaries (by putting SSS staff on a higher civil service ranking), limits on how soon SSS staff can be employed by the regulated industry, and higher fines for regulatory infractions (see Sánchez and Sanhueza, 1997 for an excellent discussion of these issues).

Our findings suggest that the workings of Chile's water regulation could also be improved by increasing the use of competition. Increasing yardstick competition would require the SSS to reveal more information to the companies about what standards are used in tariff setting and simplify the process so that company managers can compare their costs with the model company. The risk that the company might manipulate information could be reduced through greater competitive bidding of concessions and higher sanctions for misinformation. The incentive effects of yardstick competition could be lost, however, unless the dividend policy is changed. The incentive for companies to improve their efficiency under the benchmark tariff used in Chile is curtailed if government captures all the gains through dividends and taxes, especially for regional companies without the strong reputation effects of EMOS.

Another way to improve information and strengthen the incentive features of the regulation would be to require competition for all new concessions and for existing concessions when the current owner fails to live up to important obligations such as to invest or to maintain water quality over some prolonged period, such as five years. Greater consumer information and representation of consumers and other interests in the tariff setting process would also improve monitoring.

Those looking to Chile as a model might keep in mind three factors present in Chile that are not typical of water systems in most developing countries. First is the long tradition of private water rights that meant that management of the sector was shaped by a very early recognition of water as a scarce and tradable private good. A second factor is Chile's strong bureaucratic institutions and norms governing the public sector that prevented EMOS and other SOEs from operating with the degree of corruption, inefficiency and paralysis common to many water systems. Third, Chile's electoral, constitutional and judicial institutions permitted the introduction of a contract that mimicked a private concession as well as a means-tested subsidy. Finally, Santiago's relatively low cost water sources helped make it politically possible to price services at full cost recovery.

APPENDIX A: CHILE: SUBSIDY PROGRAM

In 1989 Chile created a system of subsidies for water and sewerage, which took effect in 1990. Anyone who holds a social classification card (CAS), has a household water connection and is not more than three months late in paying their water bill is eligible for the subsidy. The CAS has been in effect since 1980 and establishes eligibility for a number of targeted subsidies in addition to water and sewerage, including a flat-rate family allowance, minimum pensions, and housing assistance. Households applying for a CAS are visited by a municipal official who allocates points to the family based on the size and composition of the household, the occupation and education of the household head, the assets (car, refrigerator, land, etc.) and per capita income of the family, and the characteristics of the dwelling (such as location, number of rooms, exterior and interior materials used for construction, whether it has a water or electrical connection, etc.) Households with less than a maximum number of points are then issued a CAS and those with lower points receive higher priority for getting subsidies.

The yearly allotments for the water subsidy are based on bi-annual surveys of family income by region. The allotments are allocated so as to assure that no family in the region pays more than 5% of its income for water and sewerage.[62] Regional allotments are then divided among municipalities based on the number of CAS holders with lower points in each local area. The municipality then distributes its allotment across the list of eligible beneficiaries, starting with those with the lowest points, and puts those it cannot cover on a waiting list.[63] The subsidy is then paid directly to the water supplier, who subtracts the appropriate amount from each household's water bill and collects the remainder of the bill directly from the customer. Initially only the first 15 m^3 of consumption per month could be subsidized; this was later raised to 20 m^3. By comparison, average consumption for a poor urban family is 20–30 m^3 per month. On average the subsidy covers 60% of the water bill of eligible households, although in regions with high tariffs, up to 85% may be covered.

The number of Chilean households covered by the subsidy has gone from 5% of those eligible in 1990 to 95% in 1996 (see Table 6A.1). The expansion in coverage was largely

[62] The 5% is a standard based on the World Health Organization's recommendation that water and sewerage expenditures should represent no more than one day of a worker's income a month.

[63] Persons can lose the subsidy if they fail to pay their share of the water bill, change dwellings, etc.

Table 6A.1

Chile: coverage of the subsidy for water and sewerage, 1990–1996

Year	*Number of budgeted subsidies*	*Number of subsidies used*	*Usage rate (%)*
1990	424,625	21,824	5.1
1991	424,625	177,719	41.9
1992	441,040	315,901	71.6
1993	443,038	351,925	79.4
1994	454,038	389,712	85.8
1995	461,508	399,205	86.5
1996	466,508	442,524	94.9

Source: MIDEPLAN (1997).

the result of a public education campaign by the water companies to convince eligible consumers to apply, although an easing of the conditions for eligibility also helped. As a result of this increased coverage, the total, national cost for the water subsidy has risen from US $8000 in 1990 to US $500,000 in 1991 to US $30 million in 1996.

Only about a third of the households which received a water subsidy were in the lowest-income quintile, and 30% in the second quintile, while some benefits reached even the upper-income quintile (see Table 6A.2). This distribution by number of households is somewhat misleading, since the share of the total water subsidy which was received by the upper-income groups is probably much less than for the other groups (data are not available) and the number of people in a lower-income household is much higher. The fact that some subsidies were paid to upper-income groups is partly a result of Chile's highly unequal income distribution, where 96% of the households have incomes of less than $1260 per capita (see Table 6A.3); Chile's per capita income in 1995 was US $4805. In addition, the way the subsidies are distributed by regions means that even a household in the upper quintile could qualify under the

Table 6A.2

Chile: water and sewerage subsidy by income group, 1996

Income quintile	*Per capita household income (US $)*	*Average subsidy (CH $)*	*Number of beneficiary households*	*Percentage distribution of of beneficiaries*
I	0–250	2419	64,594	32.7
II	250–440	2370	59,541	29.6
III	440–680	2392	43,045	21.6
IV	680–1260	2676	24,220	13.6
V	>1260	2507	4,920	2.6
Total		2432	196,320	100.0

Ch $ = Chilean Pesos.
Source: MIDEPLAN (1997).

Table 6A.3

Chile: distribution of monetary income by quintiles, 1987–1996 [a] (percentages)

Quintile	*1987*	*1990*	*1992*	*1994*	*1996*
I	57.2	56.9	56.3	56.9	56.7
II	19.0	18.1	18.4	18.5	19.1
III	11.7	12.3	2.2	12.0	11.9
IV	7.9	8.2	8.5	8.2	8.2
V	4.3	4.4	4.6	4.3	4.1

[a] Monetary income includes all income to factors of production (i.e., wages, salaries, pensions, rents, dividends, interest, etc.) plus monetary transfers from the public sector.
Source: MIDEPLAN (1997).

'no-more-than-five-percent-of-income-for-water' rule in some regions. Only a third of the beneficiaries are in the lowest quintile, probably because the subsidy cannot reach indigent households (approximately 5% of all households in 1995). The subsidy is only available to those with a house connection (which requires a fixed dwelling) and who can afford to pay some part of their bill, which rules out the indigent. (As well as the non-indigent part of the 21% of Chile's population that does not have indoor plumbing, most of them in rural areas.) In addition, some poor families may be discouraged by the need to apply for the subsidy, while some municipal workers may be duped or corrupted by wealthier beneficiaries.

APPENDIX B: DEMAND ESTIMATION AND CONSUMER SURPLUS

Since the estimation of demand curve is an integral part of the cost–benefit analysis, this appendix provides more details on the assumptions used for the estimation of the demand curve and consumer surplus.

In our analysis of consumer surplus we cover two goods, water and sewerage, for which the same demand elasticity is assumed.[64] (Lack of data prevented us from decomposing the revenue into finer categories.) The price elasticity we use is −0.259.[65] The connection fee is not treated explicitly, but this should not significantly affect the results because coverage was quite high: the water market was almost 100% covered, and the sewerage market more than 80%. Continuity for the Santiago metropolitan area

[64] From total revenue and quantity of water and sewerage, respectively, which are consistently available, we derive their implicit prices.

[65] This elasticity comes from two sources. The first is an unpublished study about water demand in Santiago, according to which the price elasticity is −0.516 for households which consume 1–20 m^3 of water, −0.2 for 20–30 m^3, and −0.25 for more than 30 m^3. The second source is the consumption patterns among households, which suggests that there were 5% of households using 1–10 m^3 per month, 40% using 10–30 m^3 per month, and 55% using more than 30 m^3 per month. Then the weighted average elasticity is just $0.10 \times (-0.516) + 0.35 \times (-0.20) + 0.55 \times (-0.25) = -0.259$. Although we know that the demand elasticity for water of industrial and commercial use is −0.20, we did not use this piece of information because this category of water use accounts for only 5.8% of the total use according to the EMOS Annual Report of 1990. This elasticity is quite similar to that which is assumed in Argentina (−0.32).

was not an issue: all connections had running water for 24 hours a day. Finally, since the metering percentage has been 100%, there is no need to distinguish metered and unmetered consumers.

The total demand curve is assumed to be:

$$TQ = N(a - b\text{P})$$

where TQ is total demand, N is total population, $(a - b\text{P})$ is an individual demand curve, a is the demand when marginal price is 0, and p is the price of water and sewerage combined. Thus the total demand curve shifts outward with the expansion of population, while the individual demand curve remains stable over time.

The estimation of a and b involves a few assumptions. First, the demand elasticity around the observed demand point in 1988 is, as mentioned in the text, −0.26. Second, except for the population shifter for each year, the demand curve remains the same, whether under the factual or the counterfactual. Third, at the price in 1988, the total demand consisted of all the residents of EMOS; the uncovered population is treated as excess demand. Given the relative low price of water and sewerage that year (roughly 0.14 US dollar), this assumption is likely justified. From these assumptions, then, one may impose the following simultaneous equation system, with two equations and two unknowns, a and b:

$$-\varepsilon_d = \frac{\mathrm{d} \ln TQ}{\mathrm{d} \ln P} = -Nb\frac{P}{TQ}$$

$$\frac{Q}{\text{coverage}} = N(a - bP)$$

where Q is the number of people without water connections.

These equations can be solved by using the data for 1988, the last year of the pre-reform era. With the knowledge of demand function, we can then proceed to estimate consumer surplus associated with a price and quantity pair.

The estimation of consumer surplus then proceeds as follows.

- slope $= \dfrac{\mathrm{d}P}{\mathrm{d}TQ} = \dfrac{1}{Nb}$
- $P_{\text{intercept}} = TQ \times \text{slope} + P_t$, where $P_{\text{intercept}}$ is the interception of the demand curve with the P axis, and P_t is the observed marginal price.
- Consumer surplus $= S_1 + S_2$, where $S_1 = 0.5 \times Q \times (P_{\text{intercept}} - P_Q)$, and $S_2 = Q \times (P_Q - P)$.

Caution. The cost–benefit analysis has its limitations. While we have tried our best to take into account as many relevant details as possible in constructing our factual and counterfactual scenarios, we also had to leave out many relevant ingredients. For instance, it was hard to incorporate into the benefits of the reform the improvement of the quality of the system.[66] Also ignored are the benefits in terms of better ability to deal with a major drought.

[66] The number of pipe breaks per kilometer decreased from above 0.50 before 1990 to 0.25 in 1996, representing a significant drop.

REFERENCES

Alfaro, Raquel, 1995. *Competition and Social Extension in a Water Supply and Sanitation Company.* Unpublished report for the World Bank.

Alfaro, Raquel, 1997. *Linkages Between Municipalities and Utilities: An Experience in Overcoming Urban Poverty.* Unpublished report for the World Bank.

Alfaro, Raquel, and Caarl Bartone, 1997. *Wastewater Reclamation for Irrigation in the Santiago Metropolitan Area.* Unpublished report for the World Bank.

Baldez, Lisa and John Carey, 2000. "Budget procedure and fiscal restraint in post-transition Chile." In: Stephan Haggard and Mathew D. McCubbins, eds., *Presidents, Politics and Policy.* Cambridge: Cambridge University Press.

Brook-Cowen, Penelope, A. Dianderas and Guillermo Yepes, 1996. *Different Paths to Successful Reforms: The Case of Chile and Argentina.* Infrastructure Notes, October, Washington, DC: World Bank.

Brzovic, Francisco, 1993. *Saneamiento y la Rentabilidad Agricola.* Unpublished report prepared for the World Bank.

Constable, Pamela and Arturo Valenzuela, 1989. "Chile's return to democracy". *Foreign Affairs* 68: 89–90; 169–186.

Cox, Gary and Mathew D. McCubbins, 1997. *Political Structure and Economic Policy: The Institutional Determinants of Policy Outcomes.* Processed.

Fernanadez, Juan Francisco, 1993. "Algunos efectos en la agriculture de la contaminacación microbiológica de las agues de riego." Draft report for the SAG.

Galal, Ahmed, Leroy Jones, Pankaj Tandon and Ingo Vogelsang, 1994. *Welfare Consequences of Selling Public Enterprises: An Empirical Analysis.* New York: Oxford University Press.

Gilbert, Alan, 1993. *In Search of a Home. Rental and Shared Housing in Latin America.* London: UCL Press Ltd.

Haggard, Stephan and Mathew D. McCubbins, 2000. *Presidents, Politics and Policy.* Cambridge: Cambridge University Press.

Hearne, Robert R. and K. William Easter, 1995. *Water Allocation and Water Markets. An Analysis of Gains-from-Trade in Chile.* Technical Paper No. 315, Washington, DC: World Bank.

Jones, Leroy, P., Pankaj, Tandon and Indo Vogelsang, 1990. *Selling Public Enterprises: A Cost–Benefit Methodology.* Cambridge, MA: MIT Press.

Laffont, Jean-Jacques and Jean Tirole, 1986. "Using cost observation to regulate firms". *Journal of Political Economy* 94, Part 1: 614–641.

Laffont, Jean-Jacques and Jean Tirole, 1993. *A Theory of Incentives in Procurement and Regulation.* Cambridge, MA: MIT Press.

Levy, Brian and Pablo Spiller, eds., 1996. *Regulations, Institutions and Commitment: Comparative Studies of Telecommunications.* New York, NY: Cambridge University Press.

McCubbins, Mathew D., Roger Noll and Barry Weingast, 1987. "Administrative procedures as instruments of political control". *Journal of Law, Economics and Organization* 3(2): 243–277.

McCubbins, Mathew D., Roger Noll and Barry Weingast, 1989. "Structure and process, politics and policy: administrative arrangements and the political control of agencies". *Virginia Law Review* 75: 431–482.

Méndez, Roberto, 1990. Opinion publica y la eleccion presidencial de 1989. *Estudios Publicos*: 68–97.

North, Douglass C, 1990. *Institutions, Institutional Change, and Economic Performance.* New York: Cambridge University Press.

Panzer, John and Ricardo Paredes, 1991. "The role of economic issues in elections: the case of the 1988 Chilean presidential referendum". *Public Choice* 71: 51–59.

Poblete Juan Antonio, 1993. "Institucionalidad y control de la calidad del agua dentro de un marco de gestión integral del uso de los recursos hidráulicos: el case de Chile." Unpublished report prepared for the World Bank.

Republic of Chile, 1980. *Constitución Politica de la Republica de Chile* (Political Constitution of the Republic of Chile).

Republic of Chile, EMOS, various years. *Memoria.* Santiago: EMOS.

Republic of Chile, Ministry of Economy, Development and Reconstruction, 1990a. *Law No. 18.902: Crea la Superintendencia de Servicios Sanitarios* (Creates the Superintendency of Sanitary Services).

Republic of Chile, Ministry of Economy, Development and Reconstruction, 1990b. *Supreme Decree No. 453:*

Reglamento del Decreto con Fureza de Lay No. 70 que Establece la Fijación de Tarifas de Servicios de Agua Potable y Alcantarillado (Regulation for Decree with Force of Law No. 70 that Establishes Tariff Setting for Potable Water and Sewerage Services).

Republic of Chile, Ministry of Finance, 1981. *Law No. 18.046: Lay Sobre Sociedades Anonimas* (Law Governing Incorporated Companies).

Republic of Chile and Ministry of Finance, 1989. *Law No. 18.778: Subsidio al Pago de Consumo de Agua Potable y Servicio de Alcantarillado de Aguas Servidas* (Subsidy for Consumption of Potable Water and Sewage treatment Services).

Republic of Chile and Ministry of Finance, 1991. *Supreme Decree No. 529: Reglamento De Ley No. 18.778: Subsidio al Pago de Consumo de Agua Potable y Servicio de Alcantarillado de Aguas Servidas* (Regulation of Law No. 18.778: Subsidy for Consumption of Potable Water and Sewage treatment Services).

Republic of Chile and Ministry of Planning and Cooperation (MIDEPLAN), 1997. *Pobreza y Distribución del Ingreso en Chile, 1996. Resultados de la Encuesta de Caracterización Socioeconómica Nacional.* Santiago: MIDEPLAN.

Republic of Chile and Ministry of Public Works, 1988a. *Decree with force of law No. 70: Fijación de Tarifas de Servicios de Agua Potable y Alcantarillado* (Setting Tariffs for Potable Water and Sewerage Services).

Republic of Chile and Ministry of Public Works, 1988b. *Decree with force of law No. 382: Ley General de Servicios Sanitarios* (General Law of Sanitary Services).

Republic of Chile and Ministry of Public Works, 1990. *Supreme Decree No. 453: Reglamento del Decreto con Fuerza de Ley No. 70, de 1988, que Establece la Fijación de Tarifas de Servicios de Agua Potable y Alcantarillado* (Regulation for Decree with Force of Law No. 70 that Establishes Fixation of Tariffs for Potable Water and Sewerage Services).

Republic of Chile and Ministry of Public Works, 1991. *Supreme Decree No. 121: Reglamento de Concesiones de Servicios Publicos Sanitarios del D.F.L. MOP. No 382, de 1988* (Regulation of Concessions of Public Sanitary Services).

Republic of Chile and Superintendency of Sanitary Services, 1995. *Memoria Anual 1995* (Annual Report, 1995). Santiago.

Republic of Chile and Superintendency of Sanitary Services, 1997. *Informe Semanal de Estados de Procedimientos de Multas*. Processed.

Rios, Monica A. and Jorge A. Quiroz, 1995. "The market of water rights in Chile: major issues". *Cuadernos de Economia* 32: 317–345.

Sáez Lozano, José Luis, 1996. "Economía, política y elecciones en la democracia Chilena, 1989–1996". *Estudios Económicos* II(2): 237–296.

Sánchez, José Miguel and Ricardo Sanhueza, 1997. *Autonomy and Regulation in the Chilean Water and Sewerage Sector: An Institutional Evaluation*. Processed.

Scully, Timothy R., 1995. "Reconstituting party politics in Chile". In: Scott Mainwaring and Timothy R. Scully, eds., *Building Democratic Institutions: Party Systems in Latin America*. Stanford, CA: Stanford University Press.

Shirley, Mary and Lixin Colin Xu, 1998. "Information, incentives, and commitment". *The Journal of Law, Economics and Organization* 14(2): 358–78.

Shirley, Mary and Lixin Colin Xu, 2001. "The empirical effects of performance contracts". *Journal of Law, Economics, and Organization* 17(1): 168–200.

Shugart, Matthew S. and John M. Caarey, 1992. *Presidents and Assemblies. Constitutional Design and Electoral Dynamics*. New York, NY: Cambridge University Press.

United Nations. Economic Commission for Latin America and the Caribbean, 1991. "Una estimación de la magnitud de la pobreza en Chile, 1987". *Colección Estudios CIEPLAN* 31(March): 107–29.

Valenzuela, Arturo, 1994. "Party politics and the crisis of presidentialism in Chile: a proposal for a parliamentary form of government". In: Juan J. Linz and Arturo Valenzuela, eds., *The Failure of Presidential Democracy. Comparative Perspectives*. Baltimore, MD: The Johns Hopkins University Press.

Weyland, Kurt, 1997. "Growth with equity in Chile's new democracy". *Latin American Research Review* 32(1): 37–67.

Williamson, Oliver, 1976. "Franchise bidding for natural monopolies — in general and with respect to CATV". *Bell Journal of Economics* 7(1): 73–104.

Williamson, Oliver, 1985. *The Economic Institutions of Capitalism: Firms, Markets, Relational Contracting*. New York, NY: Free Press.

World Bank, 1986a. *President's Report on Proposed Loans with the Guarantee of the Republic of Chile to EMOS and ESVAL*. Internal Report No. P-4147-CH, January 21, 1986.

World Bank, 1986b. *Staff Appraisal Report: Chile: Santiago Water Supply and Sewerage II and Valparaiso Water Supply Reconstruction Projects*. Internal Report No. 5821-CH, January 21, 1986.

World Bank, 1992. *Project Performance Audit Report. Chile: Water Supply Project*. Internal Report of the Operations Evaluation Department No. 10464, March 19, 1992.

World Bank, 1994. *Chile: Managing Environmental Problems: Economic Analysis of Selected Issues*. Internal Report No. 13061-CH, December 19, 1994.

World Bank, 1995. *Bureaucrats in Business: The Economics and Politics of Government Ownership*. New York: Oxford University Press.

World Bank, 1997. *1997 World Development Indicators*. CD-ROM, Washington, DC: World Bank.

Thirsting for Efficiency: The Economics and
Politics of Urban Water System Reform
Mary M. Shirley (Ed.)

CHAPTER 7

REFORMING WATER SUPPLY IN ABIDJAN, CÔTE D'IVOIRE: A MILD REFORM IN A TURBULENT ENVIRONMENT

CLAUDE MÉNARD AND GEORGE R.G. CLARKE

INTRODUCTION

Especially when compared to other urban water systems in West Africa, the water supply system in Abidjan performs well. For over 30 years, despite rapid population growth, there has been enough investment to cover basic needs and, consequently, coverage is good by regional standards. Prices, although high compared to countries in Latin America and Asia, are lower than in many neighboring African countries and have been declining in real terms.[1] The strong commercial performance of the operator in Abidjan has produced sufficient revenue to subsidize the expansion and development of urban water systems throughout the country. Further, the stable contractual arrangement between the government and the private operator, SODECI, has proven adaptable enough to survive several adjustments and crises. All stakeholders — private users, government officials, international organizations and the private operator — agree, with few reservations, that the arrangement has worked well. Consequently, the model has been imitated throughout Africa and other regions of the world, although the results have sometimes been more mixed than they have been in Côte d'Ivoire.

This chapter documents the recent history of the water system in Abidjan, focusing on a reform to the contractual arrangement in 1988. We focus on three main questions. (1) What motivated reform in a system that was already performing well? (2) How, and why, did the reform affect sector performance and what additional changes might improve performance further? (3) What explains the relatively strong

[1] Chapter 8 shows that in the mid-1990s, prices were lower in Abidjan than in many other African countries.

performance of Abidjan's water system — is the success primarily due to an efficient contractual arrangement or is it due to more general characteristics of the institutional environment in Côte d'Ivoire? Given that the Ivoirian model is being transplanted to other countries, understanding the reasons for its strong performance will allow us to anticipate the likely benefits and limitations of the arrangement in other developing countries.

The primary motivation for reform was a macroeconomic crisis, which reduced the budgetary resources available for public investment. This made it difficult for the government agencies responsible for investment to increase system capacity fast enough to serve the city's rapidly expanding population. However, the crisis did not change the government's support base nor did it have a dramatic impact on service quality for existing customers and, therefore, there was little support for a major restructuring. Consequently, the crisis led to only a minor reform that shifted responsibility for investment from the government budget to sector resources and increased the private operator's responsibilities. This provides counterfactual evidence that supports one of the central hypotheses of the project: in the absence of a sector crisis or a realignment of political forces, the *desirability of reform* is weak and, therefore, reform will tend to be limited.

As the chapter shows, the reform was quite successful. Most performance indicators appear to have improved or stayed the same, tariffs have fallen, and all parties to the agreement appear satisfied with system performance. Most notably, despite a drop in investment expenditures, the private operator managed to reverse a downward trend in coverage levels in Abidjan and to maintain service quality. However, the modest nature of the reform meant that some opportunities for improvement were missed and that some problems remain, calling into question the continued viability of the arrangement. The chapter discusses the challenges facing the system in the near future and speculates about steps that might further improve sector performance.

Finally, the chapter tries to explain the success and limitations of the contractual arrangement by looking at the interaction between specific characteristics of the institutional environment in Côte d'Ivoire and the design of the contract. Notwithstanding recent events, in a region of the world plagued by political instability, Ivoirian political institutions were remarkably stable for close to 40 years.[2] In part, the success of the Ivoirian model is the result of the stability and credibility of these institutions. Although the single party system in place at the time of reform might suggest that there were few restraints that would prevent the government from behaving opportunistically, several features of the institutional environment, discussed in detail in the next section, protected against this. Because of this, and because the reform was based on a system that was already performing well, the contractual arrangement proved to have exceptional capacity to adjust even when faced with dramatic changes in the external environment. This allowed the government to maintain confidence and to guarantee the safeguards

[2] Political instability has increased in Côte d'Ivoire since the military coup of December 1999, which occurred after the field research underlying this chapter was completed. Further, notwithstanding the coup, Côte d'Ivoire was very stable for nearly 40 years (i.e., 1960–1999), especially by regional standards.

protecting the private operator during the reform process, ensuring the feasibility and sustainability of reform. Because the institutional environments of most other countries in the region are not as favorable as in Côte d'Ivoire, similar contractual arrangements might be less successful when copied in these countries.

The next section of the chapter will analyze the working conditions of the water supply system in Abidjan prior to reform. The third section describes the institutional environment in Côte d'Ivoire, while the fourth section discusses the macroeconomic crisis that motivated reform and the effect that this had on the government's support base and the water sector. The fifth section examines the reform itself. It discusses changes introduced in the preexisting governance structure, characteristics of the contractual arrangement adopted in 1988, and modalities of its implementation. The sixth section summarizes the effect that reform had on sector performance and the reasons for the changes. The seventh section discusses difficulties to overcome and lessons that can be learned from the Abidjan case and the final section concludes.

A CONTINUING PRIVATE SECTOR PARTICIPATION

Even before the 1988 reform, sector performance in Abidjan was impressive by regional standards. Access to potable water was high, almost all connections were both metered and billed, and private customers routinely paid their bills. Further, piped water was drinkable and service interruptions were infrequent. In this section of the chapter, we discuss sector organization and performance before reform. The success of Abidjan's water supply system supports the view that a *combination* of stable and well-defined institutional rules and an appropriately designed contract determines success.

Sector Organization before Reform

Private participation in urban water supply is a well-established tradition in Côte d'Ivoire, especially in Abidjan. A private enterprise, the *Société de Distribution d'Eau de Côte d'Ivoire* (SODECI), has been responsible for operation and maintenance of the Abidjan water system since before independence and has been responsible for operation and maintenance of all urban water systems in Côte d'Ivoire since 1974. Although its responsibilities have changed several times, its continuing presence demonstrates the feasibility of private sector participation in Côte d'Ivoire's urban water sector.

Partners and responsibilities

Until 1956, most water systems were managed by local public agencies under the supervision of city councils. Because the systems were often operated inefficiently and were in need of significant investment, some municipal governments turned to the private sector. In 1959, before Côte d'Ivoire became independent, an international tender for a concession contract to supply water in Abidjan was launched. A French company, SAUR, which was already active in West Africa, won the tender. However, since as a condition of the contract, the private operator had to be subject to Ivoirian laws and be majority-owned by Ivoirian shareholders, a new enterprise, SODECI, with SAUR as the main shareholder, was formed.[3] The 30-year *concession contract* that the city government signed with SODECI made SODECI responsible for operation, maintenance and investment in Abidjan (World Bank, 1977). Over the next few years, SODECI signed five additional lease contracts and three additional management contracts in other municipalities in Côte d'Ivoire (SODECI, 1996).

In 1967, SODECI's role in Abidjan was significantly reduced when responsibility for most investment was passed to the government, making the contract into a management contract with SODECI responsible for operation and maintenance of the system.[4] SODECI also continued to implement significant investment on behalf of the Ivoirian government. In 1974, SODECI's success in Abidjan encouraged the government to award it a 15-year contract for the operation and maintenance of urban water systems in all other Ivoirian cities, to improve sector performance outside of Abidjan and to equalize tariffs across urban areas. Since costs in Abidjan were considerably lower than in other areas, Abidjan ended up subsidizing urban water supply in secondary cities.

At this time, responsibility for the sector management was transferred to the central government and, from then on, local authorities played only a minor role in the sector. A government agency, the *Direction de l'Eau* (known as the *Direction Centrale de l'Hydraulique* until 1984) retained ownership of sector assets, managed and planned investment and monitored the contract with SODECI.[5] After January 1976, sector finances were managed by an autonomous agency, the *Fonds National de l'Hydraulique (FNH)* located in the *Caisse Autonome d'Amortissement (CAA)*. The FNH received part of the water tariff to service sector debt and could borrow to finance investment.

SODECI has thus been an active participant in Abidjan's water supply system for over 40 years. Indeed, no other enterprise, private or public, has been involved in the operation and maintenance of any of the urban water systems in Côte d'Ivoire since 1974. SODECI has also played a limited role in the sewerage system, although operation and maintenance of the sewerage system has never been included in the contracts for urban water supply.

Characteristics of the Water Supply System prior to Reform

Source of water

Almost all of Abidjan's water comes from an abundant aquifer, which should be able to supply sufficient water to satisfy demand in the medium term.[6] Although Abidjan is close to the ocean, the water is not saline because the aquifer is in the hilly northern part

---

[3] SAUR was the largest shareholder (46%), individual Ivoirian investors owned 45%, the Ivoirian government held 5% and private French investors held the remaining 4% (World Bank, 1977).

[4] SODECI also was responsible for renewal of electromechanical equipment for Abidjan's boreholes (World Bank, 1990).

[5] Although many agencies changed their names over the 30-year period covered by this study, we will refer to all agencies by their 1997 names for ease of exposition.

[6] World Bank (1982) notes that it is estimated that the aquifer has a safe yield of around 700,000 m^3/day.

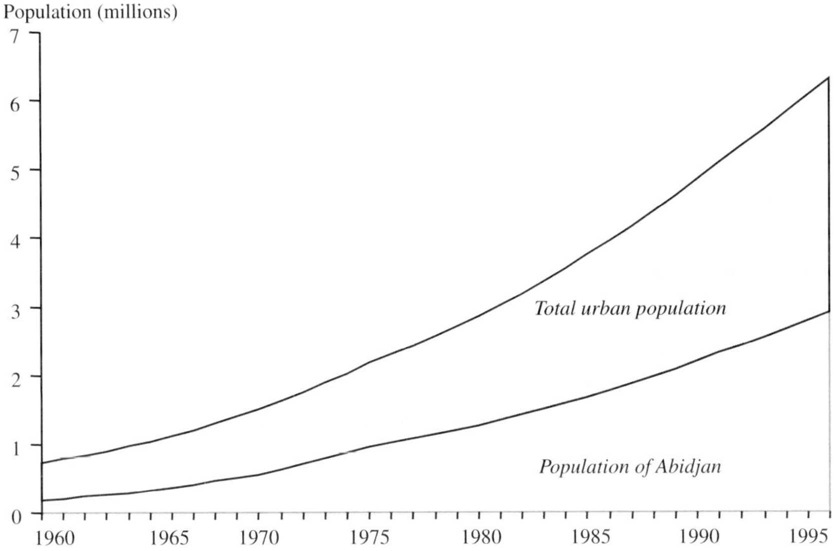

Source: World Bank (1998).

Figure 7.1. Urban population in Côte d'Ivoire, 1960–1996.

of the city. The aquifer is located about 60 m underground and is protected by a thick layer of clay (World Bank, 1989) and, consequently, the water is not highly polluted.[7]

System expansion

Partly due to a large in-flow of immigrants from neighboring countries, the population of Abidjan grew quickly following independence (see Fig. 7.1). While the population was estimated to be about 180,000 in 1960, it was estimated to be close to 3 million in 1996. Although this rapid growth put pressure on the system, the number of connections grew faster than the population for most of the 1960s and 1970s (see Fig. 7.7) and, consequently, coverage increased. Between 1960 and 1980, the number of connections increased from 3802 to 75,645, making the coverage rate in Abidjan close to 80% by 1980 (see Fig. 7.9).[8]

In large part, this was due to an ambitious investment program. During the 1970s, annual investment averaged US $40 million (World Bank, 1993), much of it financed through government borrowing and subsidies. In addition to increasing the number of connections, this also increased water produced (see Fig. 7.2) and water billed (see Fig. 7.6). Between 1960 and 1980, production in Abidjan grew from 8.9 billion m^3/year to 65.0 billion m^3/year and water billed grew from 5.0 billion m^3/year to 50.7 billion m^3/year.

[7] Although the aquifer had become polluted in low-lying areas, SODECI had abandoned these wells by the mid-1970s (World Bank, 1978).

[8] This assumes that the average connection serves 13.5 people, based on estimates by SODECI officials that each served between 12 and 15 people.

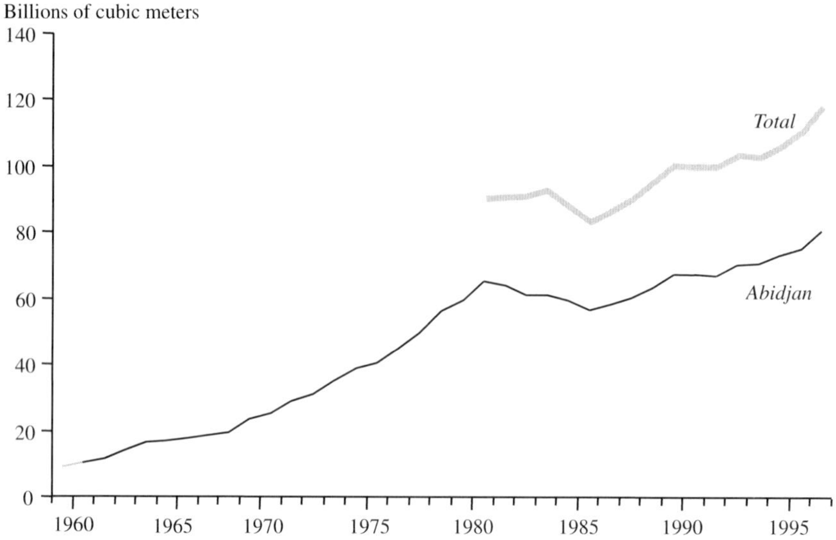

Source: SODECI.

Figure 7.2. Water produced.

Tariffs

In 1974, the tariff was CFAF 55/m^3 (\$0.25/m^3) in Abidjan and averaged CFAF 80/m^3 (\$0.36/m^3) elsewhere. After SODECI became responsible for water supply in all urban areas, a nationally uniform tariff of CFAF 89.5/m^3 (\$0.40/m^3) was imposed — subsidizing water supply in secondary cities was one of the main reasons for the 1974 reform.[9] By 1979, it was estimated that nearly 30% of the tariff in Abidjan went towards subsidizing water supply in secondary cities.[10] To compensate low-income users for the higher prices, a social tariff of CFAF 60/m^3 (\$0.27/m^3) was introduced for the first 5 m^3 of consumption per month.[11] In addition, the connection fee for small, low-income, consumers was lowered and then abolished. Starting in 1976, consumers using pipes of 15 mm and less did not have to pay a connection fee, other than a deposit of CFAF 6000 (\$25). In 1981, tariffs were increased, with the full tariff reaching CFAF 224/m^3 (\$0.78/m^3) and the social tariff reaching CFAF 128 (\$0.45/m^3). This made the average tariff CFAF 205 (\$0.71/m^3), of which CFAF 100/m^3 (0.35/m^3) went to SODECI, CFA 78/m^3 (\$0.27/m^3) to the FNH and CFAF 27/m^3 (\$0.09/m^3) to the FNA (World Bank, 1982).

[9] World Bank (1982) estimated the average incremental cost (AIC) of a cubic meter of water, which it used to as a measure of long-run marginal cost, for several cities in Côte d'Ivoire. The AIC in Abidjan was estimated at CFAF 104/m^3 (\$0.49) In comparison, the average AIC in eight secondary centers was estimated to be CFAF 271/m^3 (\$1.27). In the late 1980s, World Bank (1990) estimated that the long-run marginal cost of water was CFAF 48 (\$0.13) in Abidjan and CFAF 180 (\$0.51) in the rest of the country.

[10] CFAF 42/m^3 (\$0.20) out of the CFAF 149/m^3 (\$0.70) average tariff (World Bank, 1982).

[11] Large industrial companies paid a lower tariff until 1984, when increases in the 'industrial tariff' made it higher than the full domestic tariff.

Water and service quality

Although it is difficult to find exact data, it appears that service and water quality in Abidjan were good even before reform. World Bank (1982) reports "SODECI is efficiently managed and provides first class service to its consumers; water quality and pressure are uniformly good and unaccounted-for-water ... is very satisfactory."

Unaccounted-for water

Unaccounted-for water (UFW) in Abidjan has been low by both regional and international standards (see Ménard and Shirley, 2001).[12] Although the low level of UFW is impressive, it is not a direct result of reform. Indeed, apart from a short period between 1978 and 1980, UFW in Abidjan has been consistently less than 20% since the early 1960s (see Fig. 7.3). The success might partially reflect the relatively young age of most water systems in Côte d'Ivoire. Between 1960 and 1995, the network grew from only 176 km to 10,149 km nationwide (SAUR International, 1997). In 1974, only 34 urban areas had water systems, compared to 409 by 1995. Another feature that might have contributed to SODECI's success with UFW is that SODECI has been responsible for maintenance of the entire system. In contrast, responsibility for maintenance has been divided between two agencies, SEEG and SONEG, in Guinea, generating conflict

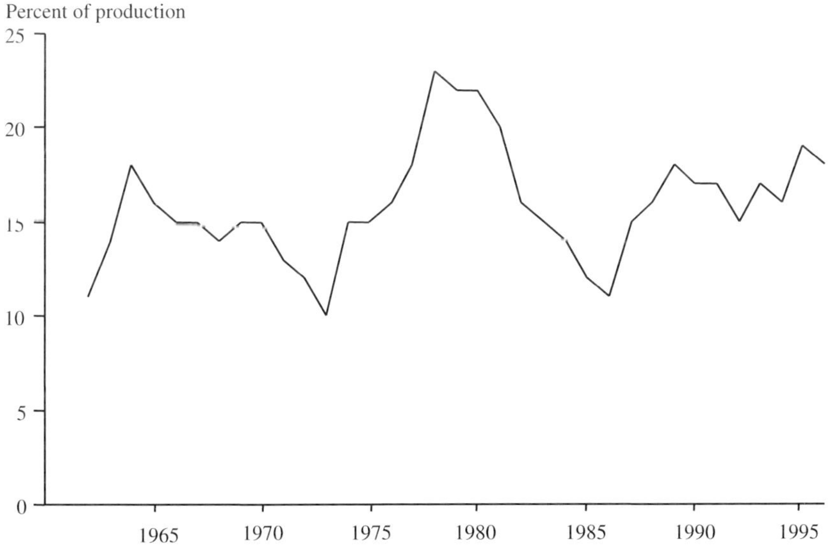

Source: SODECI.

Figure 7.3. Unaccounted-for water in Abidjan, 1960–1996.

[12] Since metering has been close to 100% since well before the 1988 reform, UFW is measured quite accurately.

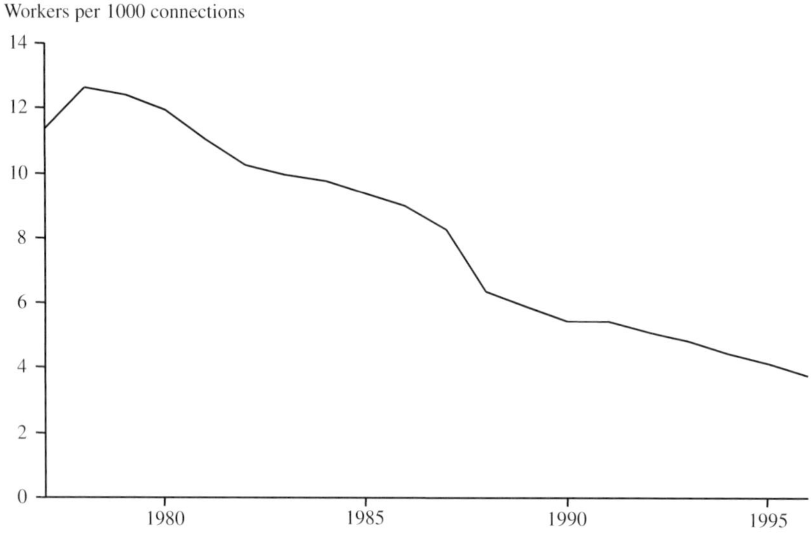

Source: SODECI.

Figure 7.4. Workers, nationally.

and ambiguities in decision-making (see Chapter 8). This emphasizes the importance of well-defined responsibilities.

Billing and collection

By the early 1980s, SODECI's commercial operations (i.e., billing and collection) were strong by regional standards. Metering was almost universal, billing, which was computerized, was executed efficiently (as demonstrated by the low level of UFW) and collection rates from the private sector were high (World Bank, 1982). In 1980, accounts receivable from the private sector accounted for about six months of private sales (see Fig. 7.12).[13] Given the delays associated with hand delivery and that no direct debit system existed, meaning that bills had to be paid at local kiosks, this was quite impressive. One factor that might account for the high collection rate was that private sector consumers were routinely cut off for nonpayment. Further, households were entitled to receive a free 'social' connection only once. This provided a strong incentive for regular payment since households that were cut off would have to pay a large reconnection fee.

The collection rate from the public sector was considerably lower. By 1980, public sector accounts receivable were over one year of public sector sales (see Fig. 7.12; see footnote 13). Most overdue public bills, at this time, were from public institutions such as universities and hospitals. Part of the problem was that SODECI was not allowed to cut off delinquent public sector customers (World Bank, 1994) and, consequently, public sector customers had little incentive to pay bills in a timely manner.

[13] Private sales are estimated as 75% of total sales. Private sales accounted for about 72% of sales (World Bank, 1977) in 1973, 78% of sales in 1987 and 74% of sales in 1992 (World Bank, 1993).

Labor productivity

Labor productivity has improved continuously since the mid-1970s and, by regional standards, was quite high by the mid-1980s. SODECI had approximately 10 workers per 1000 connections in the early 1980s (see Fig. 7.4), compared to about 24, 32, and 34 employees per 1000 connections in the public operators in Benin, Togo and Guinea, respectively. Although, SODECI initially relied heavily on expatriates for managerial and technical skills, it was slowly reducing its reliance upon them through an extensive training program for local employees.[14] By the early 1980s, there were less than 20 expatriate workers in SODECI, compared to 43 in 1978 (SODECI, various years). In part, the high productivity reflects the commercial orientation of SODECI, which dominates the company despite the small government holding.

Unconnected households

In the early 1980s, since there were few public standpipes, only about 30 in the entire city (World Bank, 1982), most unconnected households got water from private water vendors or from wells. At that time, and even now, resellers played an active role in providing water to unconnected households. Compound owners often sold water to their tenants (the cost depending upon the volume sold) and other individuals sold water to unconnected neighbors. This provided extra income to the owner, while allowing poor households access to potable water without having to pay access charges or face regular bills.[15] Water from the system was trucked to unconnected areas by professional water vendors. Although this was officially not allowed, it was tolerated as a means of providing water to unauthorized residential areas. In some low-lying, and less wealthy, areas (e.g., Marcory and Koumassi), households have wells in their compounds. Since the water from these wells is salty and can be heavily contaminated, it is not generally used for drinking except in some especially poor areas (e.g., Treichville).

Industrial enterprises were also allowed to use water from wells, but only after they had been granted permission from the government to do so. In addition, they had to pay SODECI for the use of groundwater resources and the disposal of wastewater. In practice, many enterprises were exempt from these requirements, failed to register their boreholes, or failed to pay the required fee. In the late 1980s, it was estimated that less than 20% of fees were collected from inventoried boreholes (World Bank, 1990).

---

[14] In 1974, SODECI opened a training facility that trained many SODECI employees, including a significant number of workers who later moved to the government ministries responsible for regulation.

[15] In the early 1980s, World Bank (1982) reports that 10 l of water cost about CFAF 10. This implies a total cost of CFAF 1000/m^3. At this time, the official (nonsocial) tariff was CFAF 224/m^3. The cost seemed similar in 1997. For example, in one compound in Adjamé, a connection provided water to the (extended) family of 15 persons and the owner sold water to 10 tenants. The prices were: 5 CFA for a bag, 25 CFA for a frozen bag, 25 CFA for a basin and 150 CFA for a barrel of 200 l (e.g., 750 CFA/m^3; the domestic rate at that time for an official connection was of 286 CFA/m^3).

Sewerage

Although recent studies have shown that the gains from improved water supply are far greater when sewerage is also improved (Esrey, 1996), sewerage has always been separate from, and considerably less developed than, water supply in Abidjan.[16] In 1975, there were about 34,900 connections to the sewerage system, serving about 227,000 people — about 24% of the population (World Bank, 1986). An additional 44% of the population used septic tanks and pit latrines, while the rest disposed of wastewater and sewage in unlined pits and watercourses (World Bank, 1986). There were no sewage treatment plants and, therefore, untreated sewage was dumped into the Ebrié Lagoon, around which the city was built. By the early 1970s, this dumping had overwhelmed the Lagoon's natural purification ability and large parts of the lagoon were highly polluted. As part of a long-term plan to improve the system, two World Bank supported projects attempted to increase coverage and to divert sewage to areas where natural purification was strongest. In the medium term, the government intended to by-pass the lagoon altogether by constructing an ocean outflow pipe and pumping the sewage out into the deep ocean. Two World Bank projects increased coverage of the sewerage system quite substantially in the 1970s and 1980s. By 1985, there were 60,000 connections serving an estimated 635,000 people, about 38% of Abidjan's population. In comparison, there were over 90,000 connections to the water system.

Although SODECI often had limited maintenance contracts for parts of the sewerage system, its role in the sector has been small. Before 1974, the municipal governments were formally responsible for sewerage, while SODECI had a limited maintenance contract for some parts of Abidjan's sewerage system and collected a small surcharge on water (CFAF $4.5/m^3$) to finance it (World Bank, 1974). In 1974, the sewerage subsector was reorganized, with the central government becoming more involved in the sector. A public agency, SETU, became responsible for planning and contracting new investment and a sister agency to the FNH, the *Fonds National de l'Assainissement* (*FNA*) was given responsibility for sector receipts and payments (including debt service and financing investment). Through a management contract, SODECI became responsible for the operation and (limited) maintenance of the sewerage system and continued collecting a surcharge on water.[17] Responsibility for the planning, design and construction of sewerage facilities was transferred from SETU to the *Direction de l'Eau* in April 1984 (World Bank, 1986).

INSTITUTIONAL ENDOWMENT AT THE TIME OF REFORM

A central hypothesis of this research is that the feasibility and sustainability of reform rely heavily upon the credibility and stability of government commitments. At least four aspects of the institutional environment in Côte d'Ivoire increased the credibility of

[16] Abidjan was the only city with a sewerage system in Côte d'Ivoire. A few other cities have rudimentary sewerage/drainage systems (World Bank, 1999).

[17] There was also a surcharge on property taxes, which the FNA used to finance debt service and investment related to drainage.

the government's commitment to private sector participation in the water sector. First, for nearly 40 years following independence, the Ivoirian government was stable and, consequently, there was little threat of sudden changes in policy due to the violent overthrow of the government. Further, although a single party dominated Ivoirian politics, it maintained power through a tradition of extended political consensus, which restricted arbitrary action by the executive and made sharp changes in policy difficult. Second, partly due to the tradition of extended consensus and despite the single party state, Côte d'Ivoire developed a bureaucracy and judiciary that were quite professional by regional standards. This promoted the credibility and stability of government policy. Third, the government adopted a pragmatic position in its relations with the West, leading to a strong relationship with both the donor community and the French government and increasing the confidence of foreign (especially French) investors. Finally, the government had a long-standing policy of encouraging private sector participation in the economy in general, and in the water sector in particular. Together, these four factors reduced the likelihood that the government would arbitrarily expropriate private investment and boosted confidence in the Ivoirian economy.

Credibility: Historical Background

Political stability

In sharp contrast to other West African countries, Côte d'Ivoire was politically and socially stable for close to 40 years following independence. The political institutions that developed define what has been called an 'illiberal democracy'. Although a single party, the *Parti Démocratique de la Côte d'Ivoire* (PDCI), dominated government, it ruled through a process of extended consensus and, thus, was not a classic dictatorship.[18] Further, Ivoirian institutions proved to be capable of making major adjustments, surviving a severe economic downturn in the 1980s and 1990s.

Although the Constitution continued to allow other parties to exist following independence, Côte d'Ivoire quickly became a single party state under the PDCI. However, single party rule in Côte d'Ivoire was different than in most other West African countries. Rather than eliminating opponents through arrest and exile, the PDCI progressively absorbed other political groups. Leaders of opposition groups were given influential and lucrative positions in the government or the PDCI to ensure their cooperation (e.g., the 'Political Board' elected by the Congress of the Party included leaders of most political tendencies) and, consequently, the PDCI became more ideologically and ethnically diverse over time and gained the support of most ethnic groups. Further, although the unicameral assembly was, and remained through the 1990s, dominated by the PDCI, it promoted stability by allowing local and tribal interests to be involved in the decision-making process. This diffused tension and made redistribution between regions easier (e.g., the cross-subsidization between Abidjan and secondary cities in the water sector).

[18] One plausible reason for the importance of consensus in Ivoirian politics is that no single ethnic group was large enough to impose its will on other groups. Consequently, President Houphouët-Boigny, who was from a very minor group, had to perfect the politics of inclusion (Handloff, 1991).

To summarize, notwithstanding democratization in the 1990s, a single party dominated Côte d'Ivoire for nearly forty years. However, because the party generally worked as a coalition and absorbed opponents, and due to the charismatic figure of President Houphouët-Boigny, the regime remained remarkably stable, even during major economic crises. Moreover, the regime showed a remarkable capacity to adapt (e.g., after the death of President Houphouët-Boigny). This stability and tradition of informal consensus made the government's commitment to contractual arrangements more credible and, therefore, ultimately made reform sustainable.

A pragmatic policy towards the west

Following independence, the Ivoirian government adopted a pragmatic policy towards France, the colonial power that had ruled Côte d'Ivoire before independence, and other Western countries. In return, Côte d'Ivoire received considerable support from both France and international organizations following independence.[19] These policies also encouraged French firms involved in West Africa to invest in Côte d'Ivoire and French expatriates from other former colonies to move to Côte d'Ivoire, contributing to the economic boom that followed independence.

There are several plausible reasons for the limited reaction against the colonial regime following independence. First, the French administration's policies in Côte d'Ivoire might have been less restrictive than in other West African colonies, provoking a more moderate reaction following independence.[20] In part, the less restrictive French rule reflected the French view that Côte d'Ivoire was a relatively minor territory with limited resources, poor access to the sea, and weak prospects for development. Second, the pragmatic policy was a response to political developments in Côte d'Ivoire before independence. In contrast to neighboring countries, the most important pre-independence movement in Côte d'Ivoire, the PDCI, had supporters with a strong interest in promoting capitalism and good relations with the former colonial power (Young, 1982). Because the PDCI grew out of an association of relatively wealthy African coffee and cacao planters, the *Syndicat Agricole Africain,* it initially adopted policies that would benefit the capitalist and export-dependent planters. Although the character of the PDCI changed as it absorbed other political groups, the economic success of Côte d'Ivoire following independence legitimized its policies towards foreign investment and the private sector.

Stable institutions

The credibility of the Ivoirian government was also backed by a set of stable institutions, based upon those in France. By regional standards, Côte d'Ivoire developed a relatively

---

[19] The support of international organizations became particularly obvious in the late seventies and early eighties, when a large amount of international resources were provided to help the regime confront the deteriorating economic conditions.

[20] This does not mean that there were no restrictions and, in fact, many standard colonial policies were implemented. For example, the French dominated administrative and export activities and many laws restricted local freedom (e.g., political parties specific to Côte d'Ivoire were banned).

efficient and well-trained bureaucracy. The French policy of assimilation, which encouraged the repression of local cultural traditions, resulted in the colonial administration training a local elite. Consequently, by independence, Côte d'Ivoire had a relatively large pool of well-trained individuals, with technical as well as managerial expertise. Further, in areas where local expertise was lacking, Côte d'Ivoire's pragmatic policy towards France allowed it to rely upon expatriates after independence (Young, 1982). Finally, perhaps due to the political diversity within the PDCI, the bureaucracy is more independent than in many single party regimes.

Reflecting the strong bureaucratic tradition, observers from the private operator, international organizations and the Ivoirian government generally view the public agencies involved in the water sector favorably. In particular, the *Bureau National d'Etudes Techniques et de Développement* (BNETD), which currently advises the main regulator on technical issues related to investment and the institution responsible for managing sector debt, the *Caisse Autonome d'Amortissement* (CAA) are both seen as highly competent.

In addition to a relatively professional bureaucracy, the judiciary is active and, although the Supreme Court (now called the Constitutional Council) is only a limited check on executive action, the judicial system is relatively developed. Judges are trained for two years after graduation from university and there is a tradition of judicial intervention to settle commercial disputes. Contracts can be enforced, although the cost of appealing to the courts is high in financial and, more importantly, symbolic terms.

A tradition of private sector participation in the economy

In addition to a pragmatic approach to the West, the Ivoirian government also adopted policies that encouraged private sector participation in the economy. As discussed above, these policies, which were supported by the relatively wealthy African planters who were the original supporters of the PDCI, were legitimized by the economic success that followed independence. This outlook towards the private sector, in part, explains the government's decision to allow SODECI to continue operating in Abidjan after independence.[21] The long involvement of SODECI in the water sector makes this commitment especially credible in this sector.

MOTIVATION FOR REFORM

In contrast to the other cases, where reform generally resulted from a combination of macroeconomic crisis, a realignment of political forces, and a sector crisis, a macroeconomic crisis alone was the main motivation for reform in Côte d'Ivoire. The macroeconomic crisis reduced public resources or investment, slowing expansion of the water system, and resulted in reductions in formal sector employment and wages, making high tariffs more painful. However, the overall performance of the system remained

[21] However, the government's capitalist outlook was tempered by a tradition of public intervention (Young, 1982), explaining the increases and decreases in SODECI's responsibilities over time.

strong, especially by regional standards. In the absence of either a sector-specific crisis or a major change in the government's support base, the macroeconomic crisis encouraged only a minor reform aimed at lowering tariffs and financing investment through sector revenues. Since the crisis affected the finances of the government agencies involved in the sector more than SODECI's finances, and since the government had been responsible for investment, sector problems were not blamed on the private operator. Consequently, there was little pressure to reduce its role and, in fact, its responsibilities were increased following reform.

A Major Macroeconomic Crisis[22]

After 20 years of sustained growth — GDP growth averaged over 8% per year between 1961 and 1979 — Côte d'Ivoire was hit by a series of macroeconomic shocks. Serious drops in cacao and coffee prices (which represented over 60% of Côte d'Ivoire's merchandise exports in 1979), led to a 30% decrease in the price of exports between 1977 and 1980 (Demery, 1994). This radically slowed growth, and, combined with an overly ambitious development program in the late 1970s, resulted in a large increase in the public sector deficit. This, in turn, caused a dramatic increase in public sector debt, which grew from 25% to 83% of GDP between 1975 and 1985 (see Fig. 7.5).

Source: World Bank (1998).

Figure 7.5. Public debt.

[22] Data in this section on fiscal deficits, growth, exchange rates, exports and investment are from World Bank (1998). Data on international coffee and cacao production are from Food and Agriculture Organization (1998a,b). Per capita GDP is PPP-adjusted GDP (in US dollars from World Bank, 1998). See Demery (1994) for a description of Côte d'Ivoire's structural adjustment programs, and Berthelemy and Bourguignon (1996) for an analysis of macroeconomic trends in the period under review.

To reverse the disturbing trends, the government implemented a series of structural adjustment programs (SAPs). Public investment was severely reduced (dropping over 50% between 1982 and 1984); water, electricity, and transportation tariffs were raised significantly; wages were frozen; and state-owned enterprises were closed and privatized. However, the economic problems continued, with real per capita GDP dropping 33% between 1980 and 1985. In spite of a second SAP between 1984 and 1986, and a brief spurt of growth following a temporary improvement in the cacao and coffee markets, coffee and cacao prices reversed again in the late 1980s and the economic deterioration continued.[23] Although the government responded with new measures to attract investment and liberalize the economy, the measures failed to reverse the economic decline. Further attempts at stabilization in the early 1990s failed and the economy did not start to recover until the devaluation of the CFA franc in 1994. It is in this context that reform of the water sector was proposed and implemented in the late 1980s.

Political Unrest and Democratization

The long period of economic crisis was accompanied by a significant evolution in political institutions. Since the success of the Ivoirian economy in the 1960s and 1970s contributed to the legitimacy of PDCI rule, the economy's collapse in the early 1980s led to an increase in unrest and political instability. Initially the opposition's main supporters were students and teachers, but gradually it gained support in urban areas, including from public workers whose wages were falling in real terms. Although the movement was not strong enough to encourage immediate democratization, it did encourage some basic reforms.[24]

Continuing economic problems led to further political unrest in the 1990s. In 1990, unrest and violence increased and riots and strikes by public sector workers and students forced the government to abandon major parts of its IMF and World Bank backed structural adjustment program — most notably, proposed cuts in public sector wages — and led the government to recognize opposition parties. However, despite the economic problems, elections in October 1990 confirmed the continuing popularity and influence of President Houphouët-Boigny and the PDCI.[25]

Although the reform of the water system occurred during the transition to democracy, it does not seem that this had a significant effect on the nature of reform. In the late 1980s when the reform was proposed and implemented, opposition parties were still illegal and the PDCI remained dominant, and reasonably popular, throughout the Ivoirian society (even in urban areas). Although the economic crisis had provoked unrest, no credible alternative to the PDCI had emerged (and still had not even by the

[23] The long-term downward price trend appears to have been due to an increase in worldwide production as cacao and coffee production increased 66% and 20% between 1978 and 1988.

[24] In 1980, President Houphouët-Boigny had permitted open elections within the PDCI to the National Assembly, resulting in the replacement of two-thirds of the legislature. This was expanded to local party and municipal offices in 1985. In 1987, the state-controlled broadcaster began to allow controlled political debate on one of its shows (Handloff, 1991).

[25] According to official results, President Houphouët-Boigny received 81.7% of the vote in the presidential election and the PDCI won 163 of the 175 seats in the National Assembly. The opposition claimed that the result was fraudulent.

mid-1990s) and, therefore, there was no large shift in political forces. Consequently, although the falling wages and social unrest probably motivated the government to look for ways to reduce tariffs, since sector performance was reasonable there was little pressure for wholesale sector reform.

The Effect of the Macroeconomic Crisis on Sector Performance

In most respects, the water system in Abidjan continued to perform well in the early-mid-1980s (see previous section) — quality remained high, UFW remained low, the commercial performance of the operator remained strong and productivity continued to improve. However, the macroeconomic crisis negatively impacted sector finances, especially for the public entities responsible for debt service and investment. Although this did not threaten the system with immediate collapse, it did raise questions about the sustainability of the current system and led to a modest reform designed to tackle the most pressing issues.

The most direct impact that the macroeconomic crisis had on sector performance was that it slowed demand growth. The amount of water billed in Abidjan, which had been growing at an average rate of 8% in the 1970s, stagnated in the early 1980s (see Fig. 7.6). Since most consumption was metered this reduced sector revenues. The main reason for the stagnation was a collapse in demand from large industrial and commercial consumers, which fell 30% between 1980 and 1985 (World Bank, 1990). Although the number of connections continued to grow, most new connections were 'social connections' for poor households who consumed little water. Consequently, the consumption from new connections only just offset the decline in industrial consumption and the amount of water that was billed stagnated.

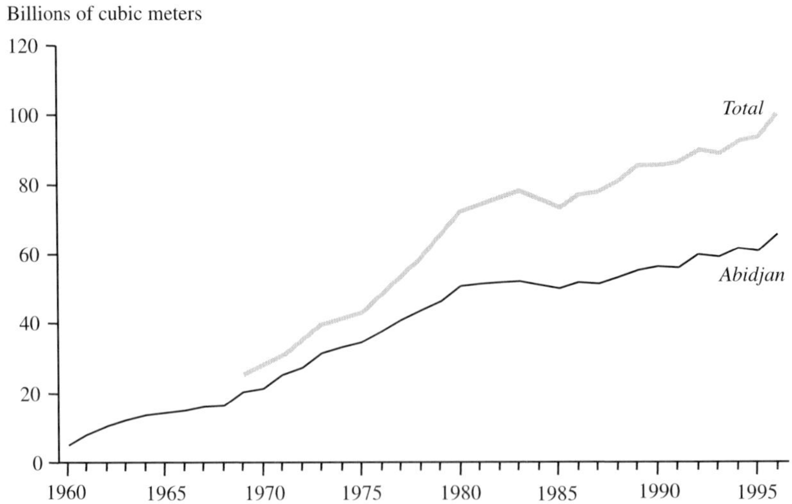

Source: SODECI.

Figure 7.6. Water billed.

To reduce the effect of the slow growth on sector finances, the government implemented a large price increase in 1984 (see Table 7.1). Although, due to concerns about social unrest, tariffs for small consumers were unchanged, prices for industrial consumers were increased dramatically — from CFAF 252 ($0.52) to CFAF 458 ($0.95)/m^3. This large price increase had a significant impact on the behavior of large users, already squeezed by the macroeconomic crisis, who started to reduce waste, recycle water, and rely upon their own wells.[26] Consequently, despite a recovery in the Ivoirian economy, total demand fell in 1984 and 1985 (see Fig. 7.6).

Although the slower growth affected the entire sector, SODECI was shielded in two ways. First, SODECI's compensation was based upon projected, not actual, revenues. Consequently, since the slower demand growth was largely unanticipated in the early years of the crisis, SODECI was partially protected from the slowdown. Second, SODECI's remuneration per cubic meter was about the same for all tariff levels (between about CFAF 150 and 180/m^3 in 1984). Consequently, since demand from small users increased in the early-mid-1980s, SODECI's additional revenues from small consumers made up for the decrease in revenue due to the drop in industrial consumption.

In contrast, remuneration per cubic meter for the government funds (i.e., the FNH and FNA) was far lower for smaller consumers than for industrial consumers (CFAF 0/m^3 for small users compared to 300/m^3 for industrial consumers). In fact, about 70% of the revenues from the 'rental fee' paid to the government funds came from a few hundred large industrial users by the time of reform (World Bank, 1990). The increase in demand from small users, therefore, failed to compensate the government for the large drop in industrial consumption. This significantly reduced the FNH's ability to service sector debt and, by 1986, it was estimated that FNH resources could cover only 60% of sector debt service (World Bank, 1990).

Through its impact on public finances, the crisis had a second, more indirect effect on the sector. As noted earlier, the crisis resulted in large deficits and a massive increase in sector debt. In response, as part of the structural adjustment program, the government was forced to reduce spending, including on public investment. Since population growth remained fast (see Fig. 7.1) and over 90% of investment was financed through government subsidies and public borrowing, this seriously threatened the continued development of the water sector.[27] Due to this, the growth in the number of connections started to lag population growth (see Fig. 7.7) and coverage fell (see Fig. 7.9). The problems in public finances also increased overdue bills from the public sector, further impacting sector finances. Between 1980 and 1985, the ratio of accounts receivable to total sales increased from about 15 months to over three years for the public sector (see Fig. 7.12). Since the public sector accounted for about 25% of total revenue, this was a significant problem.

---

[26] Although, technically, industrial users had to get permission to drill wells and had to pay an extraction fee to SODECI, legal loopholes and poor compliance meant that very few actually did.

[27] Between 1975 and 1987, 80% of investment was funded from borrowing from bilateral and multilateral agencies (52% of lending) and from commercial banks (48% of lending including lending by the *Banque Centrale des Etats d'Afrique de l'Ouest*). In addition, 13% of investment was funded from government subsidies. The remaining 7% was funded from internal cash generation (World Bank, 1990).

Table 7.1

Tariff rates in Côte d'Ivoire (in CFA francs)

	1983	*1984–September 1987*	*October 1987–January 1994*	*January 1994*[a]*–May 1996*	*May 1996–1999*[b]	*Percentage change in real terms 1987–1997*
Social (up to 18 m^3)[c]	187	187	159	159	184	–36.1
Domestic (19 to 90 m^3)	261	261	209	230	286	–25.7
Normal (91 to 300 m^3)	300	330	307	368	464	–16.5
Industrial (more than 300 m^3)[d]	252	458	350	424	532	–28.7
Administrative	261	261	261	311	390	–17.5

[a] Tariff was approved in April 1994, but was applied retrospectively from January 12, 1994 (the date of the devaluation of the CFA franc).
[b] Tariffs were in force at least until early 1999 (World Bank, 1999).
[c] Minimum consumption is 9 m^3. If consumption is less than 9 m^3, it is billed as 9 m^3. Before 1987, there was a fee for 'social' connections that resulted in the average price shown here.
[d] Tariff was reduced from CFAF 412 to CFAF 350 in 1990, as a condition for second tranche release of a World Bank loan.
Source: SODECI.

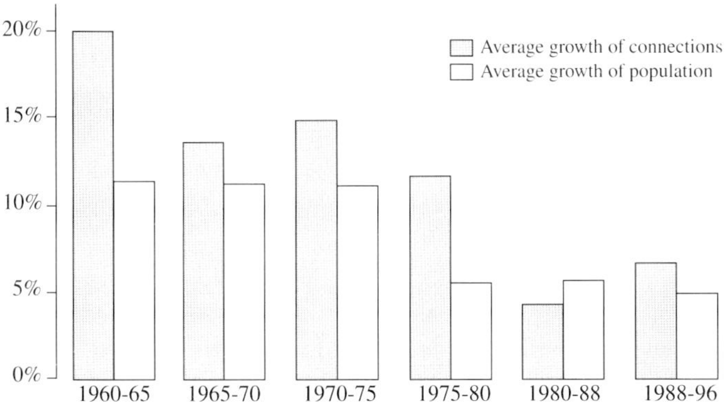

Source: SODECI (various years); World Bank (1998).

Figure 7.7. Growth of connections and population in Abidjan.

Although sector finances suffered due to the macroeconomic crisis, the crisis did not result in a major sector crisis and, consequently, failed to motivate a major sector reform. One reason for this was that the financial difficulties (i.e., the reduced resources for debt service and future investment) had only a minor impact on current consumers. Indeed, since quality remained high and tariffs for nonindustrial consumers were unaffected, sector problems were largely invisible to most nonindustrial consumers. This meant that pressure for reform was not widespread.

Further, although the private operator had been responsible for operations and maintenance, the problems did not lead to calls for reducing private sector participation in the sector. In large part, this was because SODECI did not appear to be to blame for the problems. In fact, the public entities had greater financial problems and were responsible for both areas where problems were evident — debt service and investment. Further, mistakes made by the *Direction de l'Eau* reduced confidence in the public entities involved in the sector. Despite the economic problems, the *Direction de l'Eau* continued to forecast fast growth in water demand in the 1980s and continued to invest heavily in productive capacity. Since the crunch in investment had slowed connection growth and caused a drop in the coverage rates (see Fig. 7.9), the *Direction de l'Eau* was accused of misdirecting investment towards large-scale production facilities, explaining why the responsibilities of the *Direction de l'Eau* were reduced following reform.

CONTRACTUAL ARRANGEMENT AND INSTITUTIONAL COMPLEXITIES

Although the macroeconomic crisis had a negative effect on some aspects of sector performance, in the absence of a sector crisis or a major realignment of political forces, it failed to provoke a complete reorganization of the sector. Consequently, the 1988 reform was not a radical departure from the pre-existing situation and it did not disrupt the relationship between the partners to the contract. The success of the previous arrangement made the initial reform *feasible*, while the credible commitments built by the Ivoirian government, which allowed the contractual arrangement to continue to evolve over time, ensured *sustainability*. In this section, we examine the design and implementation of the contractual arrangement that came into effect in July 1988 and discuss subsequent adjustments, focusing on the interaction between micro-institutions and the main characteristics of the contract.

Changes in Micro-Institutions

The new 20-year contract, which became effective in July 1988, affected sector organization in several ways. First, SODECI's responsibilities related to investment planning and implementation were increased, its compensation was reduced, and it was exposed to greater risk. Second, the reform brought BNETD (*Bureau National d'Etudes Techniques et de Développement*), an additional public agency, into the sector. This had a significant effect on responsibilities of the *Direction de l'Eau*, which had been responsible for both investment and monitoring the contract with SODECI before reform. Third, the macroeconomic crisis and the large government debt convinced the government and international donors that investment should be self-financed. Part of the water tariff was to be paid to a new fund, the *Fonds de Développement de l'Eau* (FDE), which would be used to finance investment. Finally, a new entity, the *Fonds National de l'Eau* (FNE), which was formed by merging the insolvent *Fonds National de l'Hydraulique* (FNH) with its solvent sister agency the *Fonds National de l'Assainissement* (FNA), became responsible for debt service.

Initial negotiations between the World Bank, the *Direction de l'Eau* and SODECI had focused on implementing a full concession contract. Under this agreement, SODECI would have assumed all responsibilities related to investment and would have become responsible for sector debt. However, after BNETD became involved in the negotiations, it questioned whether sector revenues would cover debt service requirements in the years immediately following reform. SODECI then asserted that the imbalance between its equity and sector liabilities would create considerable risk (World Bank, 1993). The idea of a full concession was therefore abandoned in favor of a contract that is close to a lease, but which gives the private operator significant control over investment.

Public agencies: who does what?

The reform resulted in a major realignment of responsibilities for the various government agencies involved in the sector (see Fig. 7.8). One significant change was that a public agency, BNETD, which had not been active in the sector before reform, assumed a pivotal role. BNETD became involved in the sector when, in January 1987, the president made it responsible for the preparation of a series of Sectoral Adjustment Loans (SECALs) from the World Bank. Since these loans included a Water Supply and Sanitation SECAL, designed to support sector reform, BNETD assumed responsibility for negotiating the new contract with SODECI.

After the negotiations with the World Bank and SODECI were complete, BNETD continued to be active in the sector. As discussed in detail below, the 1988 reform transferred responsibility for investment planning and implementation from the *Direction de*

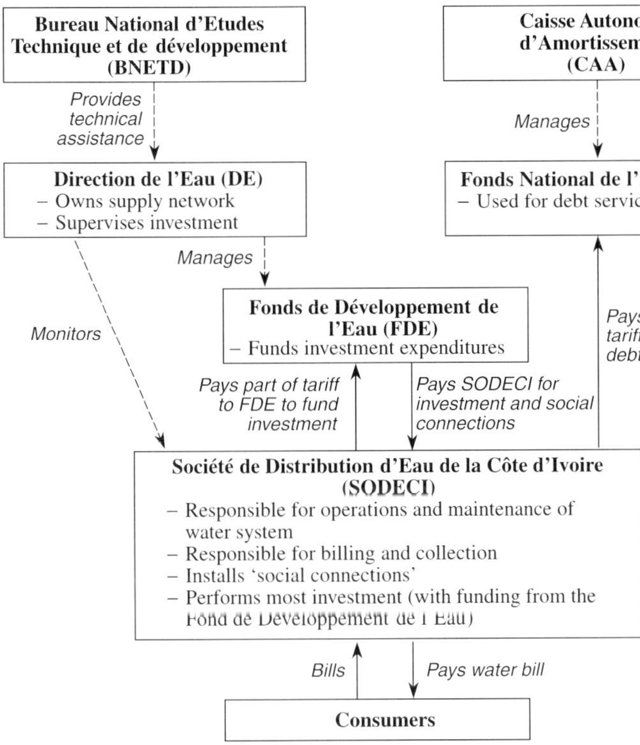

Figure 7.8. Sector organization.

l'Eau to SODECI. At the same time, BNETD assumed responsibility for supervising SODECI's investment activities. Several factors influenced the transfer of responsibility from the *Direction de l'Eau* to BNETD. First, BNETD's successful contract negotiations with SODECI convinced many observers that BNETD was competent to assume the role. Second, the powerful director of BNETD, Antoine Cesareo, who was close to President Houphouët-Boigny, used his influence to expand BNETD's role. Finally, the *Direction de l'Eau's* perceived problems managing investment (i.e., the over-investment in production facilities and under-investment in the distribution network) before reform made observers question its competence.

Conflicting micro-institutions. Although it no longer had direct responsibility for investment, the *Direction de l'Eau* remained responsible for monitoring the contract with SODECI and negotiating the price of water. This created ambiguity in its relationship with BNETD and led to tension between the two agencies that was aggravated by ill will due to the transfer and by the different skills and priorities of the two agencies.

The *Direction de l'Eau* was far smaller than BNETD (staffs of 250 and 1000, respectively, in 1998). It is part of the METT, a large ministry, and has most of its expertise in rural water supply. Consequently, a significant part of its staff is dispersed throughout the country. This provides the *Direction de l'Eau* with a detailed knowledge of the situation outside of Abidjan, but leaves it with neither the time nor the expertise to closely monitor the contract with SODECI. Consequently, it tended to adopt a hands-off approach towards SODECI.

BNETD is located in the prime minister's office and its directors have been close to the highest levels of government. This has allowed it to overrule line ministry decisions. It proposed, planned and implemented most major public works and, because of its special status (including higher salaries) attracted an excellent group of engineers, mostly trained abroad. BNETD took an active role in planning and developing sector facilities and, consequently, closely monitored and continuously interfered with SODECI's plans.

Tensions between these agencies and between the agencies and the private operator developed over time. The conflict forced SODECI, which implemented most investment, to get approval from both the *Direction de l'Eau* and BNETD, making the approval process long, inefficient and costly. This further increased tension between the three partners, making investment decisions difficult and leading international lenders to push for change. In May 1995, a formal agreement was signed that made the *Direction de l'Eau* responsible for supervising new investment and reducing BNETD to a technical consultant to the *Direction de l'Eau.* After a difficult adjustment period, all participants seemed happy with the new system.[28] This supports the hypothesis that water reform success depends on clear rules for making decisions and for solving disputes. The 1995 institutional adjustment simplified the decision-making structure, putting one well-identified agency in charge of the sector. The main weakness of the new arrangement is that the *Direction de l'Eau* has only limited technical capacity, reducing the control it has over SODECI.[29]

Investment. Two other public agencies, the *Fonds de Développement de l'Eau* (FDE) and the *Fonds National de l'Eau* are responsible for financing investment and debt service. The FDE was created to finance new investment and 'social connections' for low-income consumers. Its resources primarily come from the water tariff, through the 'rental fee'. Although it can borrow from lenders, one of the main goals of reform was to make the system largely self-sufficient.[30] In 1997, about 24% of the average tariff (CFAF 76.55) was paid into the *Fond de Développement de l'Eau.* The budget of the

---

[28] Although the 1995 adjustment clarified the relationship between the *Direction de l'Eau* and BNETD, another agency, the *Haut Commissariat à l'Hydraulique* (HCH), was introduced at about the same time. At that time, it was not clear whether the HCH would simply be a coordinating body where agencies could meet and reach agreement on sector policy, or whether it would be a full agency with regulatory power. Further, the different agencies were unsure whether the HCH's sector policy might constrain their activities. For example, if the HCH became a regulatory entity, it would presumably affect the relationship between the *Direction de l'Eau* and SODECI and might even interfere directly investment decisions.

[29] Other public agencies monitor infrastructure and water quality. Although in principle, the *Institut National d'Hygiène Publique* (INHP) monitors sanitary standards since it is both under-staffed and under-equipped, SODECI is the main monitor in practice. The *Laboratoire National d'Essais et de Métrologie Appliquée* (LANEMA) tests equipment quality, while the *Direction de l'Eau* with technical support from BNEDT, checks that infrastructure conforms to international standards. SODECI directly controls the quality of technical equipment and had implemented ISO 9000 standards by the end of 1998.

[30] Between 1987 and 1997, the FDE did not borrow any funds. The 1990 Sectoral Adjustment Loan from the World Bank was balance of payments support and, therefore was not linked to specific investments. In 1996, loans from the *Caisse Française de Développement* and a smaller amount from KfW were mostly to cover the government's debt to SODECI.

FDE, according to *Direction de l'Eau*, was CFAF 7 billion ($11.7 million) in 1997, including CFAF 4 billion ($6.7 million) for Abidjan. 'Social connections' accounted for 50% of investment, maintenance for 20% and system expansion for 30% (field interviews 1998).

An additional 14% of the tariff (CFAF 44) went to the *Fond National de l'Eau*, which services sector debt and pays for the operation and maintenance of the sewerage system. This fund is managed by the *Caisse Autonome d'Amortissement* (CAA), an agency in the Ministry of Economy and Finance that manages public debt. Because new investment was being funded through the tariff rather than through new borrowing, it was hoped after a few years that the FNE would be able to transfer resources to the FDE for investment. In practice this never occurred, due to continued nonpayment by the public sector.

Private operator

Under the 1988 contract, SODECI remained responsible for operation and maintenance of all urban water systems, including responsibility for metering, billing and collection. However, there were several significant changes that affected SODECI's responsibilities, its remuneration, and its incentives.

Responsibility for investment. Although the contract specifies that the Republic of Côte d'Ivoire retains ownership of sector assets, most responsibility for investment was transferred to SODECI. As noted earlier, this was partly due to suggestions that the *Direction de l'Eau* had inappropriately over-invested in large production facilities at the expense of the distribution network during the early years of the crisis. Formally, the government remained in charge of supervising investment and managing the *Fonds de Développement de l'Eau* (FDE), while responsibility for planning and executing investment was transferred to SODECI. Because of this, many observers call the contract a concession. However, since SODECI cannot borrow to finance investment and, therefore, does not bear investment-related risk, the contract is a *lease* under the terminology of this project. Despite the provisions that made SODECI formally responsible for planning and executing investment, BNETD actually performed these roles for the first seven years of the lease. As discussed in the next sub-section, when responsibility for supervising investment was transferred from BNETD to the *Direction de l'Eau*, SODECI's control over investment planning and execution increased substantially.[31]

By mutual agreement with the *Direction de l'Eau*, SODECI can implement investments of less than 120 million CFAF (about $220,000 in 1996) with no tender.[32] The provision was intended to cover daily maintenance of the system and the installation of 'social connections'. Proponents of the agreement argue that this is appropriate since maintenance and the small investments covered by this provision usually need to be

[31] During field interviews, all participants agreed that BNEDT played a far more active and aggressive role concerning investment planning and execution than the *Direction de l'Eau* has.

[32] Before the devaluation of the CFAF in 1994, the limit was set at CFAF 80 million (about $270,000).

implemented quickly and often require substantial coordination with operations to be implemented at least cost. Although public advertising and bidding might lower the up-front cost of these investments, they argue that the cost of delays and disruption to operations would outweigh the small financial benefit. Opponents note, however, that SODECI has used this provision to split major investments into small lots that could be realized without a tender. They emphasize that no investments were bid between 1993 and 1997 and that very few were bid at any time in the first ten years of the contract. SODECI responds that, due to the economic crisis and subsequent governmental nonpayment, there were no large investments over this period and, therefore most investment was, in fact, for maintenance and social connections. As noted by the World Bank (1999), although the evidence is not conclusive, some studies have suggested that SODECI might be making excessive profits on its infrastructure-related activities.

Although the transfer of responsibility for investment coincided with an improvement in the coverage rate in Abidjan, despite a large drop in investment expenditures, this arrangement has been criticized for several reasons. First, although SODECI controls investment, as it would in a concession contract, it does not bear the financial risk associated with investment. While this might not be a large concern when investment is funded through the tariff, if the government starts to finance significant amounts of investment through government borrowing, SODECI might have an incentive to encourage over-investment or to invest inefficiently. Since sector resources do not appear to be sufficient for funding future investment needs without additional borrowing, this is a distinct possibility. Second, the ambiguous allocation of responsibilities might distort incentives. On the one hand, SODECI's revenues depend upon its capacity to connect and collect, which gives it an incentive to invest efficiently to expand the system. On the other hand, SODECI's incentive to invest may be low since it might prefer to hold funds hostage to government receivables or to try to divert investment funds to its owners. Finally, provisions allowing SODECI to avoid bidding for small investments have reduced competition in the sector, increased information asymmetries, and might have distorted investment decisions (i.e., encouraged excessive investment in social connections and other small-scale investments).

All of these problems have been aggravated by the poor quality of the information available on investment and sector assets — a problem noted in World Bank studies since the period before the 1988 reform (see, for example, World Bank, 1999). Lack of information on investment and assets makes it difficult to assess sector needs and to evaluate current investment policy. Although public agencies often find it difficult to obtain and collect relevant information, especially in cost-plus regulatory schemes, poor information appears to be an especially significant problem in Côte d'Ivoire. One possible reason for the problem is the large number of public agencies involved in the sector and the overlapping and changing responsibilities. By making it easier for SODECI to conceal information and by making it more difficult for the regulatory agencies to develop the skills required to monitor the contract, this has increased information asymmetries. In addition, the resulting inter-agency disputes might have worsened information flows between the relevant public agencies, further aggravating the problem.

Remuneration. In addition to a change in responsibilities, the new contract reduced SODECI's remuneration. Suggesting that it might allow other companies to bid for the contract, BNETD managed to negotiate a 20% reduction in the part of the tariff paid to SODECI, allowing BNETD to allocate CFAF 28/m^3 for financing investment, while reducing water tariffs.

Rural water. In return for the cut in SODECI's remuneration, in addition to increasing SODECI's control of investment and agreeing not to put the contract out for tender, BNEDT also released SODECI from its responsibility for rural water points. Since SODECI estimated that it lost close to CFAF 2.2 billion (about 10% of its 1987 revenues) between 1982 and 1987 maintaining and operating rural water points (World Bank, 1990), this benefited SODECI greatly.

Incentives. Another change was that 'take or pay' provisions were eliminated from the new contract and SODECI's compensation was instead based upon actual collections. Before reform, SODECI's remuneration had been based upon the *Direction de l'Eau*'s forecasts of water sales and, therefore, when actual sales were lower than projected, such as during the crisis, SODECI was entitled to compensation. World Bank (1990) estimated that US $11 million had to be paid to SODECI due to the 'take or pay' provisions before reform. In part, the threat of bidding seems to have induced SODECI to agree to this change. Since SODECI also effectively controlled the portion of the tariff that was paid to the FDE for investment, this arrangement maintained SODECI's incentive to collect and bill existing customers and to connect new customers.

This complicated division of responsibilities between several public agencies and the private operator resulted in shifting responsibilities and some confusion over the roles of these agencies. The lack of clear rules for conflict resolution created a vacuum that increased SODECI's discretionary power. Binding rules that usually accompany a lease contract (e.g., rules regarding investments) were difficult to implement and gave way to conflicts between ministries, relaxing the constraints that the contract formally imposed on the operator. This severely hampered reform and increased the risk that the private operator would capture rents.

EFFECT OF REFORM ON SECTOR PERFORMANCE

Notwithstanding institutional problems and difficulties due to the ambiguous contractual arrangement, the urban water sector continued to perform impressively in the 1990s. Although the system was already functioning well, the 1988 reform and the subsequent adjustments managed to reverse some of the problems related to the macroeconomic crisis, while maintaining most other performance measures. Prices fell in real terms and, despite a drop in investment expenditures when the system became self-financing, the coverage rate in Abidjan improved. Some problems, however, remain — in particular, the tariff failed to provide adequate resources to finance investment and nonpayment by the public sector became a serious problem.

The reform also demonstrated the strength of the contractual arrangement. Despite the adverse conditions under which the reform was implemented, the arrangement per se was not questioned. Safeguards were considered strong enough to guarantee the sustainability of the system, and, although the credibility of the government might have been shaken, it was not perceived as a threat to the contract. In fact, the arrangement's ability to absorb major shocks reinforced the idea that the system would be a good model for developing countries inside and outside the region. In this section of the chapter, we carefully evaluate the recent performance of the system, focusing on those areas where performance changed.[33]

Continued Strong Performance

As discussed previously, SODECI was already performing well prior to reform. Many performance measures — UFW, water quality, productivity, and billing and collection — continued to perform well following reform. In this subsection, we briefly discuss these measures, noting how the new contract affected them.

UFW and water quality

UFW remained low following the 1988 reform, remaining between 15 and 20% over the post-reform period. Although the quality control process is relatively complex, and data are sometimes inconsistent, water quality does not appear to be a problem. Under the contract, SODECI is required to meet the World Health Organization's (WHO's) water standards and according to SODECI close to 99% of its water met these standards in 1997 (World Bank, 1999). A 1996 study, based on data collected from January 1993 to December 1995, confirms this for Abidjan (Balla, 1996).[34]

Productivity

The steady improvement in productivity that was evident before 1988 continued after reform (see Fig. 7.4). In addition, when the new contract was implemented, SODECI reduced its workforce from 1665 employees to 1349 employees, resulting in an additional one-time improvement. In part, this was due to the change in SODECI's responsibilities under the new contract — 86 employees were laid off due to suspension of SODECI's involvement in rural water supply. However, presumably in response to the lower compensation under the new contract, 164 additional employees were laid off and 66 positions that were vacated by retirements, dismissals, resignations and deaths were not filled. By the end of 1996, there were 3.8 employees per thousand connections (1294 employees for 344,855 connections nationwide), which is good by regional, or even international, standards. In comparison, SEEG and SONEG in Guinea had around 17 employees per 1000 connections and SDE in Senegal had around 6 employees per

[33] Our case study focuses on Abidjan, which accounts for about 68% of the water produced by SODECI. However, some data (e.g., number of employees) are not available by city and, therefore, we sometimes present national data. In general, the situation in secondary cities is usually not as favorable as the situation in Abidjan.

[34] The equipment in some neighborhoods, including many of the poorest (e.g., Plateau, Adjamé, Anoukoua-Kouté, Riviera Nord), is less adequate.

1000 connections in mid-1997.[35] SODECI has also continued to decrease the number of expatriate employees — from 19 in 1987 to 7 in 1996.

Commercial performance

Metering, which was close to 100% before reform, remained at this level following reform. In addition, as demonstrated by the continued low levels of UFW, the billing rate also remained high. Although private collection was good before reform, the ratio of accounts receivable to total sales for the private sector improved after reform, falling from 7 months in 1987 to under 4 months by 1996 (see Fig. 7.12). Indeed, SODECI reports that its collection rate from the private sector (i.e., amount collected divided by amount billed) was about 97–98% (SAUR International, 1997). Reform improved SODECI's incentives to collect in two ways. First, before reform, SODECI's compensation was based upon the FDE's projections of total sales, not on actual sales. After reform, SODECI's compensation was based directly upon collections. Second, by self-financing investment from the water tariff rather than through government borrowing and subsidies, system expansion started to depend directly upon SODECI's success with collecting. Since SODECI effectively controlled the *Fond de Développement de l'Eau* (FDE) this effect was magnified.

Sewerage

As in the past, sewerage was not included in the 1988 sector reform. However, over this period, the World Bank did finance a large sewerage expansion project that connected 174,000 additional Abidjan inhabitants (World Bank, 1996). In addition, an ocean outfall pipe was constructed, ending the practice of disposing of waste from the system into the lagoon.[36] When SODECI's management contract for operation and limited maintenance of the sewerage system expired in 1987, it was put out for tender. Although SODECI won the bid, bidding resulted in a 35% reduction in SODECI's compensation. When SODECI stopped passing the rental fee to the FNE in 1992, the government was unable to pay SODECI's management fee, which was paid from FNE resources. SODECI, however, continued to perform its responsibilities.[37]

The Effect of Reform on Coverage and Prices

System expansion

The number of connections in Abidjan increased from 105,806 to 179,202 between 1988 and 1996. Fig. 7.9 shows an estimate of the connection rate assuming 13.5 people

[35] Outsourcing is marginal and has not changed significantly over the period studied. According to SODECI, it represents about 10% of turnover and is mostly related to services such as cleaning and security (meeting with B.A. Ebah, May 6, 1998).

[36] However, waste from non-connected industrial plants and households might still end up in the lagoon.

[37] Although the contract expired in 1997, the government was still negotiating a new lease contract for sector operations and maintenance with SODECI in mid-1999.

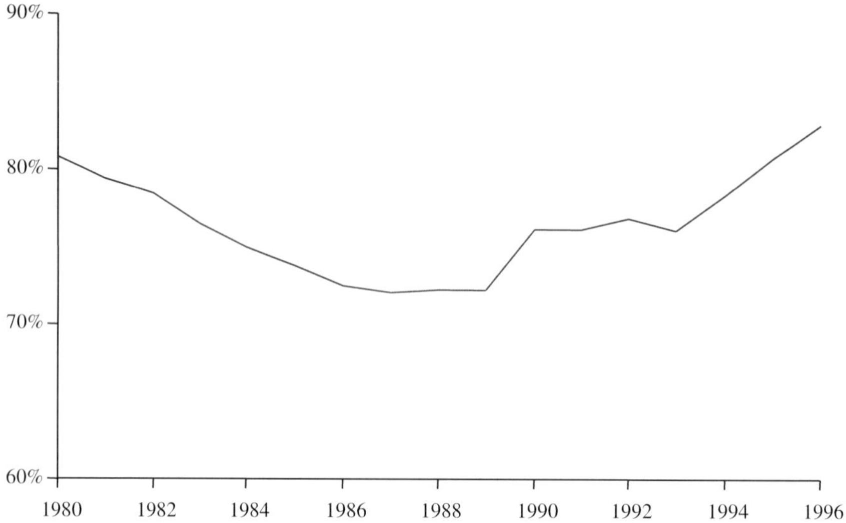

Note: Assumes 13.5 people per connection (see footnote 8).
Source: Connections (SODECI); population (World Bank 1998).

Figure 7.9. Estimated connection rate in Abidjan.

per connection. As noted earlier, the slow expansion of the system, combined with the fast growth of the city, led to a decline in coverage in the period before reform. After the reform, this trend reversed and coverage improved. The improvement accelerated in the mid-1990s, following the devaluation of the CFA franc and the recovery of the Ivoirian economy.

The increase in coverage occurred despite a significant decrease in investment, due to the shift to self-financing and the government's failure to pay its water bill. Indeed, while the *Direction de l'Eau* invested $40 million/year in the 1970s, SODECI spent only $7 million/year between 1988 and 1992 (World Bank, 1993). However, the reversal also reflects that almost all investment between 1988 and 1998 was in the distribution network, primarily financing social connections for the low-income population. There was no significant investment in production for over ten years (e.g., no new boreholes were drilled) and, consequently, according to SODECI, the system was operating at over 90% of total capacity by 1998 (field interviews in 1997 and 1998). Since then, SODECI has used a loan from the *Caisse Française de Développement* (CFD) to finance the construction of new facilities in Abidjan (World Bank, 1999).

The faster increase in connections might partly be because SODECI was more efficient than the *Direction de l'Eau* was. However, other explanatory factors related to contract design might also be important. First, SODECI has greater incentives to connect new customers, rather than invest in other facilities, because expanding distribution has a direct effect on revenues. Consequently, it has operated production facilities close to capacity. Second, under the contract, SODECI has to go through a competitive bid process for large investments (over CFAF 120 million), but can implement small investments (e.g., new connections) without bidding. This gives SODECI, which received 18% of revenues from investment-related activities in 1996 (SODECI, 1996), an incentive to focus on small investments such as new connections as much as possible. Third, although the tariff system, under which small customers who consume little water pay lower tariffs than customers who consume more, might seem to discourage SODECI from connecting small customers, this does not seem to be a major problem. In large part, this is because SODECI's compensation is similar for all tariff levels. Although the 'social tariff' is CFAF 184/m^3, while the 'industrial tariff' is CFAF 532/m^3, SODECI's remuneration is CFAF 144/m^3 for the 'social tariff' and only CFAF 198/m^3 for the 'industrial tariff'. In contrast, the government funds (the FNE and FDE) receive far more revenue from the industrial tariff (CFAF 312/m^3 compared to CFAF 24/m^3 from the social tariff).[38] Finally, SODECI appears to have concentrated more investment in Abidjan and less in secondary centers than the public agencies did before reform. Although the average growth of connections in Abidjan increased from 4.2% in 1980–1988 to 6.6% in 1988–1996, the average growth in secondary centers fell from 8.3% in 1980–1988 to 5.5% in 1988–1996. Since the marginal cost of water is lower in Abidjan than in secondary centers, directing investment towards the capital would make sense from a commercial standpoint.

Prices

As noted in the previous section, the renegotiation of the concession contract in 1988 led to a significant reduction in tariff rates, especially for large users (see Fig. 7.10). The 'domestic' tariff was reduced by 20% and the 'industrial' tariff was reduced by 23% (see Table 7.1). Since then prices have continued to fall. In field interviews, most Ivoirian consumers seem to accept prices in the mid-1990s — an additional strong indication of the acceptance is that SODECI reported a private collection rate of 97–98% in 1997 (SAUR International, 1997). In 1997, the average price of water was US $0.54/m^3 (CFAF 323.5), compared to the pre-reform price of US $0.85/m^3 (CFAF 296).[39]

The contract stipulates that prices should be adjusted every six months according to a fixed formula based on inflation, the cost of labor and certain inputs, and import duties and taxes, and renegotiated every five years. In practice, these procedures have not been followed, with prices not being adjusted either according to the formula every six months or at the first five-year review (World Bank, 1999). There have been, however, several ad-hoc adjustments. In 1990, as a condition for the release of the second tranche of a World Bank loan, the industrial tariff was reduced from CFAF 412 to CFAF 350. In addition, since the devaluation of the CFA franc significantly increased SODECI's

[38] VAT is CFAF 16/m^3 and CFAF 22/m^3 for the social and industrial tariffs, respectively.

[39] Access charges for small consumers are also subsidized — there is no charge for connection pipes less than 15 mm in diameter for private noncommercial connections. Low-income consumers do have to pay CFAF 6934 for installation of a meter, CFAF 2080 for registering and pay CFAF 16,500 in advance for water consumption. In contrast, large consumers (with access pipes between 20 and 40 mm in diameter) have to pay, on average, CFAF 90,000 for connection pipes and additional charges for meters and advances for future consumption. The additional charges vary from between CFAF 60,177 for pipes 20 mm in diameter to 346,480 for pipes 40 mm in diameter.

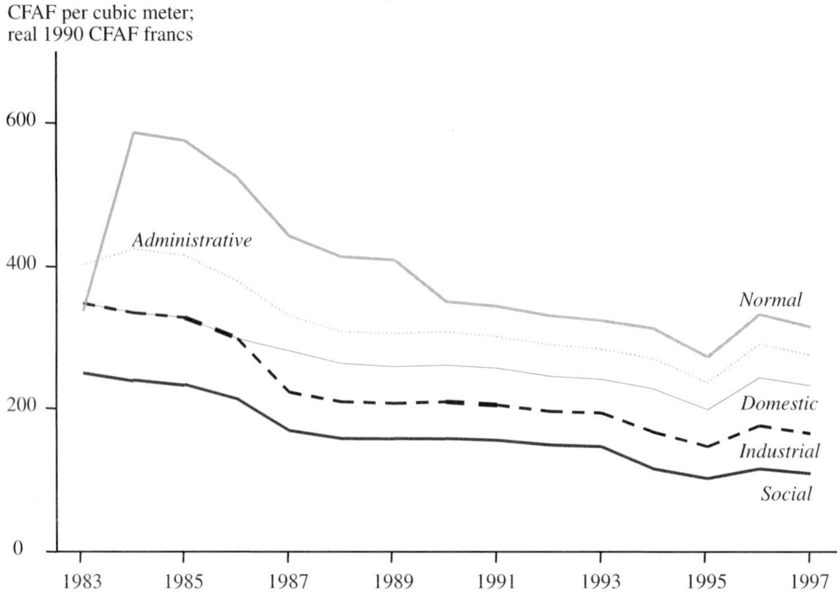

Source: SODECI and authors' calculations.

Figure 7.10. Tariff levels.

costs for imported goods, a tariff increase was negotiated following the devaluation in December 1993 (see Table 7.1). Since it would have been difficult for political reasons to adjust prices the full 25% that would have been needed to fully compensate SODECI, the adjustment was only partial. However, after further negotiations and in return for a loan from the *Caisse Française de Développement* (CFD), the government agreed to an additional price increase in May 1996 (see Table 7.1).

Despite the price increases, SODECI's profits, after an initial drop at the time of reform, appear to have recovered (see Fig. 7.11). At least, in part, SODECI has managed to cope with falling prices by reducing costs. Between 1987 and 1996, operating expenditures (expenditures on labor and intermediate inputs) fell from close to CFAF 17,668 million to CFAF 15,662 in 1990 prices. This occurred despite a 30% increase in water production, from 90 billion m^3 in 1987 to 118 billion m^3 in 1996. Although expenditures on intermediate inputs remained at about CFAF 12,300 million (in 1990 prices), labor expenses fell 36% from CFAF 5329 million to CFAF 3369 million (in 1990 prices).

Challenges that Remain

Nonpayment by the public sector

Although nonpayment by the public sector became a problem before reform, public sector arrears continued to grow and, by 1994, they were equal to over five and a half

Source: Authors' calculations.

Figure 7.11. Net profit of SODECI.

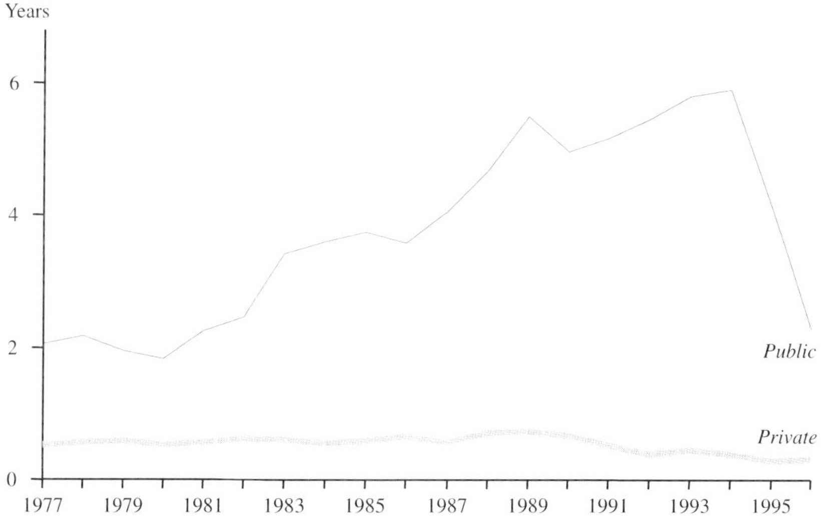

Source: SODECI

Figure 7.12. Public and private accounts receivable turnover period: SODECI.

years of public sector sales (see Fig. 7.12). Although there was a marked improvement after 1994, this is primarily due to a loan that the *Caisse Française de Développement* gave the Ivoirian government to pay its water bill. One of the main reasons for the public sector's poor performance record is that SODECI is unable to cut off public sector entities for nonpayment and, therefore, can do little when public agencies fail to pay.

Since the public sector accounts for about 25% of total revenues, nonpayment by the public sector is a significant problem and has been a continuing source of tension between the partners to the contract. In addition to affecting SODECI, since SODECI was only obliged to transfer the 'rental fee' to the FNE and FDE for bills that were paid, nonpayment also reduced the resources available for investment and debt service. By the end of 1991, public sector arrears had reached CFAF 23 billion ($73 million). In January 1992, SODECI applied to the prime minister's office for, and was granted, a waiver from paying the 'rental fee' to the FNE. SODECI resumed payments in 1995, when the government started to repay its arrears (see Fig. 7.12). In other terms, SODECI received some protection from nonpayment of their bills by governmental bureaus and public agencies, with the government transferring the burden to the agencies in charge of investment and debt service, leading to tension with the *Caisse Autonome d'Amortissement* (CAA).

Sustainability of investment

Although the private operator managed to keep increasing connections, it did this by neglecting other investment. Consequently, production was hitting capacity limits by the end of 1997. It now seems unlikely that the rental fee will provide sufficient funds for investment and future investment will probably require financing from additional borrowing. The fiscal crisis has meant that the government is unable, or unwilling, to finance system expansion (and has even delayed paying its own bills — see Fig. 7.12). However, SODECI cannot borrow to finance investment under the current contract. A special arrangement, reached between the government and the *Caisse Française de Développement* (CFD) in 1996, allowed an exception to this general rule. Conditional on a price adjustment and the payment of overdue public sector bills, the CFD loaned the government CFAF 12 billion and provided an additional CFAF 4.2 billion loan to SODECI for investment. About half of the loan was for new production units and about half for social connections. However, the loans will be serviced through the *Fond de Développement de l'Eau*, making the government ultimately responsible for repayment. Consequently, although SODECI does not bear any risk, it will have direct access to the investment funds and will not have to compete for works contracts.

Cost-plus price system

In practice, the decline in real tariffs does not appear to have impacted SODECI's profits. Consequently, it seems possible that the contractual arrangement suffered from the usual bias of a cost-plus system (i.e., the operator had little incentive to reveal cost information to the regulator or to reduce costs). In theory, this should be mitigated by the fact that prices are set based upon expected costs, not actual costs. Further, since prices are supposed to be set for five-year periods, SODECI could keep any additional cost savings it implements over this period until the next five-year review. However, as discussed above, prices renegotiations have been ad-hoc and the six-month adjustment formula has not been applied. Since the operator could lose cost savings at any point through ad-hoc adjustments, its incentives to cut costs have been reduced.

LESSONS FROM A STYLIZED CASE

In the previous section, we provided evidence that supports the view that the 1988 reform was largely successful. The strong performance of the sector was maintained and even improved in several important ways, despite the continuing economic crisis and the large increase in population. Given the difficult environment in Côte d'Ivoire during the 1980s and 1990s, the main question is why has this system been so successful? Is the contractual arrangement particularly well designed or is success due to institutional factors specific to Côte d'Ivoire? Our tentative answer is that the institutional endowment, with its long standing acceptance of private sector participation and the guarantees it provided to investors, has allowed a well-designed system to function. The good fit between stable institutional rules providing guarantees to the operator and a contract providing adequate incentives is likely the foundation of this success.

Despite this positive assessment, some problems with the initial contractual arrangement persisted after the reform. This report identifies at least five factors that have hampered success and that may threaten the sector's long-run performance, namely:

(1) The separation of sewerage and water supply and the failure to expand the former.
(2) The ambiguous and constantly changing roles of different public agencies.
(3) The public sector's failure to pay its water bill, which has threatened investment.
(4) The contract, which mixes aspects of concessions and leases, allows the private operator to perform most investment without competition or financial risk.
(5) The government's failure to introduce competition has potentially reduced the operator's incentives to operate efficiently and has made it more difficult for the public agencies to collect relevant information.

The first three points are mostly related to failures or problems in the institutional setting, while the last two concern the design of the contract. In this section of the chapter, we discuss these issues and suggest some ways to improve sector performance.

Institutional Bugs

The institutional decision to separate water and sewerage has led to uneven sector development. While 80% of households in Abidjan had access to safe drinking water by 1996, only about 35 percent had connections to the sewerage system. Although the aquifer, which is located at a depth of over 60 m and is protected by a thick layer of clay, has not become polluted, the situation appears to be getting worse. Judging from the record in water, increased private sector participation would improve sector development.

Regulation and monitoring of the private operator has been made more difficult by the unclear and constantly changing roles of different public agencies. Between

1987 and 1995, the *Direction de l'Eau* and BNETD competed for responsibility over sector investment. Although the situation was clarified with the May 1995 Convention, which firmly established the leadership of the *Direction de l'Eau*, the simultaneous creation of the *Haut Conseil à l'Hydraulique* led to the emergence of a new, potentially competing, agency. Disputes over bureaucratic territory have made the private operator's position difficult, slowed the decision-making process, and given SODECI and its main shareholder (SAUR) a strong incentive to go over these agencies directly to the president. Moreover, conflict between the agencies has meant that the regulators have failed to develop coherent competencies or share information, increasing information asymmetries and hurting the quality of regulation.

A third problem, rooted in the institutional environment, is the government's failure to pay its water bill. This restricted funds at the time when the system was shifting towards self-financing investment and led SODECI to withhold funds for debt service. Although a loan from the *Caisse Française de Développement* reduced this problem significantly in 1996, unless public sector discipline improves, it could easily occur again. Allowing the private operator to cut off public sector agencies that fail to pay their bills is certainly useful. However, private operators, which remain heavily dependent upon the government, have often been unwilling to do this. For example, in Guinea, the private operator has not cut off any central government agencies for nonpayment even though the contract explicitly allows it to do so (Chapter 8). In a country such as Côte d'Ivoire, where recent price adjustments have been made through a process of negotiation rather than by following strict formulas, this could be a potential problem.

Several methods to improve discipline have been tried in other sectors. First, under some circumstances, international donors can provide guarantees for government payment of utility bills. This essentially shifts the risk of nonpayment onto international donors, who have greater leverage over the government. Second, some part of export revenues might be tied to government payment. These methods are not, however, perfect. A private operator might be unwilling to call a guarantee for much the same reasons that it is often unwilling to cut water off to a government agency. In the end, the most important thing is for the government to recognize the effect that nonpayment has on sector development and to act accordingly. This raises the difficult issue of building government's credibility with regard to guaranteeing full property rights, including the right for private operators to be paid.

Contractual Failures: Lease versus Concession

The fourth problem is the fundamental ambiguity of a contract that mixes characteristics of a lease and a concession. Since the private operator does not bear any investment-related risk, the arrangement is a lease in the terminology of this project. However, despite not bearing any investment-related risk, SODECI plans and implements most investment, allowing SODECI to virtually eliminate any competition over investment projects, creating opportunities for SODECI to act opportunistically, and further aggravating information asymmetries.

As well as affecting competition over investment, the contract has also had a significant impact on the quantity of investment. Under the 1988 arrangement, SODECI was not able to borrow to finance investment and was not responsible for debt service. Since the fiscal crisis and the large increase in government debt reduced the government's willingness to borrow, the only funds available for investment were generated through water tariffs. Even if the government had continued to pay its water bill and the crisis had not affected industrial consumption, it would have been difficult to finance all investment needs in this way. Although SODECI managed to increase both the number of connections and the coverage rate in Abidjan, investment in productive capacity was ignored. Although recent loans from the CFD have allowed additional investment in production facilities, this is likely to continue to be a problem in the future.

The government's continuing financial difficulties make it unclear whether the necessary financial resources should be raised through the government, by borrowing from abroad, or by allowing SODECI to obtain direct access to financial markets. The final solution, which would pass investment risk to SODECI, would result in the contract becoming a true concession. As noted in the previous section, a recent loan by the CFD is a step in this direction, although it partially protects SODECI against risks. Considering the stability of the Ivoirian regime[40] and the long tradition of private management of utilities, particularly in the water system, and the country's strong rating by private risk assessment agencies, passage to a full concession may be both desirable and feasible. But what are the respective merits and limits of concession and lease in an environment like that of Côte d'Ivoire?

The fundamental advantage of a *concession* is that it makes the firm fully responsible for investment, management and debt for the duration of the contract. Since profits depend upon its capacity to collect bills, the operator has a strong incentive to extend the network and to meter and bill users. However, unless certain conditions are fulfilled, concessions can be inefficient. *First*, the contract should be awarded through competitive tender, something that can be difficult in developing countries. Further, competition is likely to be even less effective for contract renewals, since the incumbent has a significant informational advantage. In Africa, because of the small number of firms that have been actively involved in the urban water sector, it is even more difficult to ensure that bidding is competitive. *Second*, the contractor's obligations (e.g., investment and network development) should be well specified, and easy to implement and monitor. Again, this can be difficult in developing countries, since many factors that affect the optimal evolution of the system are unpredictable (e.g., unplanned urban developments). *Third*, because water systems are close to a natural monopoly, there is a risk that the operator would continuously pressure for renegotiations, once the contract has been allocated. Hence, dispute resolution mechanisms must be well defined and efficient, which requires especially sophisticated and credible institutions. *Fourth*, the regulator monitoring the concession needs to have access to and the technical ability to assess

[40] As noted previously, although there has been considerable political instability since the military coup in December 1999, Côte d'Ivoire was remarkably stable for the 40 years prior to this.

relevant information (e.g., on costs and maintenance). Therefore, the contract needs to have adequate information revealing schemes and the regulator needs to have the technical and managerial skills to assess this information. *Fifth*, the firm might over-use equipment and under-invest in maintenance particularly towards the end of the contract, if there is a risk that it will not have the contract in the next period. The only way to avoid this bias is to have a credible and efficient regulator, with sufficient enforcement power. This requires highly qualified, and independent, civil servants. *Finally*, the most important problem is that large fixed investments, which cannot be re-deployed to other uses, are required. These investments are risky in countries with limited capital markets and unstable institutional environments. Although, as discussed earlier, Côte d'Ivoire has stronger institutions than many other developing countries, many observers believed that the risks were still too high to implement a full concession, especially since the contract includes systems in secondary centers.

The main advantage of a lease is that it is more attractive to private operators, who might otherwise be reluctant to get involved in risky environments. If the contract is well designed (i.e., if there are adequate incentives to develop the water system and to perform efficiently), private operators might be able to significantly improve the management of the system. Adequate incentive schemes should base the operator's revenues on bills collected, and should provide the operator with motivation to reduce unaccounted-for water (UFW) and to extend the network. If the firm can capture part of the benefits of cost reduction, the contract might also encourage them to reduce costs. This is the solution adopted in Abidjan and it surely explains part of the success of the contract.

However, there are also problems with lease contracts. *First*, since the operator is not responsible for investment, the incentive to overuse physical assets is high. *Second*, because firms are not responsible for nor involved in the management of the debt, there is a bias towards investment. This bias is aggravated by the affiliation that many large water companies have with construction companies (e.g., SAUR is a subsidiary of the construction group BOUYGUES). *Third*, unless the contract is specific, since the operator's income is based on bills collected, there is a strong incentive for the leasing company to develop the network only in the most profitable areas. If there are public health externalities or equity issues, leases might result in under-investment, especially in poor areas. Requiring the operator to charge the same price throughout the country or charge poor residents a lower 'social tariff' will magnify this risk. *Fourth*, for the reasons mentioned above, there is a risk that competition from bidding will be limited, especially for contract renewals.

Several factors alleviate the difficulties listed for both concession and lease contracts. *First*, since water systems rely on technologies that are well established, inputs are easily identifiable and costs are easier to assess. Therefore, a relatively competent regulator can estimate the validity of claims made by the operator more easily than in other infrastructure sectors. For example, in Côte d'Ivoire, the experienced engineers at BNETD could determine whether the operator needed to change a pump and, based upon international prices, how much they should charge. *Second*, since there are few operators on the international market for water supply services, and they are fierce competitors, there are strong reputation effects. If the operator wishes to expand, and to gain other contracts elsewhere, they would want to avoid problems with existing contracts.

The last problem in Côte d'Ivoire, as in many other countries, is the failure to introduce competition into the sector. There was no bidding for contract renewal in 1988 and bidding has been eliminated for most investment. Although the macroeconomic problems meant that most new investment was for social connections, allowing the operator almost exclusive access to the *Fonds de Développement de l'Eau* is likely to increase costs in favor of the operator.[41] Competition for investment could be increased by splitting the water system into autonomous subsystems for the different cities and allow bidding over these contracts.

In addition to foregoing efficiency gains, the lack of competition also makes it difficult to assess whether water could be provided at lower cost. Although SODECI appears to perform well by regional standards, the price of water is high compared to the prices observed in Latin America and Asia (see footnote 1) Although international comparisons are difficult due to both institutional and physical differences between countries, this suggests that gains might be possible. Because SODECI has a monopoly on urban water distribution in Côte d'Ivoire, there are no other companies to compare its performance within Côte d'Ivoire. Splitting the contract into several separate contracts for different cities would help to develop the bidding process and would allow yardstick competition.

Yardstick competition allows the operator to cut its own costs, without affecting the revenues that it receives (i.e., the regulator bases the price that the operator charges on the performance of *other* utilities). Although this is an attractive idea, it has proven difficult to implement even in developed countries (see Cowan et al., 2000). For example, in Côte d'Ivoire, Abidjan is very different from most other major cities. It is far larger than any other city, gets its water from an aquifer rather than a dammed river, and receives more rain than cities in the interior of the country. Consequently, most estimates of the long-run marginal cost of water are far lower for Abidjan than for smaller secondary systems. It would be difficult, therefore, to split the contract and allow yardstick competition, without abandoning the government's commitment to having the same tariffs throughout the country.

Rather than introducing yardstick competition, another solution would be to set prices based on a *price-cap* system. The system adopted in 1988 was a simple cost-plus system, where SODECI negotiates tariff levels with the government every five years.

[41] According to our interviews, the standard procedures in mid-1998 were the following. For new investments (extensions etc.) SODECI prepares plans that are thereafter submitted to the *Direction de l'Eau*. Once approved, these plans are either realized by SODECI (if the amount is less than CFAF 120 million, which has been systematically the case recently) or through bidding (if more than CFAF 120 million), under the supervision of the BNETD. For maintenance of the system and of related equipment (i.e., replacing a pump) the *Direction de l'Eau* assesses whether it is justified and it approves (or rejects) SODECI's proposal. However, this is often an ex-post formality since there are often emergencies. Finally, the management of social connections is largely left to SODECI, although the *Direction de l'Eau* decides the criteria for qualifying for the subsidized connections. The *Direction de l'Eau* checks that connections have actually been installed and that the beneficiaries meet the criteria ex-post.

Prices are determined based upon SODECI's costs and the amount necessary to cover sector debt and to pay for required investment. Since the cost-plus system guarantees profitability, the private operator has less incentive to reduce costs or reveal information to the regulator. In contrast, under a price cap, since the operator can keep cost-savings for a fixed period, it has a stronger incentive to continuously improve efficiency. However, price caps can be politically sensitive, since the regulator is unable to cut prices during the period between reviews even if the operator is making large profits. A price cap might also encourage the operator to develop the network only for the more profitable sections of Abidjan and to avoid developing systems in poorer areas and in secondary cities, since this would keep costs low. Although the contract might try to correct this in other ways (e.g., specifying the number of new connections), this will increase complexity and transactions costs. Further, the government's failure to follow the tariff adjustment procedures specified in the current contract suggests that a change to a price-cap system might not be credible, especially if the government suspected that SODECI was making high profits.

To conclude our discussion, there is no optimal contract; there are contracts that fit more or less with the institutional endowments. In the case of Abidjan, the lease contract with significant responsibilities for the operator with regard to the choice of investments has worked well. On the other hand, there are many indications suggesting that the operator should have assumed more financial risk. Similarly, the high price of water in Abidjan suggests that there is room for gains; the cost-plus system should be revised accordingly, in order to provide the operator greater incentive to reduce both prices and costs.

CONCLUSION

This case substantiates two important hypotheses. First, in general, in the absence of a change in the government's support base or a sector-specific crisis, a macroeconomic crisis will not make substantial changes in a specific sector *desirable*. The modest 1988 reform was motivated by a major economic crisis, which reduced the resources available for investment. However, the continued strong performance of private operator and the absence of a significant sector-specific crisis meant that the reform resulted in minor adjustments to the existing contract rather than radical change. The increased responsibility given to the private operator following reform reflects that the problems resulting from the crisis were primarily concentrated in the public sector entities involved in the sector. Second, the case study illustrates the importance of the long-run stability and credibility of the government with respect to the contract and the guarantees of associated rights. Continuity of these commitments made continued private sector participation *sustainable*.

The reform has, in general, been successful, especially when sector performance is compared to the performance of systems in other cities in the region. Coverage has been maintained at a high level despite rapid population growth, water and service quality have been excellent and prices have declined in real terms. Although the success is partly due to specific characteristics of the contract and to private sector participation in the sector, other cities in developing countries have adopted similar contracts, some with considerably less success.[42]

As noted in World Bank (1999) to a large degree, SODECI and the government's good relations have been based upon using negotiation and flexibility to achieve outcomes that are acceptable to both parties rather than on specific contract provisions. Indeed, some contract provisions have been ignored (e.g., the automatic six-monthly adjustment of tariffs based upon a fixed formula). At a general level, success in Côte d'Ivoire has been due to the government's consistent support for private sector participation in the sector and the institutions that have guaranteed the property rights of the operator. More specifically, the existence of strong micro-institutions with sufficient human capital has allowed the government to supervise the private operator and monitor the contractual arrangement well, at least by regional standards.

Nevertheless, this chapter has identified some flaws in the arrangement. The quality of information available on the private operator and investment — a well-known bias in cost-plus systems — has been a consistent problem. Moreover, the contractual arrangement adopted in 1988 gave the operator, SODECI, considerable influence over investment decisions, without requiring it to bear related risks. Another problem is that, under the new arrangement, all aspects of competition were seriously weakened. The concession was not open to competitive bidding at the time of reform and SODECI has managed to almost eliminate bidding for new investment. Since bidding is one way to reduce information asymmetries, and given that the information asymmetries between the private operator and the regulating agency are large, this has likely slowed sector development. Other aspects of the contractual arrangement have also had a negative impact on the information available to the public sector agencies. Unclear and overlapping responsibilities for the two main public agencies involved in the sector, which appear to have been recently resolved, led to an extended power struggle. Consequently, the agencies failed to develop the relevant skills required to monitor the contract and failed to fully coordinate their decisions and share information. This resulted in an awkward situation for the private operator and slowed sector development.

In summary, sector performance in Abidjan has been impressive, due in no small part to the strong performance of SODECI. This has been supported by the government's commitment to private sector participation and an institutional framework that protects the private operator. However, the resolution of some remaining problems could further improve sector performance and guarantee these gains into the twenty-first century.

REFERENCES

Balla, Cisse Mahamadou, 1996. *Qualité de l'Eau d'Adduction Publique de la Ville d'Abidjan: Bilan de Trois Années d'Analyse de Routine du Laboratoire de l'Institut National d'Hygiène Publique*. Processed.

Berthelemy, Jean-Claude and François Bourguignon, 1996. *Growth and Crisis in Côte d'Ivoire*. Washington, DC: World Bank.

Cowan, Simon, George Clarke and Ana-Maria Zuluaga, 2000. *Competition and Regulation in Urban Water Supply: The Case of England and Wales*. Processed.

[42] For example, see Chapter 8 for a discussion of a similar lease contract in Guinea.

Demery, Lionel, 1994. "Côte d'Ivoire: fettered adjustment". In: Ishrat Husain and Rashid Faruqee, eds., *Adjustment in Africa: Lessons from Country Case Studies*. Washington, DC: World Bank.

Esrey, Steven A., 1996. "Water, waste, and well-being: a multicountry study". *American Journal of Epidemiology* 143(6): 608–623.

Food and Agriculture Organization, 1998a. *Production Yearbook, 1998*. Rome: Food and Agriculture Organization.

Food and Agriculture Organization, 1998b. *Trade Yearbook, 1998*. Rome: Food and Agriculture Organization.

Handloff, Robert E., 1991. *Côte d'Ivoire: A Country Study*. Washington, DC: Superintendent of Documents.

Ménard, Claude and Mary Shirley, 2001. *Reforming Contractual Arrangements: Lessons from Water Systems in Six Developing Countries*. Processed.

SAUR International, 1997. *World Bank and SAUR International Experience Sharing on Private Participation in Public Services. Presentation to the World Bank by SAUR International on December 11, 1997, World Bank, Washington DC*. Washington, DC: SAUR International.

SODECI, various years. *Rapport Annuel*. Abidjan: Société de Distribution de la Côte d'Ivoire.

SODECI, 1996. *SODECI: Institutional Framework, Organization and Results in the Management of Water Supply and Sewerage Services*. Abidjan: Société de Distribution de la Côte d'Ivoire.

World Bank, 1974. *Staff Appraisal Report: Abidjan Sewerage and Drainage Project*. Washington, DC: World Bank.

World Bank, 1977. *Staff Appraisal Report: Secondary Centers Water Supply Project*. Washington, DC: World Bank.

World Bank, 1978. *Report and Recommendation of the President. Second Abidjan Sewerage and Drainage Project*. Washington, DC: World Bank.

World Bank, 1982. *Staff Appraisal Report. Second Water Supply Project*. Washington, DC: World Bank.

World Bank, 1986. *Project Completion Report. Second Abidjan Sewerage and Drainage Project*. Washington, DC: World Bank.

World Bank, 1989. *Staff Appraisal Report. Abidjan Environmental Protections Project*. Washington, DC: World Bank.

World Bank, 1990. *Report and Recommendation of the President of the IBRD: Water Supply and Sanitation Sector Adjustment Program*. Washington, DC: World Bank.

World Bank, 1993. *Project Completion Report. Water Supply and Sanitation Sector Adjustment Loan*. Washington, DC: World Bank.

World Bank, 1994. *Project Audit Report. Water Supply and Sanitation Sector Adjustment Loan*. Washington, DC: World Bank.

World Bank, 1996. *Project Completion Report. Abidjan Environmental Protections Project*. Washington, DC: World Bank.

World Bank, 1998. *World Development Indicators, 1998*. Oxford: Oxford University Press.

World Bank, 1999. *Strengthening the Framework for Private Investment in Infrastructure: The Case of Côte d'Ivoire. Sectoral Report: Water and Waste Water*. Washington, DC: World Bank.

Young, Crawford, 1982. *Ideology and Development in Africa*. New Haven, CT: Yale University Press.

Thirsting for Efficiency: The Economics and
Politics of Urban Water System Reform
Mary M. Shirley (Ed.)

CHAPTER 8

A TRANSITORY REGIME: WATER SUPPLY IN CONAKRY, GUINEA

CLAUDE MÉNARD AND GEORGE R.G. CLARKE

INTRODUCTION

The performance of Conakry's water system, which had been deteriorating since independence, had become critical by the mid-1980s. The public agency in charge of the sector, the *Enterprise Nationale de Distribution de l'Eau Guinéenne* (DEG) was inefficient, overstaffed and virtually insolvent. Less than 40% of Conakry's population had access to piped water — low even by regional standards — and service was intermittent, at best, for the lucky few with connections. By 1983, residents had to line up at neighbors' connections and standpipes for hours, hoping for service. Many residents used polluted well water as their primary source of drinking water and even more relied upon it as a secondary source when the piped system was not operating. Consequently, waterborne diseases, including cholera and diarrhea, had become a major problem. Further, after disappointing results from previous investment projects, international donors were becoming frustrated. The poor, and deteriorating, performance of the sector convinced many observers that significant reform was necessary.

In 1989, after several years of discussions and delays, the government signed a lease contract with a majority privately owned enterprise that covered operations and maintenance of all urban water systems in Guinea. A newly created public enterprise was given ownership of sector assets and responsibility for investment. At the same time, a large World Bank-led project, the *Second Water Supply Project*, was approved, allowing for expansion and rehabilitation of the system.

In several ways, the reform was quite remarkable, requiring significant commitment from private investors in an institutional setting that imposed few limits on arbitrary government action. Further, informal constraints, such as traditions of regulatory or bureaucratic independence, were also weak. The World Bank, which was heavily involved in the design and implementation of the contract, had to implement new internal procedures to handle the arrangement and it was not clear to the partners how well the system would work.

Although DEG's disastrous performance made the reform desirable, sector-specific crises have often been insufficient motivation for private participation. In Guinea, four additional factors encouraged substantial change. First, DEG's poor management and performance, rather than less tractable problems such as scarcity or expensive raw water, were the main reasons for the sector's problems. Consequently, the introduction of private sector participation could be expected to improve the situation. Second, the military government that came to power following the death of President Sekou Touré did not rely heavily upon support from the urban elite who were used to receiving free water. This allowed the government to implement large price increases, allowing tariffs to cover operating expenses and provide a fair return on capital. Third, privatization was consistent with the government's philosophy of liberalizing and privatizing the economy, which was demonstrated by the structural adjustment program that it was implementing in response to the macroeconomic crisis that hit Guinea in the mid-1980s. Finally, bilateral and multilateral donors, who were pushing for reform, had significant leverage, since in addition to financing the structural adjustment program they also financed most sector investment.

Although the weak institutional framework did not prevent the private sector from becoming involved in the sector, it affected the success of reform. In practice, no independent body existed that was able to enforce the contracts between the various parties involved in the sector and the government. Consequently, many useful contract provisions were ineffective (e.g., provisions requiring that the government pay its water bill, allowing for international arbitration in the case of dispute between the partners, and enforcing performance targets). Further, in the absence of a neutral party that could mediate disputes and enforce its decisions, disagreements between the different parties had to be settled at the highest levels of government. These problems slowed sector development and poisoned relations between the public enterprise that owns sector assets, the private operator and the government. For example, one of the main reasons that the price of water became high was that the collection rate remained low due to the government's refusal to pay its water bill and to the weak judicial system that made it difficult to collect from private consumers who refused to pay.

Notwithstanding these problems, the bottom line from this study is that all parties have benefited from reform. On aggregate, increased coverage and improved quality more than compensated consumers for higher prices. Further, government subsidies were reduced and the foreign investors made modest, but positive, profits. Under the circumstances, it is unlikely that the public operator, DEG, would have had greater success than the private operator had, especially given its disastrous record before reform and the poor compliance with provisions in the public enterprise's contract. This shows that even in a weak institutional environment, where contracts are hard to enforce and political interference is common, private sector participation can improve sector performance.

This chapter discusses the mechanisms that makes progress possible, even in difficult institutional environments, and identifies factors that inhibit the potentially positive effects of reform. The next section describes the situation in Guinea before reform. The third section describes the sector crisis at the time of reform. The fourth section examines the political economic circumstances that affected the decision and the design of reform. The fifth section analyzes the contractual arrangement and its implementation. The sixth section evaluates the outcome. The seventh section goes beyond standard performance measures and presents results from a detailed cost–benefit analysis. The final section concludes and discusses options that might improve sector performance.

GUINEA AT THE TIME OF REFORM

In many ways, the collapse of the water system mirrored a collapse in Guinea's economy, and many of the problems that affected DEG affected the entire public sector. By the time that reform was enacted in 1989, Guinea, with a per capita GDP of US $430, was one of the poorest countries in the world (World Bank, 1990d). Further, life expectancy at birth (43 years) and the infant mortality rate (143 deaths per 1000 births) were both below the average for Sub-Saharan Africa (51 years and 108 deaths per 1000 births). This was not due to a lack of natural resources. In fact, due to its vast mineral wealth and its fertile land, Guinea had been seen as one of the most promising colonies in French West Africa before independence. Instead, Guinea's problems were primarily the result of years of economic mismanagement by the dictatorial regime led by President Sekou Touré.

Before independence, due to vigorous campaigning by Sekou Touré and the *Parti Démocratique de Guinée* (PDG), which dominated pre-independence politics, Guinea rejected membership of the French Community proposed by General de Gaulle. This prompted an immediate withdrawal of French expatriates and a swift end to technical assistance, budgetary aid, and trade preferences. Since the PDG held the majority of seats in the regional assembly, Sekou Touré automatically became President. He used his position to set up a one-party state, introducing 'African Socialism' and imposing a dictatorship based on personal and clan ties. President Touré's fear of being overthrown — known as the permanent plot — resulted in the arrest, persecution, and torture of any real, or imagined, opponents.

The Touré regime's policies led to state domination of the economy and urbanization. Following independence, most enterprises were nationalized and cooperative farms were set up in rural areas. To subsidize urban consumption, the government suppressed private trade of agricultural products and forced farmers to sell their produce at prices that by 1984 were lower than the estimated cost of production (Arulpragasam and Sahn, 1997). The long-term results were predictable. People living in rural areas withdrew from the state by turning to barter, subsistence farming, smuggling and the black market (Azarya and Chazan, 1987). Meanwhile, agricultural production and exports fell dramatically.[1]

These policies also encouraged fast population growth in urban areas. Between

[1] Whereas agricultural products made up 71% of total exports in 1957, they made up less than 1% of total exports by 1981 (Arulpragasam and Sahn, 1997).

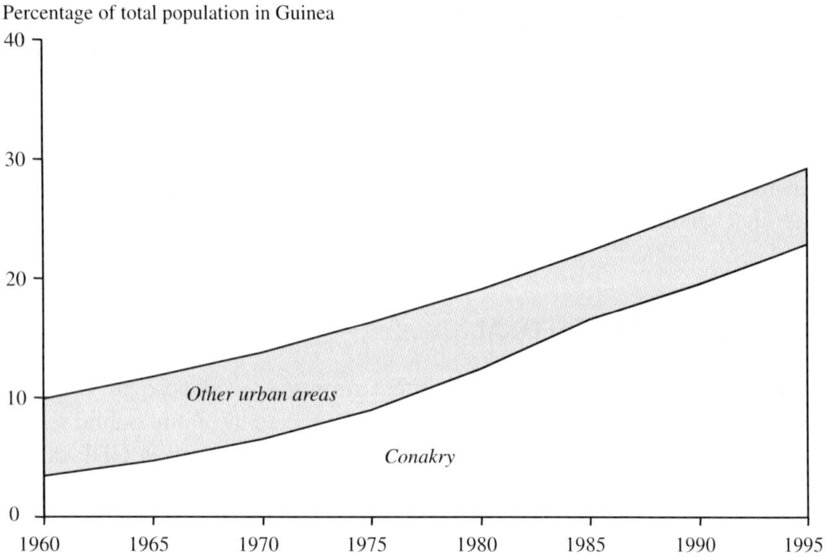

Source: United Nations (1996) and World Bank (1997b).

Figure 8.1. Population living in Conakry.

1960 and 1985, Conakry grew from around 39,000 people (3.5% of the population) to over 800,000, almost 16.5% of the population (see Fig. 8.1). Although, the estimated growth rate slowed in the late 1980s and early 1990s, it remained fast due to an influx of refugees from neighboring Sierra Leone and Liberia. Rapid population growth, combined with poor urban planning and inadequate investment, quickly overwhelmed Conakry's infrastructure, including its water system.[2]

Despite the strong performance of the partly foreign-owned and run mining sector, the Guinean economy rapidly deteriorated. Rapid expansion of the civil service led to a huge public sector wage bill, even though salaries were falling in real terms.[3] Although wages were supplemented with benefits such as access to goods at official prices, transportation allowances, and access to free piped water, it became nearly impossible to support a family on a civil servant's salary. Consequently, civil servants turned to corruption and moonlighting to make ends meet.[4]

In addition to increasing government size, these policies also increased the influence of the PDG. Price controls, rationing and control of production greatly increased the

[2] A Center for Privatization study conducted for USAID discusses Conakry's infrastructure in the mid-1980s (Marston et al., 1986).

[3] Although the First Republic did not produce accurate budgets, it is clear that public finances were in disarray. In the 1970s, public sector employment grew by 7% per year and although civil service wages were frozen between 1965 and 1980, they accounted for between 50 and 60% of government expenditures by the mid-1980s (De Merode, 1994).

[4] By 1984, the average salary was estimated to cover only about 25% of the cost of living, even after including the rice rations and other commodities used to supplement salaries (Marston et al., 1986).

potential for patronage, making links to the PDG a way of gaining access to jobs, rations at official prices, education and connections to the water system.[5] Foreign exchange obtained from the sale of cash crops was used to subsidize the consumption of civil servants and party officials, while low subsidized prices for food crops benefited urban consumers. Since both the civil service and the PDG were concentrated in Conakry, President Touré became reliant upon urban areas for support. Although protests in 1978 resulted in modest liberalization, the pace was predictably slow since many of the reforms (e.g., privatization, civil service reductions and price liberalization) would harm his urban supporters.[6]

On March 24, 1984, during preparations for the upcoming meeting of the *Organization for African Unity* (OAU) in Conakry, President Touré had a heart attack, dying two days later. President Touré's notorious fear of plots meant that there was no clear successor and, therefore, a lengthy power struggle seemed likely. The process, however, was cut short on April 3, when several top army officers, led by Colonel (now General) Lansana Conté, seized power. The new government quickly consolidated power, sweeping away the remnants of the Touré regime — the *Parti Démocratique de Guinée (PDG)* and the unions were outlawed and political prisoners were released. After years of repression, the new regime received widespread support from the public, signaled by huge and festive demonstrations. Because of this goodwill, and because it did not depend upon urban areas for support, the new military government could consider policies to liberalize the economy that had been rejected or delayed by the Touré regime.

A BADLY NEEDED REFORM: CIRCUMSTANCES IN THE WATER SECTOR

In 1983, while President Touré was still president, a group of international experts was invited into Guinea to evaluate Conakry's water system and to suggest ways of improving sector performance.[7] By this time, sector problems were critical. Despite loans from international donors, coverage was low and many non-connected residents drank polluted water from wells. Because of this, waterborne diseases were the main cause of death for infants and children and there were periodic cholera epidemics. The public enterprise responsible for the sector, the *Enterprise Nationale de Distribution de l'Eau Guinéenne* (DEG), was poorly managed, overstaffed, and practically insolvent. Although the report suggested several options for reform, and despite pressure from international donors and the system's deteriorating performance, the government did not

[5] Arulpragasam and Sahn (1997) note that PDG officials "exercised control over everything from employment in state enterprises to access to foreign exchange at official rates."

[6] After demonstrations by market women throughout the country in 1978, the Touré regime started to tentatively implement reform (see Young, 1982). By 1983, several enterprises were given operational autonomy (e.g., over employment and pricing), 15% of public sector workers were released (although the policy of guaranteeing employment for graduates was continued) and several business units of public enterprises were disbanded. The government, however, refused to implement other reforms, including realigning the exchange rate, since access to hard currency and imported goods at subsidized rates were among the benefits enjoyed by high-level civil servants and PDG party members.

[7] The group, coordinated by SONEDE (Tunisia) and *Lyonnaise des Eaux* (France), submitted their report in March 1985.

enact reform until 1989. In this section of the chapter, we describe the situation prior to reform and discuss reasons for the sector's poor performance.

Sector Performance Prior to Reform

Sector organization

Prior to reform, a public agency, the *Enterprise Nationale de Distribution de l'Eau Guinéenne* (DEG) was responsible for operations, maintenance and investment in Conakry and other urban areas. Theoretically, DEG was an autonomous agency within the Ministry of Natural Resources and Environment (MNRE) and, under its statutes, its board contained representative from several ministries. However, in practice, its board never met and the MNRE treated it as it would any other department (Triche, 1990). Consequently, DEG had no autonomy and suffered from the same problems that plagued the rest of the civil service during President Touré's regime. For example, the policy of guaranteeing employment to university graduates meant that DEG, like the rest of the civil service, was extremely overstaffed.

Water resources

Despite the sector's problems, scarcity has never been a major problem in Guinea, which, due to its abundant rainfall, has earned the reputation of being the "water tower of Africa".[8] Although water from the Conakry aquifer is unsuitable for drinking because it is highly polluted, due to the use of pit latrines, and saline, due to proximity to the ocean, a safe source of abundant fresh water is available. The reservoir at Grandes Chutes, which is located about 60 km from the city, could produce over 500,000 m^3/day (World Bank, 1989). In comparison, DEG's average production in Conakry was only 44,000 m^3/day in 1985 and total production remained at only 68,000 m^3/day in 1996 (World Bank, 1987, 1998). Further, because Grandes Chutes is located 233 m above sea level, the system can be gravity-fed, reducing the cost of water. Although potential production was high, the transmission pipeline from the Grandes Chutes dam, which could carry only 45,000 m^3/day, limited actual productive capacity.

Low connection rate

Despite the abundant water supply, coverage was low before reform — we estimate that less than 40% of the population of the city had access to piped water in 1989.[9] According to data from SEEG, there were 3100 *legal unbilled* connections for deputies,

---

[8] Average annual rainfall in Conakry is about 4.2 m (World Bank, 1990c).

[9] In addition to private connections, there were 40 operational standpipes in 1989 (World Bank, 1989, p. 3). Assuming 27 people per connection (see footnote 42), 1000 people per standpipe (see footnote 42) and population from United Nations (1996), this implies a coverage rate of 38% in 1989, omitting people who obtained water from illegal connections. This does not take into account that service interruptions meant that people with connections had to use water from wells as a secondary water source.

senior civil servants and DEG employees and 10,200 *billed* connections in 1989 for close to one million people. In comparison, Abidjan in Côte d'Ivoire, which was about twice the size of Conakry, had over 111,505 connections (SODECI, 1990). Most private connections were a single tap inside a lot or compound — only a small wealthy minority had running water inside their house — usually serving many households (Durany and Morel à l'Huissier, 1994). According to field interviews, whether a household had a connection depended primarily on personal links (e.g., tribal and family) and fidelity to the Sekou Touré regime. For this reason, a household's capacity to provide water to neighbors rapidly became part of its power, prestige, and control.

Although it is difficult to estimate the exact number, many Conakry residents had *illegal* connections to the system. In 1992, three years after the private operator took control of the company, Castilia (1993) estimated that there were 4000 *illegal* connections in Conakry, compared to 13,371 legal connections.[10] Since a private operator, operating along commercial lines, presumably has greater reason to reduce the number of illegal connections, it is likely that there were significantly more illegal connections before reform.

In part, the low connection rate was the result of rapid urban growth due, in part, to policies that encouraged urbanization. Problems related to fast growth were exacerbated by poor urban planning, something that remains a problem today. Service delivery and system expansion were difficult and costly because the poor road network made it difficult to lay pipes and because a relatively small number of potential customers were spread over a wide area. Consequently, even if maintenance had been adequate, Conakry's infrastructure would have been strained beyond its capacity.

Low tariffs

Before June 1986, when it was increased to GF $60/m^3$ ($\$0.12/m^3$), the water tariff was GF $10/m^3$ ($\$0.02/m^3$). Even after the increase, the tariff was considerably lower than prices in other West African countries (see Fig. 8.2) and far lower than estimates of long-run marginal cost. Based upon DEG's actual expenses and estimates of the optimal level of operating and maintenance costs, World Bank (1987) estimated that the average incremental cost of water (AIC) was about US $\$0.25/m^3$. However, even this might have been somewhat low, given that in 1989, World Bank (1989) estimated that the AIC in Conakry was US $\$0.82/m^3$.[11] Because of this, it would have been difficult for even an efficient operator to operate and maintain the system at the tariff levels in place before reform. Since, as we discuss below, DEG was overstaffed, inefficient, and hopeless at billing and collection, there was little money available for either maintenance or system expansion.

[10] Similarly, Durany and Morel à l'Huissier (1994) estimated that based on the number of household connections SEEG reported that only about one-third of households should have had private connections in 1992, compared to the 45% they observed.

[11] These estimates should be treated with caution due to the large differences in estimates and to the poor state of DEG's accounts.

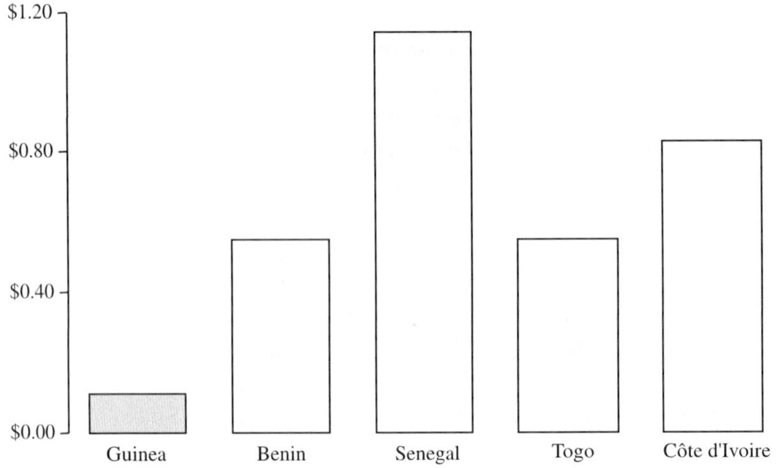

Source: World Bank files.

Figure 8.2. Price of water in 1988.

Low labor productivity

By 1984, DEG had 504 employees nationally, a ratio of 34 employees per 1000 connections (World Bank, 1987).[12] Even compared to other West African public water utilities this was high — there were 32 and 24 employees per 1000 connections, respectively, in the public utilities in Togo and Benin. The private operator in Côte d'Ivoire, SODECI, had only 9.8 employees per 1000 connections at this time (SODECI, 1986). Further, Guinea's severe financial problems led to a serious decline in civil service salaries, including salaries at DEG, in the 1970s and 1980s. Because of this, DEG employees had little or no incentive to do their job and many would show up only to collect their salaries, which were often paid in cash.

Low billing and collection rates

In principle, water distribution was metered and consumers were charged according to consumption but, in fact, metering was rare. Despite a $12 million IDA loan in 1979, which included a component intended to increase the number of connections and the extent of metering, only 5% of connections had a working meter in 1984.[13] Few private customers willingly paid for water and many were not billed at all — the 1985

[12] In 1984, there were 11,167 connections in Conakry and 3644 connections in the rest of the country (World Bank, 1987). Between 1978 and 1984, DEG's workforce increased from 280 to 504 employees (World Bank, 1978), mirroring similar growth in civil service employment.

[13] In 1979, when the project was appraised, 95% of connections were supposed to be metered by 1984. However, the Project Completion Report noted, "about 1000 connections were built but no meter could be installed since those supplied under this project were diverted by persons unknown from their intended use" (World Bank, 1987).

consultant report estimated that less than 12% of private users were billed in 1982. For the few customers that were billed, the highly political nature of water prevented DEG from collecting. For example, in 1987, DEG billed customers for approximately GF 800 million, but collected only about GF 100 million (World Bank, 1989). Since the private billing and collection rates were so low, DEG relied upon infrequent payments and subsidies from the government. However, even this was unreliable, and nonpayment by the government often led to conflict with donors.[14]

Troubled financial state

Since DEG did not comply with international accounting standards, did not have audited accounts for most of the 1980s, and maintained poor records and data collection standards, it is difficult to assess DEG's financial performance before reform.[15] However, given that DEG maintained poor commercial habits and that tariffs were set below long run marginal cost, it is not surprising that DEG appears to have had serious financial problems by the time of reform. One indication of this is that DEG owed over US $4 million in unpaid interest and was over US $14 million in debt by 1984. Another is that, even according to DEG's own accounts, it was losing large amounts of money. In 1988, DEG's total operating revenues (i.e., amount billed) were about GF 938 million ($2.6 million in 1996 dollars) and its total loss (i.e., retained earnings) was GF 805 million ($2.0 million in 1996 dollars). This probably understates DEG's actual losses, since actual collections were significantly lower than billed water sales.

Poor service and water quality

In addition to affecting the coverage rate, DEG's troubled financial state had a large effect on water and service quality. Although it is difficult to find reliable data — DEG did not collect information on service interruptions, water pressure or the time it took to repair breaks — by all accounts, water and service quality were poor. Since water was often delivered through poorly maintained lead pipes and was not isolated from the environment, it was heavily polluted.[16] Consumers would have to run the water several minutes before it was 'clear' enough to appear drinkable. Constant service interruptions meant that even households with piped water had to use polluted water from wells as secondary source of drinking water.[17] In addition, this meant that residents would have to store water for long periods, often in unhygienic conditions (i.e., open buckets in an expanding city with almost no sewerage).

[14] For example, in the late 1980s, the World Bank waived provisions for a structural adjustment loan that had required the administration settle its water and electricity bills promptly (World Bank, 1990a).

[15] The 1985 consultants' report concluded that DEG's poor accounting practices and the non-availability of most relevant data made it impossible to perform an audit. Similarly, the World Bank (1989) concluded in 1989 "the shortcomings of DEG's accounting systems are such that they largely preclude attempts to observe trends and base forecasts."

[16] Most pipes were either only a few centimeters underground or above the ground.

[17] In 1992, about 50% of people with access to piped water reported using well water as their primary alternate source of drinking water when the system was not working (Durany and Morel à l'Huissier, 1994).

Unaccounted-for water (UFW)

Poor maintenance due to DEG's weak financial position and the large number of illegal connection also resulted in high levels of unaccounted-for water (UFW). Although the lack of metering makes it difficult to estimate UFW, most estimates were very high. According to the 1985 audit report, UFW was at least 60% in 1983 — high compared to standards for well-developed systems (between 10 and 20%) or to other water systems in West Africa (see Fig. 8.10).[18]

Alternate water sources

In 1992, almost one-fifth of households reported that they got piped water from a neighbor's connection.[19] Alternative sources, such as bottled water, were almost nonexistent. Before reform, state control prevented private initiatives to supply water. But even by 1993, nearly ten years after President Touré's death, less than 1% of households purchased water from vendors and less than 5% purchased water from connected neighbors (Durany and Morel à l'Huissier, 1994). Many households used *rainwater* for uses other than drinking (e.g., for washing dishes and bathing), especially during the rainy season. However, it was often stored in unsanitary conditions. Most non-connected residents relied upon neighbors' connections or water from wells. In 1992, 29% of Conakry residents relied upon well water as their primary source of drinking water and, as noted above, many others used it as a secondary source when the system was not working. Because the sewerage system was also undeveloped, well water was heavily polluted.[20]

Sewerage

Although the sewerage system was, and remains, underdeveloped, it has always been separate from the water supply system.[21] Consequently, the 1989 reform did not have a significant effect on the sector, other than through its indirect effect on water supply.

[18] However, one year later, the World Bank (1987) estimated, more optimistically, that UFW was 39% and, in 1989, the World Bank (1989) estimated that UFW was about 35%. Based upon the experience since the private operator completed metering in 1995, the higher estimates seem more likely.

[19] Durany and Morel à l'Huissier (1994) found that 18% of households reported that they got their water from a neighbor's connection in 1992, including 5% that paid for it. This survey was conducted before extensive metering.

[20] Although data from the 1980s are not available, it seems plausible that contamination of well water was a serious problem at this time. In 1994, a study of 57 wells found that, on average, there were over 30,000 fecal coliforms/100 ml of well water (Gélinas et al., 1996). In comparison, the bacteriological WHO (World Health Organization) norm is zero fecal coliforms per 100 ml of water and an alternate system classifies any water with over 1000 fecal coliforms/100 ml as "seriously pathogenic" (Feachem, 1980).

[21] Durany and Morel à l'Huissier (1994) found that only about 5–10% of households had latrines connected to the sewerage system in 1992. This is broadly consistent with a study financed by Luxembourg and Japan which estimated that only about 9% of Conakry residents were connected to sewerage (World Bank, 1997a).

In 1992, 89% of households relied upon primitive sewage facilities in their courtyard. 21% of these facilities were simple unlined pits, 39% were pits lined with cement, and 29% consisted of two separate pits (Durany and Morel à l'Huissier, 1994). Sludge from these pits can leak into the phreatic layer, from which households draw well water for daily activities such as washing dishes and bathing and can be close to leaking pipes that provide water to households and public fountains.

POLITICAL CIRCUMSTANCE LEADING TO REFORM

Given the weak institutional framework and the dire situation of the water system, the reform ended up being quite radical, especially in a sector that is politically sensitive and where full privatization is rare. Operational control of the system was transferred from a state-owned enterprise that was only nominally independent of the government ministry it reported to, to a private-sector company that was owned and controlled by foreigners. In addition, significant price increases were planned (over 500% in real terms over a five-year period) and even greater increases were implemented (over 600%). Despite the collapsing system, however, the government took a long time to implement reform. Before reform was finally enacted, six years passed and a change of government took place. In this section of the chapter, we discuss why the reform occurred, why it took the form it did, and why it took so long to enact.

Desirability of Reform

Although the crisis in the water sector made reform highly *desirable*, in many other cases sector-specific crises have failed to motivate significant reform. For example, over the past decade, some parts of downtown Mexico City have sunk two meters due to the overexploitation of groundwater resources. Despite this, the government implemented only a minor reform, which failed to reduce costs or curb waste (see Chapter 5). Although problems in Conakry, which were primarily the result of the poor performance of the public operator, were far more tractable than the problems in Mexico City, poor performance alone has often failed to motivate significant reform.

Theoretically, the government could have reformed the sector in many different ways: however, its practical options were constrained. The main problem was that if sector performance were to be improved additional funds would be needed to allow the operator to cover its costs and to increase investment. The 1985 consultant's report estimated that, at that time, US $73 million of investment was needed to increase production capacity and repair and extend the distribution network. Although in theory these funds could be obtained in several different ways — for example, through additional borrowing, through increased subsidies or through higher tariffs — in practice the government had only limited options. Given that the government's finances remained weak due to the disastrous economic policies of the previous regime, the government could not finance the needed investment without help from donors. However, donors led by the World Bank demanded "full cost recovery" (i.e., setting tariffs to cover operating costs and support sector debt service) (Triche, 1990). To get access to donor funds, the government would therefore have to increase prices significantly.[22]

The need for support from donors and international organizations, due to the pressure that the macroeconomic difficulties put on public finances, limited the government's options in other ways. Most importantly, the poor performance of the *First Conakry Water Supply and Sanitation Project* convinced the World Bank that significant institutional reform, including some private sector participation, was necessary (World Bank, 1987). The government, therefore, found that it would have to accept this if it were to receive aid.

Other factors also encouraged private sector participation. The government recognized that reforming DEG without increasing private participation would have been extremely difficult. Even with technical assistance from abroad, entrenched interests and existing corruption would make some tasks, notably bill collection, difficult if DEG remained in charge of the sector. The decision to allow private sector participation also reinforced the decision to allow price increases. Since the government was, and remains, delinquent in paying its bill, a private operator would have been wary about relying on government subsidies, especially given the poor state of public finances and the weak institutional environment. Consistent with this, the part of the price subsidy that went to the private operator for the first few years following reform was funded through a World Bank loan, rather than from general government revenues.

An additional factor that encouraged private sector participation was that it was consistent with the governments' philosophy of liberalization and privatization. The structural adjustment program that the government implemented in response to the macroeconomic crisis greatly increased the private sector's role in the economy. In this respect, increased private participation in water was part of a larger program to increase the importance of the private sector in the economy.

Feasibility of Reform

By restricting possible reform options, these constraints threatened the *feasibility* of reform. Price increases and private sector participation would have a significant effect on two powerful groups — the urban elite who were used to not paying the full cost of water and workers who might lose their jobs. Reform would only be possible to the extent that the government could either ignore these groups without threatening its support base or compensate them for their losses.

In practice, neither the urban elite nor public workers were vital supporters of the new military government. The clearest demonstration of this is that, in response to the macroeconomic crisis, the government enacted a structural adjustment program that harmed these groups significantly. Within days of assuming power, the military government announced that it intended to speed up the liberalization process. The focus

[22] The focus on cost recovery is reflected in the contract between the private operator and the government (the *contrat d'exploitation*); neither this contract nor the list of requirements defining SEEG's responsibilities (*cahier des charges*) mention targets on connections or UFW.

of the proposed reform package was improving the performance of the agricultural sector, reducing the size of the public sector and liberalizing prices and exchange rates.[23] The stress on agricultural production, which would initially benefit rural areas, was a marked contrast to the Touré regime's policy of favoring its urban constituents.[24]

The new government began initiating its agenda several months after coming to power. Prices were liberalized, the overvalued exchange rate was realigned, the state-owned marketing boards and stores were closed and many state-owned enterprises were privatized or liquidated. Although in the medium term it was hoped that structural adjustment would benefit all Guineans, it was clear that these reforms would cause significant short-term pain for many urban residents. This would be especially acute for higher-level officials who had better access to goods at official prices — the same group who had access to free or low-cost water.[25]

The structural adjustment program, which resulted in many layoffs, also harmed public sector employees. Employment in state-owned enterprises fell from 17,111 in early 1986 to 2000 by the end of 1989 and civil service employment fell from 70,989 in early 1986 to 51,000 by the end of 1990.[26] These massive layoffs were painful for many Conakry residents, since 24% of employed workers (and 30% of employed male workers) in Conakry worked in the public sector in 1990 (Arulpragasam and Sahn, 1997) and the poorly developed private sector was not able to quickly absorb laid-off workers.[27]

In contrast, urban water reform would not greatly affect the government's main supporters. The army, who were the most important supporters of the government, would be largely unaffected by reform.[28] Further, rural consumers, who were expected to be the main beneficiaries of structural adjustment, did not have access to piped water and, therefore, reform would mainly affect them through its effect on government finances.

---

[23] These were consistent with the policies that the IMF and the World Bank were promoting at that time (see Clapp, 1994; World Bank, 1994).

[24] The government's policy of promoting agricultural and rural development was made clear to the population. President Conté made several announcements stressing that one of the new regime's main goals was to boost agricultural production. In his 1985 New Year speech, he stressed the importance of the agricultural sector and noted steps that could improve performance (Clapp, 1996). On April 3, 1985, in a speech on the first anniversary of the coup, President Conté announced that the government's main priorities were boosting agricultural performance and improving communications (Economist Intelligence Unit, Second Quarter, 1985). This is consistent with documents produced for the 1987 Consultative Group Meeting with international donors, which note that improving agricultural production through the liberalization of the sector was one of the key objectives of the Economic and Financial Recovery Program (Republic of Guinea, 1987).

[25] Arulpragasam and Sahn (1997) note, before structural adjustment, that over half of high-level officials in Conakry purchased their entire rice ration at subsidized official prices, compared to only 15% of low-level officials and 16% of small scale merchants.

[26] Data are from World Bank (1990b) and De Merode (1994).

[27] Based upon a survey conducted in 1990–1991, Mills and Sahn (1997) estimated that nearly half of laid off civil servants had not found a job in either the formal or the informal sectors after four years.

[28] Even if the army did not pay its water bill, it would take an extremely brave private operator to agree to cut water off to army barracks or the Ministry of Defense in a weak institutional environment. In addition, the army was also largely unaffected by the structural adjustment program since it was shielded from cuts in public employment and wages (Economic Intelligence Unit, Quarter 1, 1985).

Although water sector reform would be painful for many DEG employees and customers, it is important to note that some members of these groups would benefit from reform. First, the government's, and DEG's, weak financial position meant that employees could often not rely upon receiving wages in a timely manner. DEG employees who found jobs in the new private enterprise might therefore benefit from reform. Similarly, although current DEG customers would have to pay higher prices, they would benefit from improved service and water quality. Consequently, not all DEG employees and current customers could be counted as losers.

The government also took steps to reduce opposition from current DEG customers and employees. First, the government, with support from the World Bank, agreed to subsidize prices during the early years of reform. Although price increases were implemented immediately, they were far smaller than would be needed to cover sector costs. Second, the government, again with World Bank support, compensated DEG employees who were not selected for jobs at SEEG or SONEG. Of 504 DEG employees, 40 were hired by SONEG, 250 by SEEG, and 30 qualified for other civil service positions (Triche, 1990). Additional persons retired or moved to other jobs in the formal sector. Remaining employees were offered basic equipment and special training, organized and monitored by the 'Compagnie Générale des Eaux', to help them create and develop cooperatives specializing in public works. Notwithstanding severe tension, this made reform acceptable to most employees.

Contractual Options

Within the general framework of increased prices and private sector participation, the government had several options. The 1985 consultants' report considered possibilities ranging from a management contract to full privatization. In the end, the government decided on a mixed scheme with a public enterprise owning and developing infrastructure and a mixed company, dominated by the private sector, in charge of operating the system. It chose this rather than full privatization for several reasons. First, it was thought that private enterprises might not be willing to invest in the development of infrastructure. Although the current government was committed to reform, the political situation was unstable: several coups had been attempted and civil unrest was pressuring the government to become more democratic. Further, although Guinea was trying to increase the competence and independence of the judiciary, the judiciary was weak and heavily dependent upon the executive.[29] Because of this, and because of the weak system of property rights, the judiciary had a poor record of enforcing private contracts.[30] Under these circumstances, investment in long-lived and non-transferable assets was thought too risky for many private firms. Finally, foreign ownership of assets would

[29] The government is currently developing a new system of institutions, laws protecting rights and contract laws, largely inspired by the French system and it intends to put an administration in place to fully implement these laws. However, even now, the system is mainly theoretical and largely incomplete.

[30] A 1996 bill, approved by the government, that would have allowed SEEG to sue consumers who do not pay their bill, was rejected by parliament. This might not be surprising since, in interviews, several sources noted that it is thought that most deputies do not pay their utility bills.

have been controversial and might have led to greater unrest. For these reasons, the government adopted a scheme that would involve little private investment.

The Long Delay before the Reform Was Enacted

Although radical reform eventually occurred, it was not implemented until six years after the initial studies and five years after the change in government. Overall, it is not surprising that the Touré regime did not implement significant reform. Unlike the military government that replaced it, the Touré regime was highly dependent upon the two groups that would be harmed by reform. Any reform that resulted in increased prices, especially of the magnitude actually implemented, and layoffs would harm the regime's main support base and, therefore, would have been hardly feasible. In contrast, the five-year lag following the change of government reflects the slow pace of structural adjustment and, more generally, political instability in the early years of the new regime.[31]

Consequently, sensitive reforms that would harm urban residents, such as realigning the exchange rate and civil service employment reductions, were approached cautiously (see Clapp, 1994, 1996). After years of repression and shifting policies under the Touré regime, it was going to take some time for adjustment to result in increased agricultural production and improved welfare in rural areas. Since President Touré was careful about making sure that rivals could not build effective power bases while he remained president, the new military leaders were unable to build support before taking power. Consequently, while the new leaders were forging a coalition, they remained heavily reliant on the military and on goodwill conferred to them for ending President Touré's more repressive policies. Since the benefits of structural adjustment would take a reasonable time to materialize, it is not surprising that the new government was careful about implementing reforms that could initiate protests.

CONTRACT DESIGN AND IMPLEMENTATION OF REFORM

Water reform was finally adopted in 1989. World Bank requirements meant that the private partner had to be selected through international bidding and six companies initially expressed interest: SONEDE (Tunisia), Thames (UK), Lyonnaise des Eaux (France), SAUR (France), SAGER and Compagnie Générale des Eaux (re-named VIVENDI, France). Two bidders dropped out during the process and the four remaining companies combined to form two competing consortia. A joint venture led by SAUR, with the participation of Compagnie Générale des Eaux, submitted the lower bid for tariffs that they would charge consumers, which was 30% below the consultant evaluation of the tariff level that would ensure cost recovery. SAUR quickly confirmed its leadership in the newly formed company with the help of experts from Côte d'Ivoire, where SAUR had been successfully involved in the water sector for nearly 30 years

[31] A major coup attempt less than a year after President Conté came to power, demonstrated that the new government could not ignore the welfare of Conakry residents or the army.

(see Chapter 7). New contracts were formally signed in 1989 and were supported by the World Bank coordinated *Second Water Supply Project*.[32]

Sector Organization Following Reform

As part of the reform, two enterprises were created: Société Nationale des Eaux de Guinée (SONEG) and Société d'Exploitation des Eaux de Guinée (SEEG). The reform involved a set of three contracts: one between the government and SONEG (*Contrat plan*); one between SONEG and SEEG (the lease contract); and one between SEEG and its international shareholders (a technical assistance contract). In addition, there were contracts between SEEG and its customers, i.e., users of the water system.

Fig. 8.3 shows sector organization and the pattern of ownership of the two enterprises, SONEG and SEEG. SONEG was a small (44 employees in 1997), entirely state-owned government agency, which reported to a board of directors and to the Ministry of Natural Resources and Energy. Formally, the relationship between SONEG and the government was monitored through a contract. At the time of reform, the contract defined rules governing the determination of tariffs (see below) and committed the government to

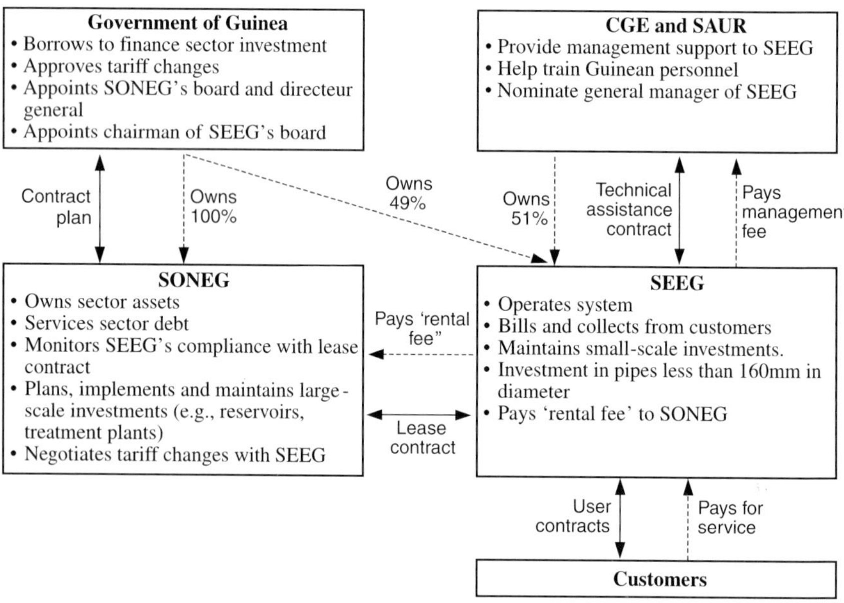

Figure 8.3. Institutional arrangement in Guinea.

[32] This $102.6 million project included $40 million from the World Bank, $23 million from the African Development Bank, and $8.6 million from the Government of Guinea and SONEG. In practice, the individual partners' contributions varied slightly from the original estimates (see World Bank, 1997a). Actual expenditures on investment components were considerably higher than planned — $84.9 million on expansion and $9.0 million on rehabilitation (World Bank, 1997a). The World Bank loan, from the International Development Agency (IDA), was on standard IDA terms (50 years at a 0.75% interest rate).

pay its bill regularly. It was the only contract that defined specific performance targets, including the increase in connections from an estimated 16,400 in 1989 to 20,200 in 1992.

The lease contract defined the responsibilities of SONEG and SEEG. SONEG owned sector assets and was responsible for planning, projecting and managing infrastructure, sector accounting, and managing funds from the 'rental fee' and donors to pay for new investment and service sector debt. Since the rental fees paid to it by SEEG were supposed to cover its operating and debt service expenses, it was financially independent of the government. It was also responsible for supervising most large-scale investment (e.g., reservoirs and transmission pipelines) and for the construction and maintenance of the primary distribution network (i.e., pipes more than 160 mm in diameter). One significant restriction was that SONEG could not finance investment through borrowing; this was the responsibility of the government. Although SONEG was formally an independent agency, its board was under direct government control, with members appointed by different ministries.[33] In addition, the government also appointed its staff, including its Directeur Général. SONEG was further constrained by its reliance upon foreign suppliers (including SAUR, SEEG's leading shareholder).

SEEG, which was 49% state-owned and 51% privately owned, had a ten-year lease contract with SONEG. SEEG was in charge of the distribution and commercialization of water, including (1) building and maintaining the secondary and tertiary distribution networks (i.e., pipes under 160 mm in diameter), (2) metering, billing and collecting, and (3) paying the rental fee to SONEG. As we noted above, the private owners were two French companies, SAUR (a subsidiary of Groupe BOUYGUES, a French company heavily involved in public works), which had the leadership in this arrangement, and Compagnie Générale des Eaux (CGE, now Vivendi). In addition to the lease contract, the two French partners signed a 'technical assistance contract' with SEEG to provide managerial support, technical assistance and financial expertise, and to help train Guinean personnel.[34] As their contribution to SEEG, SAUR and CGE provided 51% of the initial US $3 million of capital. For its contribution, the government donated equipment and infrastructure from DEG and, through SONEG, took responsibility for accumulated sector debt.

Overlapping responsibilities

Notwithstanding the presence of five government-appointed representatives on its eleven-member board, SEEG acted like a private company. The private owners were responsible for nominating the general manager, while the government was responsible for choosing the chairman of the board. Several factors increased SEEG's autonomy. First, the foreign management's close ties to SEEG's international owners significantly reduced government influence. In addition, the contractual arrangement, which put

[33] At the time of reform, ten different ministries appointed the eleven members of the board.

[34] This contract between SEEG and the foreign owners specified conditions under which the foreign companies provided home-office support for day-to-day management problems, selection of expatriate staff and audits of procedures. Remuneration was set at 2% of SEEG's revenues (World Bank, 1989).

SEEG in charge of bill collection, gave SEEG direct control over the financial resources it needed to operate.

The division of investment responsibilities between the two agencies led to significant problems, with the two agencies blaming each other for delays and disagreements. The different priorities and constraints that the two agencies faced were part of the problem. Because SONEG tended to be more concerned with social and political goals, while SEEG had commercial goals, they often disagreed on priorities concerning network expansion. Further, SONEG, as a public agency, had to follow public procedures when making investments, (e.g., awarding large investment contracts through bidding). SONEG's processes were often slow and costly and provided opportunities for public interference, corruption, and delay. In contrast, SEEG, as a private company, had much greater flexibility and, consequently, SEEG tried to go over SONEG's head, developing infrastructure beyond its contractual responsibilities. Although the contract explicitly prohibited SEEG from doing this, once it was done the situation was irreversible. When the Caisse Française de Développement provided SEEG with substantial aid to develop new infrastructure (without bidding and ignoring the clause that makes SONEG responsible for such projects), SONEG could do little. Further, despite SEEG's role in investment, SEEG assumed almost no investment-related risk — loans for infrastructure were the responsibility of the government.

Institutional impediments

Implementation of the contract between SONEG and SEEG was made more difficult by low administrative capability and the absence of an independent judiciary. This meant that no impartial body was able to constrain arbitrary action by the government and credibly enforce contracts provisions. For example, recognizing that SEEG could apply pressure to SONEG by refusing to pass SONEG the rental fee, the contract allowed SONEG to seize a deposit paid by SEEG if this occurred. However, political interference by the government prevented SONEG from doing this. For example, when the government stopped paying its water bill (see Fig. 8.6), SEEG stopped passing SONEG the rental fee on the money it was collecting from the private sector, formally a breach of contract. In turn, this threatened SONEG's ability to pay employees, reimburse international lenders and continue with investment. However, when SONEG threatened to impose the penalty, SEEG, backed by its international owners, argued that the delay was due to nonpayment by the government and government representatives forced SONEG to back down. Ultimately, under pressure from international donors who made settlement a condition for credit effectiveness for the *Third Water Supply and Sanitation Project*, SEEG and SONEG significantly reduced their cross-debts on September 26, 1996.[35]

This episode demonstrates that contract provisions designed to prevent certain government actions will not be useful unless an independent body can enforce the contract

[35] Before September 1996, cross debts were about GF 4 billion. This was reduced to GF 660 million that the government owed to SEEG and GF 608 million that SEEG owed to SONEG. Once the World Bank approved the *Third Water Supply and Sanitation Project*, cross debts began to accumulate again.

and prevent the government from taking retaliatory action. Although the lease contract explicitly allowed SEEG to cut water to government entities that were more than four months late with payments, SEEG was unwilling to actually cut off water to *any* agencies of the central government. In part, SEEG's unwillingness reflected its concerns about its relationship with the government, since the central government has significant power over SEEG.[36] For example, at that time, SEEG was trying to get the government to pass a law that would allow it to force collection for nonpayment by private parties and to approve price increases larger than had been originally planned.

Dispute resolution

Another effect that weak institutions had on outcomes was the apparent failure of several mechanisms directed at dispute resolution. First, the government, which owned 49% of SEEG, had representatives on the boards of both SEEG and SONEG. It was hoped that this would improve coordination between the two enterprises. In fact, it did the opposite by introducing ambiguity about whether the contract was open to repeated renegotiation. Second, the arrangement allowed the two parties to appeal to the judiciary in Guinea and, ultimately to an international institution, the *Centre International pour le Règlement des Différends Relatifs aux Investissements (CIRDI)*, in cases of major dispute. The clause was intended to reassure foreign investors concerned about the lack of institutional constraints on arbitrary government action (e.g., the weak judiciary). In practice, however, the clause was almost purely symbolic. It was not possible to solve everyday disputes and disagreements through an international forum and, since turning to international arbitration would suggest major failure to foreign investors and international organizations, it was difficult to use for even major disputes. Further, in the absence of an independent judiciary, it was unclear how CIRDI could enforce it decisions, if it ruled against the government.

In practice, therefore, problems were largely solved through informal negotiations between top managers — making the persons in charge crucial — or through political interference (e.g., when SONEG tried to impose the penalty for not paying it the rental fee). Political interference, however, was costly. First, it introduced bias and could result in delayed and unpredictable outcomes. Second, as noted above, these problems poisoned relations between SEEG and SONEG, making it more difficult for the two companies to work together.

Regulation: Some Major Inefficiencies

Water reform is more likely to succeed if an efficient and credible regulatory entity with the capacity to force the partners to fulfill their commitments is developed. This did not happen in Guinea. In the absence of an independent regulator, the efficient running of the system depended on the set of contracts and their implementation. SONEG, who

[36] In contrast, SEEG was willing to cut water to local government agencies, which had significantly less power over SEEG (e.g., they play no part in price reviews).

was officially responsible for monitoring SEEG, was not able to do this due to its lack of autonomy and to information asymmetries between SONEG and SEEG. SONEG's lack of autonomy meant that major decisions and conflicts went to the cabinet of the Ministry of Natural Resources and even to the President of the Republic. Consequently, the government had significant discretionary power that was only partly counterbalanced by donors and the powerful international companies that owned SEEG.

This discretionary power created uncertainty for SEEG, since SONEG's plans and their implementation depended upon political decisions. Since SONEG planned, contracted, and monitored the development of the primary network, SEEG had to rely upon SONEG to do this before it could develop the secondary network. However, there was the risk that the government might divert SONEG's rental fee to fund other non-sector related spending, meaning that needed investment in the primary network was not performed. Further, the bureaucratic process for SONEG's investment program meant that the government could use bureaucratic delays to improve its bargaining position or disguise other motives. This encouraged SEEG to make investments that should have been made by SONEG (e.g., in the primary network) and to skirt rules (e.g., by splitting large investment into small parts to avoid public control).

However, weak regulation could also work in SEEG's favor. For example, since the distribution network included both lead pipes installed in the 1950's and newly installed PCV pipes, maintenance costs were difficult to observe. Since SEEG was responsible for maintaining the secondary network, it could potentially overstate these costs. An additional problem was that SEEG also bid on construction contracts and, since it did not produce separate accounts for its separate functions (e.g., those that should be regulated and those that were not), it was difficult to assess the profitability of the lease contract. These informational asymmetries were aggravated by SEEG's refusal to provide SONEG with the reports and information, including separate accounts, which the contract required.

Prices

As noted above, one of the major goals of reform was to make the sector financially self-sufficient within six years. Since, as noted earlier, prices were thought to be below long-run marginal cost, large tariff increases were necessary. Recognizing that this would be difficult for political reasons, the government agreed to heavily subsidize prices for six years following reform through a World Bank credit. Consumer tariffs were immediately increased to 150 GF/m^3 in 1989 (about \$0.25/m^3) to cover local currency costs, including the local currency operating costs of SONEG. However, the foreign currency costs (e.g., for imported equipment and supplies) were covered by proceeds from a World Bank loan and the government continued to service part of sector debt.[37] These subsidies slowly declined until 1995, by which time the consumer tariff

[37] The World Bank contributed US $16.9 million to subsidize the scheme (World Bank, 1998). The subsidy was described as 'support for the rehabilitation of sector operations' in World Bank documents and was intended to support the institutional building that was needed to implement the lease contract.

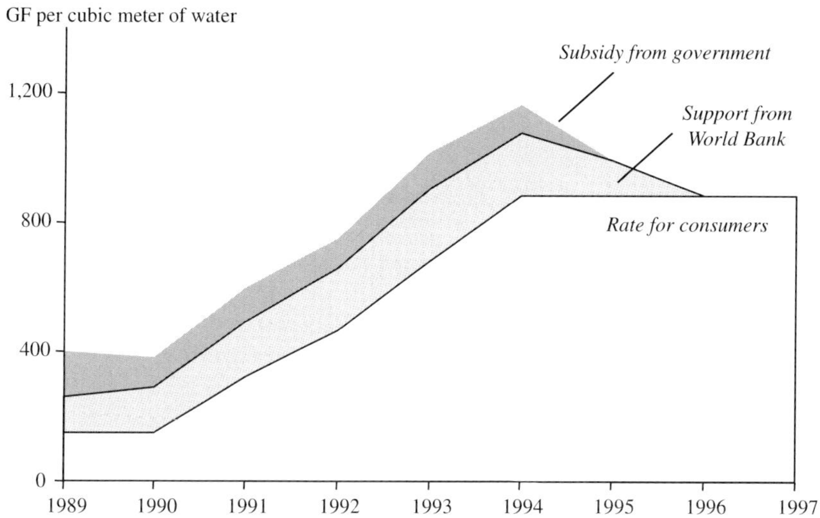

Source: SONEG.

Figure 8.4. Consumer rates and subsidies from World Bank and government: Guinea.

was supposed to cover all costs (see Fig. 8.4). It was thought that this would imply an average price to consumers of 660 GF/m^3 in 1995, which would fall to 566 GF/m^3 in 1998 (Triche, 1990). In practice, prices have been considerably higher.

Prices were regularly revised, in accordance with a cost index formula. In addition, SEEG had to renegotiate its rate with SONEG and the government every four years, based on an informal cost plus formula.[38] The renegotiations were intended to give SEEG a short-term incentive to cut costs, since it could keep cost reductions between negotiations, while allowing consumers to benefit in the medium term. Although there was initially concern that the government would not increase tariffs enough to "ensure sound sector development" (World Bank, 1989), this was not a problem in practice. In fact, prices were allowed to increase far more quickly than originally planned. By 1996, the average tariff had reached 880 GF/m^3, compared to an expected price of GNF 660/m^3. The difference between actual and projected tariffs was not the result of higher than anticipated inflation. Fig. 8.5 shows that the price grew faster in real terms than originally projected.[39]

At the end of 1997, the minimum bimonthly payment for service was GF 13,500 (about US $13). This fixed payment included payment for the first 20 cubic meters of water and had to be paid in full whether the household consumed the whole 20 m^3 or not. The price of water between 20 and 60 cubic meters was GF 850/m^3 and the price above 60 cubic meters was GF 925/m^3. In addition, it cost GF 90,000 (about US $90

[38] A clause of the lease contract also opened the possibility of renegotiation if there were significant variations in taxes.

[39] Projected price is deflated by 1988 forecasts of the CPI from World Bank files and the actual price is deflated by the actual CPI.

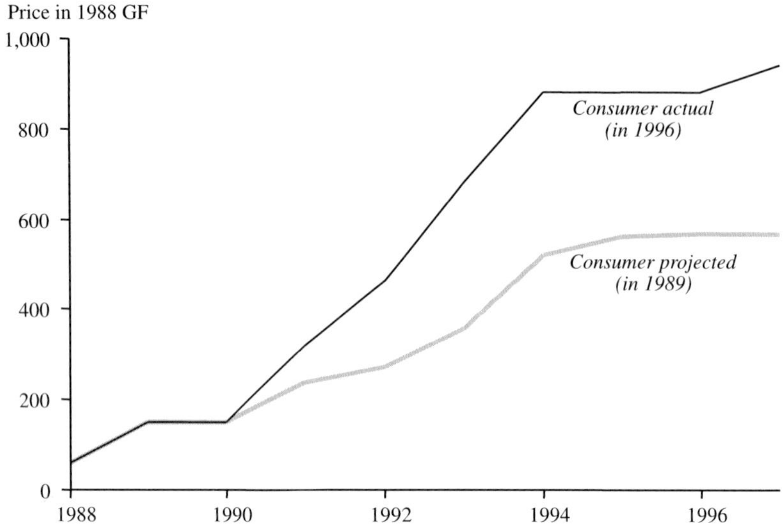

Source: SONEG (actual) and World Bank (1989, projected).

Figure 8.5. Actual and projected water prices in Guinea.

in 1996) to be connected to the system. Given that the average income of a high public servant was only about GF 150,000 (about US $150 in 1996) a month, these amounts were substantial. In field interviews, even relatively wealthy persons claimed that the high prices made it difficult for them to afford connections, especially due to the social custom of providing free water to family and neighbors. Possible reasons for the high price are discussed in the next section.

EFFECT OF REFORM ON SECTOR PERFORMANCE

Despite the problems with conflict resolution and overlapping responsibilities, there is no doubt that reform resulted in a significant improvement in sector performance. However, given the disastrous performance before reform, the results were not as dramatic as was originally hoped and significant scope for future improvement remains. In this section, we analyze the mixed results and suggest some explanations.

Improved Performance Along Many Dimensions

Increased investment

Although it was hoped that the sector could become self-supporting in the medium term through increases in tariffs, funds for investment and rehabilitation were needed immediately. Consequently, international donors led by the World Bank agreed to finance a large investment project, the *Second Water Supply Project*, that had four main investment components.

(1) US \$58 million to expand the Conakry water system, including increasing the capacity of the pipeline between Grandes Chutes and Conakry, the addition of a new treatment plant and extension of the distribution network (including 15,000 new connections).
(2) US \$4 million to support SONEG, including technical assistance, consultants, training and equipment.
(3) US \$4 million to rehabilitate existing facilities.
(4) US \$1 million to provide consultants' service for studies of secondary centers and to design a training program for staff laid-off from DEG.

Although the new pipeline from Grande Chutes dramatically increased potential water production, from 54,000 m^3/day in 1988 to 100,000 m^3/day by the end of 1993 (World Bank, 1998), actual production has been far lower. Average daily production was 60,345 m^3/day in Conakry in 1996, while average billed consumption was only 30,255 m^3/day, reflecting the high levels of UFW.[40] Consequently, if the rate of UFW could be reduced to around 20%, current productive capacity should adequately service needs in the near-term even if the number of connections increases significantly.

Increased connection rate

Although the number of legal connections increased after reform, the increase was slower than anticipated and coverage remains low.[41] By 1997, there were about 31,000 connections in Guinea, including about 25,000 connections in Conakry (population of about 1.7 million).[42] In comparison, there were close to 180,000 connections for 2.7 million inhabitants in Abidjan, Côte d'Ivoire (SODECI, 1996). The low connection rate reflected economic, technical and institutional factors. First, as discussed below, many residents could not, or would not, pay for water due its high price. In 1994, it was estimated that nearly 12,000 connections were inactive due to nonpayment (Brook Cowen, 1996). Second, Conakry's chaotic, unplanned development, which was aggravated by the large and sudden inflow of refugees from civil wars in neighboring Liberia and Sierra Leone, made network expansion technically difficult. Finally, disagreements between SEEG and SONEG, which were responsible for complementary portions of investment, also slowed system expansion.

[40] Average daily production and billing do not vary greatly over the year — in 1996, the maximum average daily production and billed consumption for any two-month period was 61,693 m^3 (April–May) and 31,962 m^3 (June–July).

[41] Only 11,000 of 15,000 planned connections had actually been implemented by mid-1997.

[42] According to a World Bank–UNDP survey from 1992, approximately 27 people used each connection including family, neighbors, etc. (Durany and Morel à l'Huissier, 1994). This might overestimate the number of people with access to piped water — the technical director of SONEG suggested that only about 15 persons used each connection in 1997. Using the higher number (27 people per connection), an estimate of 1000 people per standpipe based upon the average number of people per standpipe in 1992 (Durany and Morel à l'Huissier, 1994) and population figures from United Nations (1996), the coverage rate was 48% in 1997.

Increased number of standpipes

Although the number of *standpipes* increased following reform, they remain uncommon. There were about 130 standpipes in Conakry in 1997, including some that were inactive, resulting in 9050 non-connected residents per standpipe. During field interviews, SEEG officials stated that SEEG did not want to manage standpipes, arguing that the widespread belief that water from standpipes should be free damaged their public relations. Consequently, although SEEG installed standpipes when asked by a *Commune* (municipal government), it would not manage them. Instead, the *Commune* and SEEG jointly selected a manager, who collected user charges that he paid to SEEG. SEEG then returned a share to the *Commune* to cover access charges and employee benefits. In practice, since managers were often unable to resist pressure for free water from neighbors and were sometimes corrupt, they collected little money from users leaving the municipal government to foot the bill. If the *Commune* did not pay, SEEG could, and did, cut the standpipes off.

The low connection rate meant that water-related health problems remained a major issue due to the large number of customers who consumed unsafe water from illegal connections and contaminated wells. Because of this, diseases related to unsafe water and inadequate sewerage remained the main source of mortality (in decreasing order: malaria, diarrhea, hepatitis A, poliomyelitis, and skin diseases).[43]

Improved water and service quality

Most people agreed that water and service quality improved dramatically following reform. For example, the manager at the local *Coca-Cola* bottler noted that although water was muddy and discolored for several years after reform, regular tests confirmed that it met the standards imposed by their international headquarters by 1997. Further, the *Organization for Consumer Protection* in Conakry rated water quality as excellent and noted that it can be consumed as delivered (World Bank, 1998). Similarly, a 1994 study that measured chemical and bacteriological contamination in piped and well water, concluded that piped water in Conakry complied with WHO norms for drinking water (Gélinas et al., 1996).

The reform also improved customer service. To make it easier to file complaints, apply for new connections, notify SEEG about needed repairs and pay bills, SEEG divided Conakry into three districts, each with its own director. In addition, SEEG invested in light equipment (e.g., motorcycles) so that they could collect bills more efficiently and respond more quickly to repair requests. Further, SEEG's detailed computerized map of the network and final connections made it easier to locate disruptions. Delays for new connections were reasonable, although consumers complained about SEEG's policy of waiting until a large group of consumers paid deposits before expanding the network. Similarly, although some people complained about delays with repairs, most agreed that this was not a major problem.

[43] Periodic cholera epidemics also occurred. For example, a cholera epidemic in 1994 resulted in 675 deaths, including 330 in Conakry (World Bank, 1997a).

Rapid metering

Following reform, SEEG rapidly metered households, businesses, and the administration. Before reform, despite metering projects funded by the World Bank and other international donors, only about 5% of customers had working meters. By 1996, 98% of private customers and 100% of administration connections were metered.

Metering significantly reduced the amount of water billed to the government. In 1990, the first full year of private operation, estimated government consumption was close to 7.5 million m^3 of water. In 1996, measured government consumption was about 3.6 million m^3. Most of the decline was probably because government consumption was overestimated before metering and because responsibility for standpipes was transferred from the central to municipal governments.[44] However, since government consumption continued to decline even after public metering was completed in 1995, it appears that the government also reduced its consumption.[45] Before 1989, DEG billed the government centrally and, therefore, individual agency within the government had little reason to use water efficiently even when they were metered. Following reform, each agency became responsible for its own water bill giving them reason to use water more sparingly (e.g., reporting breaks and leaks more quickly). In practice, however, nonpayment weakened these incentives.

Improved bill collection from private consumers

Although bill collection from private consumers improved significantly after reform, it remained low compared to other privately operated systems (e.g., the private collection rate is 98% in Abidjan, Côte d'Ivoire). Further, as the price to consumers increased, the collection rate fell — from 75% of the amount billed in 1989–1990 to under 50% in 1991–1992. It recovered in 1993 and remained at around 60% through 1996 (see Fig. 8.6). SEEG could, and did, cut off customers who did not pay their bills for three consecutive months.[46]

Higher labor and total factor productivity (TFP)

Another improvement was that labor productivity increased at the time of reform (see Fig. 8.7). The rapid improvement was primarily due to a reduction in the number of workers immediately following reform. After this, connections per employee failed to increase significantly and actually dropped between 1994 and 1996. The increase in 1997 was due to a large increase in the number of connections that year (23,435 in

[44] For the first few years following reform, the central government was responsible for running and paying for consumption from standpipes and water from standpipes was free.

[45] Government consumption declined from 4,771,000 m^3 in 1995 to 3,551,000 m^3 in 1996.

[46] Field interviews with households, businesses and local administrations confirm that SEEG did this consistently, although its implementation may have raised sporadic protests. SEEG also discontinued the practice of allowing high-level civil servants and deputies to have unbilled connections (although this does not mean that they actually paid their bills).

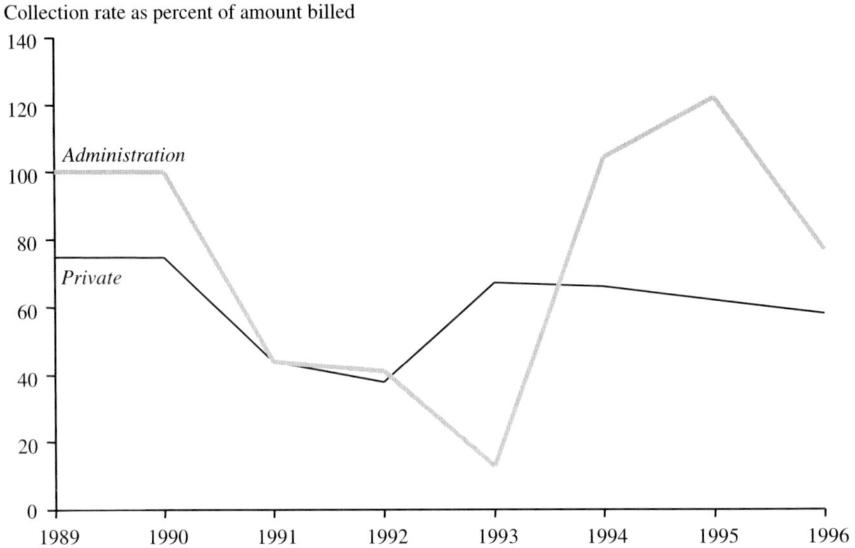

Note: Collection rate can exceed 100 percent if unpaid bills from previous years are paid in current year.
Source: SEEG.

Figure 8.6. Collection rate.

1996 compared to about 31,000 in 1997). A second measure of productivity, output per employee, followed a similar pattern.[47] Total factor productivity also increased at the time of reform (see Fig. 8.8). However, after this, TFP slowly declined (although it remained considerably higher than before reform). The most noticeable decline was a large drop in 1993, due to the large increase in fixed capital due to investment in the *Grandes Chutes* pipeline. Since a large part of productive capacity remained unused by the end of 1997, due to the slow development of the distribution network, total factor productivity should increase as the distribution network expands.

Improved financial situation

Given the improvements in billing, collection and productivity and the large increases in tariffs, it is not surprising that the sector's financial situation improved dramatically following reform, despite a drop in subsidies. Whereas DEG was losing large amounts before reform, SEEG quickly became profitable (see Fig. 8.9). By 1996, SEEG's profits reached GF 3.2 billion ($3.2 million in 1996 dollars). The increase in SEEG's accounting profits after 1993 coincided with a decrease in the 'rental fee' that SEEG paid to SONEG. Although the average consumer rate remained at GF 880 between 1994 and 1996, the share of the tariff that SEEG paid to SONEG fell from GF 527 in 1994 to GF 370 in 1996.

[47] The difference in the two indicators is primarily due to the large increase in 'miscellaneous' revenues. Although water sales increased relatively slowly following reform, revenues from construction and other works related to the water sector increased more quickly. Consequently, output per employee continued to improve for several years after reform.

Note: Number of employees includes employees at both SEEG and SONEG.
Source: SEEG.

Figure 8.7. Indicators of labor productivity for SEEG and SONEG.

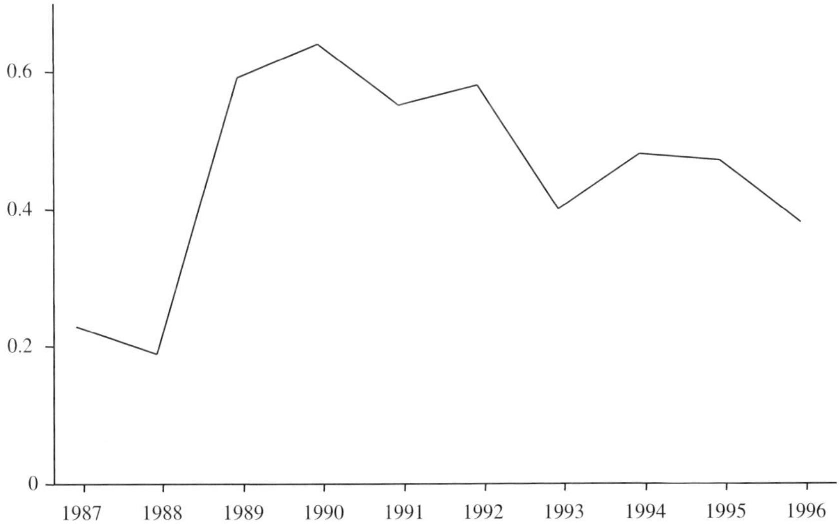

Note: Measures of TFP include inputs from both SEEG and SONEG.
Source: Authors' calculations.

Figure 8.8. Total factor productivity for SEEG and SONEG.

In 1989–1990, SONEG lost about \$2.8 million (see Fig. 8.9). After slowly improving through 1995, SONEG's operating profit became negative (−\$4.1 million) once more in 1996. The decline was due to the sharp drop in the 'rental fee' that SEEG paid to

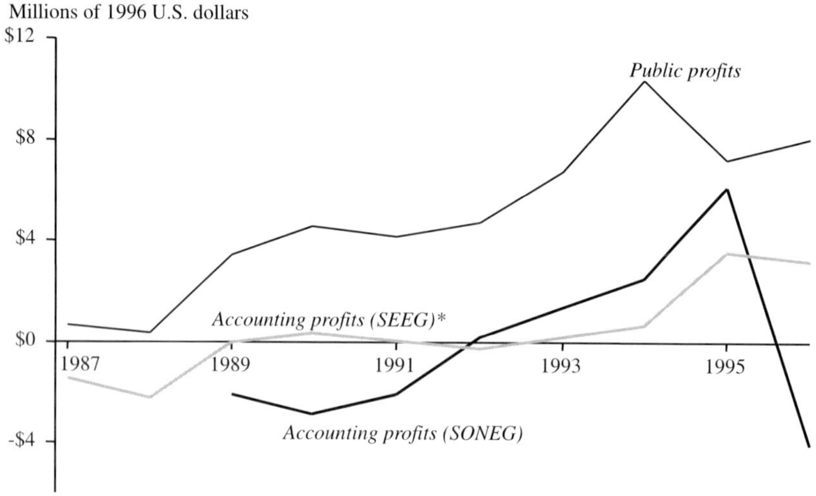

* Private profits for DEG before privatization.
Source: Authors' calculations.

Figure 8.9. Private and public profits.

SONEG and the end of the subsidy that the government paid SONEG for debt service (GF 88/m^3 in 1994). Although SONEG appeared profitable in 1995, this was due to a large increase in non-operating income (from GF 396,000 in 1994 to GF 9.6 billion in 1995). Without this, SONEG would have shown losses of close to $3.7 million in 1995. Fig. 8.9 also shows 'public profits' since reform, which include returns to debt-holders (i.e., interest payments) and the government (i.e., net taxes), and omits the opportunity cost of both SEEG's and SONEG's working capital.[48] Public profits, which were only slightly greater than zero under public ownership, increased following reform.

Remaining Challenges

Although sector performance improved along many dimensions, some problems remained. First, as noted in the previous section, many changes were modest, especially considering the poor initial performance of the sector. Second, other performance indicators, including unaccounted-for water (UFW) and public collection rates, failed to improve. Finally, prices increased dramatically, resulting in many complaints from consumers. As noted earlier, although it is difficult to compare prices across cases, tariffs are high in Guinea compared to the other countries in this study. In this subsection, we discuss these remaining challenges.

[48] Depreciation is also included to avoid the use of accountants' rates. See Jones et al. (1990) and Galal et al. (1994) for a full description.

Unaccounted-for water

By 1996, when 95% of connections were metered, UFW stood at about 48% (World Bank, 1998). Although due to the highly variable estimates of UFW before metering was complete it is difficult to say whether UFW improved, it remained high compared to either international standards (between 10 and 20%) or to other systems in West Africa (see Fig. 8.10). Several physical characteristics of the system contributed to this. First, in sections of the city where old pipes were buried only a few inches underground, it was easy to connect to the system illegally and pipes could be easily broken (e.g., by heavy vehicles). Second, overlapping lots and interlaced households made it difficult to interrupt water supply and control connections. Finally, poor maintenance and ancient infrastructure contributed to the problem.

However, institutional factors also contributed. Officials from SEEG noted that it was difficult to prosecute persons with illegal connections and, therefore, the company could do little to deter this behavior other than cutting illegal connections off. In addition, the contractual arrangement gave SEEG, which was responsible for a substantial share of maintenance, little reason to reduce UFW. Since the expansion of the Grande Chutes pipeline meant that potential production outstripped both consumption and actual production, SEEG could increase connections without reducing UFW. Moreover, SEEG paid the rental fee to SONEG based on bills collected, not on water produced and delivered by SONEG. Since SEEG did not pay for raw water and did not lose sales due to UFW, it had little incentive to take costly actions to reduce it.

Some experts argue that reducing UFW is not that important in Conakry, considering

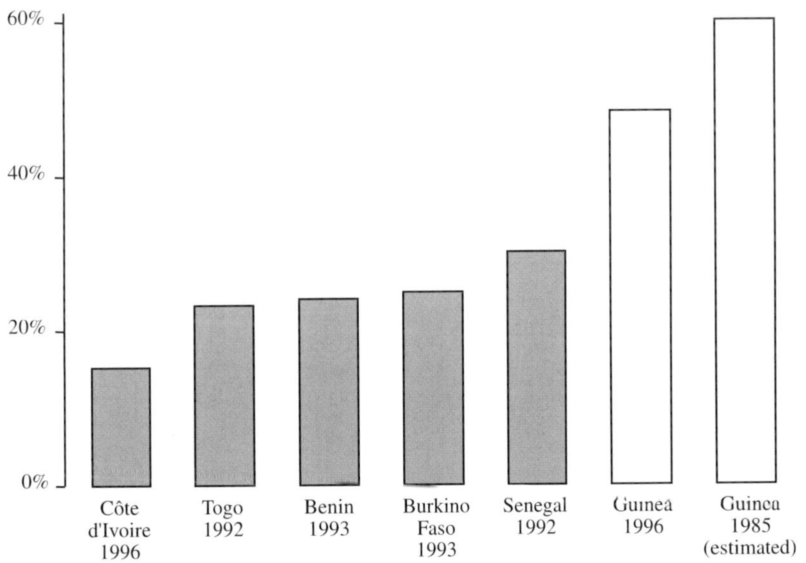

Source: World Bank files.

Figure 8.10. Unaccounted-for water.

the considerable water resources available and the high cost of a significant rehabilitation program. Although SEEG's accounts were not detailed enough for precise estimates, since the system is gravity-fed and most costs (e.g., provisioning and debt service) do not depend upon the quantity of water provided, short run marginal costs were probably quite low. For this reason, it was probably not efficient in strict economic terms to spend significant amounts to reduce UFW. Although with hindsight it might be argued that the productive capacity added under the Second Water Project could have been delayed if UFW had been reduced, this was not clear when the project was proposed. UFW was thought to be considerably lower before metering was complete and, therefore, consumption was thought to be close to productive capacity.[49]

Despite these arguments, the high rate of unaccounted for water might be costly in other ways. First, UFW can cause pollution and disease, especially when drainage is poor. Water obtained through illegal connections is of poor quality, especially when stored in unsanitary conditions, and leaky pipes can affect water quality through the entire system. Second, it is hard to convince people of the long run economic cost of water and, therefore, the necessity of paying for it when they know that large amounts are simply being wasted. Finally, the high rate of UFW reflected other problems in the system — most notably, the poor quality of the infrastructure and the inability, or unwillingness, of the population to pay for water.

Public collection rate

Another problem was the low collection rate from the public sector. Although the private operator was not to blame — the collection rate was low before reform and a private operator can do little to force an unwilling government to pay in a weak institutional environment — this was clearly a problem. For the first two years of the lease, under donor pressure, the government paid its bill regularly (see Fig. 8.6). However, in 1991 the government collection rate fell to less than 50%, and then dropped further, to close to 10%, in 1993. Since the government accounted for about 30% of sales — and more before 1996 — this was a significant problem for the private operator. Further, large fixed costs were passed on to a small number of private consumers who actually paid their bills, increasing prices significantly (see below).

The high price of water

The most common complaint during field interviews was that the price of water was too high.[50] In general, it is difficult to compare prices across countries. First, tariff schemes vary greatly across countries — for example, it is hard to compare prices in countries

[49] Before reform, productive capacity was about 54,000 m^3/day (World Bank, 1989). In 1996, by which time metering was complete, billed consumption was only about 30,255 m^3/day. This would have left some space for UFW and seasonal variation in consumption, which is quite low in Guinea, without increasing productive capacity.

[50] Similarly, the Organization for Consumer Protection wrote "living standards in Guinea make it impossible to pay the price charged by SEEG" (World Bank, 1998).

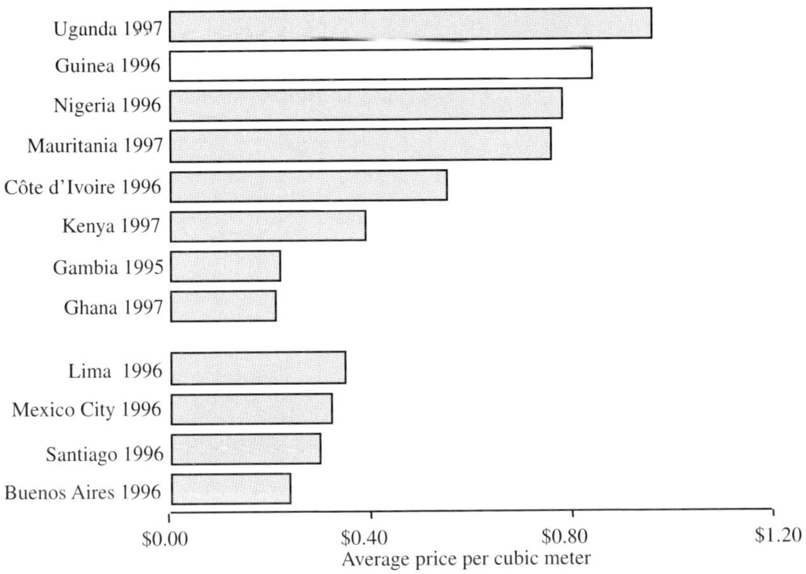

Source: Ménard and Shirley (2001) and World Bank files.

Figure 8.11. Average price of water billed in Africa and Latin America.

that are fully metered with those where customers pay lump-sum tariffs. Second, the cost of providing water also varies greatly between countries. For example, whereas the water system in Conakry is gravity-fed, water in Mexico City has to be pumped from a source that is 140 km away and 1000 m below the city. On the other hand, Conakry's system is far smaller and less dense than the systems in the other cases, increasing costs in Conakry. Finally, water systems are often heavily subsidized, often in nontransparent ways, making comparisons even more difficult. Unlike systems in many other developing countries, Conakry's system provides sufficient revenues to cover operations and maintenance and service sector debt. Finally, since prices are adjusted infrequently, and exchange rates in Africa are often unstable, it is difficult to compare prices in US dollars at a single point in time.

With these provisos, however, prices in Guinea appeared high, compared to prices in the other cases in this study (see Fig. 8.11). Further, prices in Guinea were higher than average, although not completely out-of-line with, other African countries (see Fig. 8.11). Finally, the tariff for low-income consumers in Guinea was high compared to tariffs for similar consumers in other countries. For example, although the average tariff in the mid-1990s was higher in Uganda than in Guinea ($0.96 vs. $0.84), the metered tariff for domestic users was lower in Uganda than the social tariff for low-income users in Guinea ($0.57 vs. $0.64).[51]

It is hard to assess why prices were higher in Guinea than in other African and Latin American countries. Fig. 8.12 shows the division of both revenues from water sales and

[51] Data for Uganda are for 1995 from Dinar and Subramanian (1997).

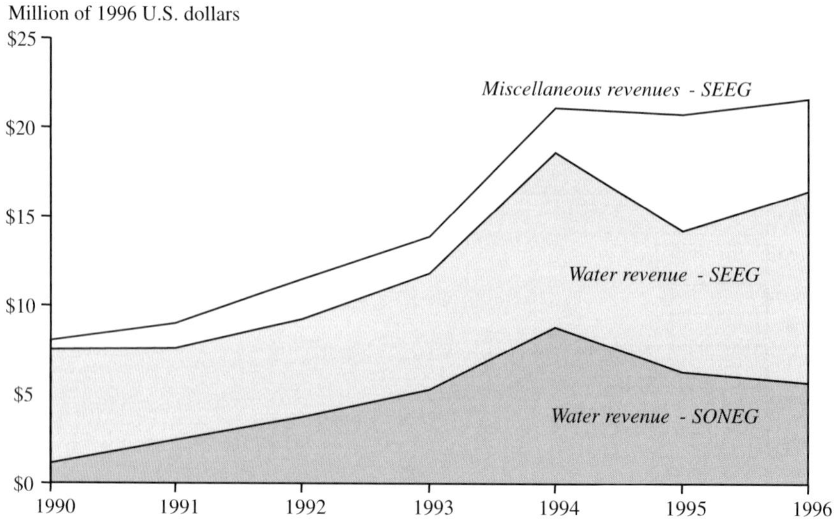

Note: Revenue from water sales includes connection charges.
Source: SONEG.

Figure 8.12. Revenues from water sales and 'other' sources.

total revenues for the post-reform period. SEEG received a greater share of both total revenues and water revenues than SONEG (about 65% of water revenues and about 74% of total revenues in 1996). Since SONEG financed most of its activities through grants and loans, this is not surprising (see Clarke et al., 2000).

Expenditures on wages and intermediate inputs for SEEG and SONEG combined consumed between 48% and 62% of total revenue between 1989 and 1996 (see Fig. 8.13). Since SEEG was far larger than SONEG (500 employees at SEEG compared to 44 at SONEG in 1997), expenditures on wages and intermediate inputs were primarily expenditures by SEEG. In 1996, SEEG accounted for 90% of total wage expenditures and 87% of expenditures on intermediate inputs. According to field interviews, the salaries for the five expatriate managers at SEEG accounted for about 15% of expenditures on wages and salaries in 1994–1995 (i.e., only about 1.6% of total revenues). Consequently, they were clearly not the main cause of the high price of water.

By 1996, interest payments and repayment of principle consumed, respectively, about 17% and 19% of revenues from water sales (see Fig. 8.14). We compare interest payments and the repayment of principle with revenues from water sales, rather than total revenues, because these payments were made by SONEG, which receives revenue only from water. Between 1990 and 1996, interest payments on sector debt increased from about \$2.0 to \$2.8 million (in 1996 US \$).[52] Since 1996, principal repayment on loans from the *Second Water Supply Project* has increased and new loans have been

[52] The estimate for 1990 includes debt service expenditures by the government, since it subsidized debt service payment at this time (see below).

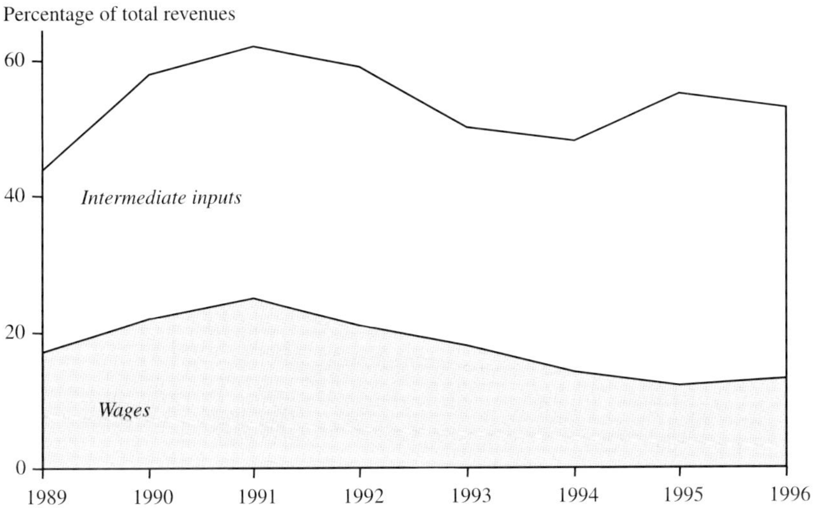

Note: Includes wages, intermediate inputs and revenues for SEEG and SONEG.
Source: SONEG and authors' calculations.

Figure 8.13. Wages and intermediate inputs.

taken out under the *Third Water Supply and Sanitation Project*. Consequently, interest payments and principal repayment were projected to increase significantly after 1996 — SONEG projected that interest and principal repayment would increase by 226% by 2005. Hence, without large increases in coverage or prices, the share of revenue consumed by debt service is likely to grow.

As discussed in the previous section, both public and private collection rates, although higher than they were before privatization, were consistently low. Therefore, provisions for bad debt consumed a large portion of revenue. In 1996, provisioning for bad loans cost SEEG and SONEG about $3.0 million (in 1996 US $) — about 18.4% of revenue from water sales (see Fig. 8.15). In comparison provisions for bad debt accounted for only about 2.8% of SODECI's pre-tax revenues in Côte d'Ivoire. Improving billing, through both improving government payment and making it easier to collect from nonpaying private customers, would allow substantial price reductions.

Given that the government accounted for a significant part of the nonpayment problem, one way to ensure lower prices would have been for the government to pay its water bill in a timely manner. Similarly, legal reform that would make it easier for SEEG to prosecute people who connect illegally or who refuse to pay their water bill would have also reduced financial pressure on the two companies. A bill to that effect, which was presented by the government under pressure from international donors, was killed by the parliament in 1996. This event illustrates the importance of political pressure on pricing and collection issues, even when the private sector is involved in the sector.

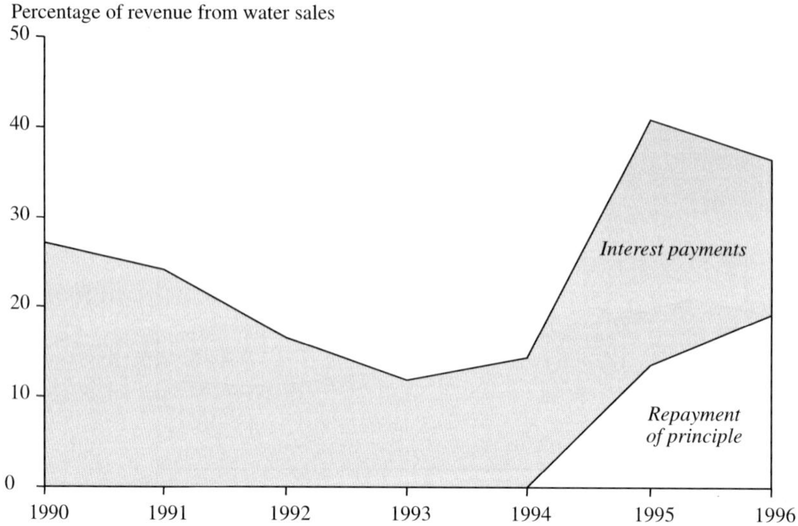

Note: Water sales include revenues from water sales and connection charges (i.e., excludes 'miscellaneous' revenues for SEEG. Between 1990 and 1995, interest payments include expenditures on debt service by the government.
Source: SONEG and authors' calculations.

Figure 8.14. Interest payments and debt service by SONEG.

WELFARE EFFECTS OF REFORM

This section presents a brief description of results from a cost–benefit analysis based upon the methodology in Jones et al. (1990) and Galal et al. (1994). In the analysis, we compare the actual performance of the sector following reform (the 'actual' scenario) with expected performance if the reform had not taken place (the 'counterfactual' scenario) for the ten years following reform.[53] Clarke et al. (2000) present a complete description of the welfare analysis, including data deficiencies, other problems and a sensitivity analysis.

We believe that the results in this section present a conservative estimate of the benefits of the reform for two reasons. First, we are not able to assess the potentially large benefits to consumers from improved service and water quality since it is difficult to value these gains. Second, the counterfactual relies heavily upon DEG's pre-privatization accounts. Since the unaudited accounts probably overstate DEG's success, we are likely to overestimate performance in the counterfactual scenario and, therefore, underestimate the gains from reform. For example, since we are using DEG's

[53] We analyze performance for ten years after the reform for comparability with the other case studies and because the original lease contract was only for ten years, providing a natural break for the analysis. Because of this, there is a two-year 'projection period' for which we use preliminary and estimated data (1997 and 1998). In practice, the discount factors applied after 1998 would make future years relatively unimportant. The projection period is described in Clarke et al. (2000), which also presents results making alternative assumptions about projections.

Note: Includes revenues and provisions by both SEEG and SONEG. Revenues from water sales include connection charges (i.e., they exclude 'miscellaneous' revenues).
Source: SONEG and authors' calculations.

Figure 8.15. Provisions for bad debts.

profit and loss statement, which records transactions as billed, we implicitly assume that DEG would have actually collected (or provisioned for) the amount billed. However, DEG had a poor collection record and provisioning appears inadequate. Since we rely upon these figures to construct the counterfactual, and because we omit the gains from improved quality, the estimated gains from reform should be seen as a lower bound on its likely effect.

We estimate consumer surplus for three categories of demand: water supplied to private consumers, water supplied to the government, and connection fees.[54] Although, under ideal circumstances, we would like a finer breakdown for demand, this is impossible because neither revenue nor volume of water billed was broken down more finely. Since neither SEEG nor DEG was responsible for sewage disposal, we do not include sewage disposal in this partial equilibrium analysis. In the analysis, we assume a high price elasticity of demand (−0.6), although, in practice, the results are similar for a large range of assumptions about demand and price elasticities (see Clarke et al., 2000). Following Galal et al. (1994), we subtract connections fees from consumer surplus for water

[54] For a final category of 'miscellaneous' revenues, we do not calculate consumer surplus. Before 1989, 'miscellaneous' revenue is 'works invoiced' in DEG's accounts. After 1989, most 'miscellaneous' revenue probably comes from the construction of water infrastructure performed by SEEG on behalf of SONEG. If SEEG and SONEG were a single company, the company would not receive revenue from 'construction' (other than for the connection fee) and, therefore, this would not appear as separate source of revenue. For comparability with the counterfactual, therefore, we do not calculate consumer surplus for this category. Further, since this is a 'catch-all' category, it would be difficult to make assumptions regarding excess demand and price elasticity of demand.

consumption. The implicit assumption is that consumers gain utility from water usage, not from the connection itself. Hence, potential consumers compare the net present value of consumer surplus from usage with the connection fee and the stream of tariff payments.

Differences between Actual and Counterfactual Scenarios

In this case study, the counterfactual scenario differs from the actual scenario in the following ways. First, we assume that the World Bank-supported *Second Water Project* would not have been approved and that, therefore, investment would have remained low. Second, we assume that labor productivity would remain poor. Third, we assume that the large price increases that followed reform would not have been implemented. Finally, given the above assumptions, we assume that the government would have increased subsidies to cover DEG's operating losses on a cash-flow basis. This final assumption is necessary given the second and third assumptions.

Investment

The assumption that the World Bank would not have approved the *Second Water Project* without some degree of private sector participation is supported by World Bank documents and discussions with World Bank staff involved in the project. Since most investment was supported through donor funding, we assume, therefore, that DEG would only perform sufficient investment to maintain assets (i.e., that investment would be equal to real depreciation). Because of this, there is no output growth in the counterfactual scenario. This is consistent with two observations. First, since average daily production in Conakry was about 45,000 m^3/day and maximum average production was estimated to be 54,000 m^3, little system expansion would be possible in Conakry without upgrading productive capacity.[55] However, the expansion of the production capacity would not have been possible at the observed level of investment in the late 1980s.[56] Second, between the end of the First Water Project (1984) and the start of the Second Water Project (1989), there was little or no system expansion.[57] Since assumed investment under the counterfactual is slightly lower than the observed average investment in 1987 and 1988, the assumption of no expansion seems generous.[58]

---

[55] Pre-reform data are for 1984 — the last year before reform that reliable data are available (World Bank, 1987). Maximum capacity was 60,000 m^3/day, but this could not be maintained as an average level of production (World Bank, 1989).

[56] World Bank (1989) reports that the estimated cost of the transmission pipes from Grandes Chutes to Conakry and the additional treatment plant to treat the water was US $38.8 million (of US $57.9 million in total investments in Conakry). However, actual costs were probably higher since total expenditures on planned investments in Conakry were US $84.9 million (rather than US $57.9 million). Total investment by DEG averaged $4.7 million in 1987 and 1988 and much of this would be required for maintenance, etc.

[57] Between 1984 and 1988, in Conakry, registered connections fell from 11,167 to about 10,200 and active standpipes fell from 112 to 40, while (nationally) estimated consumption fell from 12.83 million m^3/year to 10.89 million m^3/year (World Bank, 1987, 1989).

[58] This is also consistent with assumptions in World Bank documents about sector growth if the Second Water Supply Project had not occurred.

Productivity

In the welfare analysis, we assume that the improved productivity observed following reform was a result of the reform. This seems reasonable, since there is little evidence of increased productivity before reform. Although poor data quality makes it difficult to assess factor productivity accurately prior to 1989, estimated consumption and the number of connections were falling and the number of staff was steady in the mid-1980s, suggesting labor productivity was falling.

Prices

We assume that the large price increases observed following reform were contingent on reform. This is reasonable for several reasons. First, given DEG's poor collection performance when prices were low, it seems unlikely that DEG would have been able to collect billed amounts if prices were raised. Second, it seems unlikely that consumers would have borne the price increases without corresponding improvements in quality, which would have required additional investment. Finally, in the actual scenario, the private operator has an incentive to push the government and SONEG for price increases. This motivation would have been missing under public ownership. In fact, most increases in the 1970s and 1980s appear to have been implemented only under donor pressure. Consequently, under the counterfactual scenario of continued public ownership, with no additional donor funds for investments, we assume that prices would have stayed constant in real terms at the 1988 tariff rate.

Subsidies

Before reform, even after the large subsidies DEG received from the government, DEG was losing large amounts of money. These large losses would have been unsustainable without additional borrowing or an infusion of capital from the government (i.e., an increase in the subsidy). DEG's financial position was too weak to borrow significant amounts from private capital markets and donors such as the World Bank generally do not lend money to finance operating losses. Therefore, we assume that the government would have increased the subsidy to make the company's cash flow close to zero. In the counterfactual scenario, therefore, subsidies increase from $3.15 million in 1988 to $4.4 million in 1996 (in 1996 US $). This assumption means that subsidies are slightly lower in the counterfactual scenario than in the actual scenario through 1992.[59] After 1992, subsidies stabilize at about $4.4 million in the counterfactual scenario, whereas they drop to zero in the actual scenario.[60]

[59] As noted previously, the government subsidized sector operations in the actual scenario by paying a declining share of debt service and using proceeds from a World Bank loan to subsidize part of the tariff for five years following reform.

[60] The level of subsidies envisioned under the counterfactual scenario would have been feasible for the government in the medium term. In 1987, the subsidy was equivalent to about 1% of government consumption. Under the counterfactual scenario, this subsidy increases to only 1.6% of government consumption by 1996.

Table 8.1

Winners and losers from reform in first ten years of reform

	Total gain (millions 1996 US dollars)	*Per capita gain (1996 US dollars)*	*Gain (as percent of 1988 output)*
Total	33.2	6.12	126.6
Total domestic	29.3	5.41	111.9
Government	9.8	1.81	37.4
Fiscal effect	17.6	3.25	67.2
Government consumer surplus	−7.8	−1.44	−29.8
Consumers	19.5	3.60	74.5
Foreign buyers	3.9	0.71	14.7

Source: Authors' calculations.

Welfare Impact

Table 8.1 presents results from the welfare analysis. Under the scenario described above, the total welfare gain was over $33 million (in 1996 dollars), with most of this accruing to domestic parties. Even ignoring the presumably large gains due to improvements in service and water quality, private consumers benefited considerably from reform. Although the large price increase might have reduced the utility of connected customers, especially those disconnected for nonpayment, this appears to have been more than offset by gains accruing to new consumers.

The government benefited considerably in fiscal terms, although the increased price of water meant that the government lost consumer surplus. The reason that the government lost consumer surplus, while private consumers gained is that private consumption increased considerably between 1989 and 1996, while government consumption dropped. That is, since government consumption was not rationed before reform, government consumers did not benefit from expanded coverage. It is important to note that the loss in consumer surplus to the government is calculated based upon the assumption that the government pays its water bill in time and in full. Since this has not been the case, the 'true' price of water to government is likely to be less than the assumed price in this study. Foreign buyers benefited modestly from reform, although their gain was small compared to the gain that accrued to consumers and to the government.

In Clarke et al. (2000), we present additional results making other assumptions in the actual and counterfactual scenarios (e.g., about elasticities and excess demand). In practice, the results were not highly sensitive to the different assumptions and were similar to the results here, at least qualitatively.

CONCLUSION

Our cost–benefit analysis suggests that overall both consumers and the government gained from reform. The most notable achievement is that government subsidies were sharply reduced, with revenues covering the operational expenditures of both SEEG and SONEG and, for the most part, debt service. However, consumers also benefited. Although the increase in the number of connections was slower than anticipated and, therefore, coverage remained low, it increased faster than it would have under continued public ownership. Further, the quality of both water and service improved considerably — safe drinking water is now available 24 hours a day, 7 days a week. In summary, there is little doubt that the situation was better than it would have been under any reasonable counterfactual with continued public ownership.

However, there were some problems — water became expensive, the number of connections increased slowly, and conflicts developed between SEEG and SONEG. Our analysis strongly suggests that the fundamental cause of these problems is *the lack of strong and stable institutions*, the effect of which is amplified by a set of contracts *without adequate mechanisms for solving disputes among the parties involved.* Although progress was made towards strengthening parliament and the judiciary, there were few effective checks and balances on the executive. In particular, the judiciary lacked expertise and constitutional support, particularly in implementing contracts and protecting property rights. Consequently, there was no independent authority to monitor the lease contract effectively or to make sure that the partners met their goals and obligations. Because of this, the private operator did not have adequate incentives, disputes were not resolved, and regulation was weak. In what follows, we summarize the main problems and we indicate suggestions for improving the contractual arrangement.

High Prices

As noted in Ménard and Shirley (see Chapter 2), prices in Guinea were far higher than in any of the other case studies in this project as well as in neighboring countries, notwithstanding abundant water resources. There were several possible reasons for this. First, as noted by several experts during field interviews, the system was smaller than the other systems, meaning that it might not have benefited from economics of scale. Moreover, because connections were spread over a large area, the network was not very dense, increasing the capital cost of providing service. However, this was not the only reason for high prices — operating costs were also higher than in the other case studies.[61] This is somewhat surprising since the system was gravity-fed and raw water was readily available.

Another factor that led to high prices was that although SEEG improved collection and billing, the improvements were smaller than originally hoped. Collection rates for both the administration and the private sector were low and the high rate of UFW suggested that theft and illegal connections were still significant problems. In 1996, less than 33% of the water that was produced was actually paid for, making higher prices necessary for those consumers who did pay. The most important action that the govern-

[61] Operating costs per connection and per cubic meter of water produced are both higher than in the other case studies.

ment could take would be to pay its own water bill. However, the government could also encourage parliament to pass the measure allowing collection from non-payers — the earlier rejection of this bill almost certainly contributed to the low collection rate.

The final cause was weak regulation. World Bank (1998) reports that audits revealed that misapplication of the formulas that adjust prices in response to cost changes resulted in overvalued tariffs. Because of this, and the informal price negotiations between SEEG and the government, the private operator received far greater compensation than originally anticipated.[62] Although this was due, in part, to a lack of administrative capacity at SONEG, SEEG failed to provide the information to SONEG that is required under the contract. This suggests that steps should have been taken to strengthen SONEG and to increase SEEG's compliance with reporting requirements. However, as discussed below, other steps to reduce information asymmetries may have also been helpful.

Low Connection Rate

The second problem, the slow expansion of the system, is the result of both the high price and institutional problems. World Bank (1997a) reports that although 75% of households expressed a strong interest in individual connections, only 10% of households were willing to pay the full cost of being connected. Further, most were not willing to pay the average consumer rate of GF 880/m^3. However, if the connection fee was subsidized, so that the price was GF 40,000 (about $37 in 1997) and they paid the social tariff (GF 680/m^3), most households indicated that they would be willing to connect. This suggests that a more vigorous 'social connection' program, similar to the program in Côte d'Ivoire, could have successfully increased the connection rate. Consistent with this, support for 'social connections' by bilateral donors appears to have successfully increased the number of connections in 1997 and 1998. Between 1989 and 1996, only about 10,000 connections were added nationally. In comparison, nearly 8000 connections were added in 1997 alone.

However, this was not the only reason for the slow system expansion. In addition, disagreements between SEEG and SONEG made it difficult for the two parties to coordinate their investment plans. In the absence of a credible dispute resolution mechanism, these disagreements seriously damaged the relationship between the two entities and took a significant amount of time and effort to resolve. Although several apparently useful clauses intended to reduce this were included in the contract, in the absence of a credible body to interpret and enforce them, the clauses were almost impossible to implement.

Coordination and Disputes

Several steps could have been taken to reduce coordination problems and resulting disagreements between SONEG and SEEG. Many local officials believed that imposing

[62] In Guinea, the lease contractor rate — the part of the tariff paid to SEEG — was supposed to be 214 GF/m^3 (in current GF) in 1993. In practice, it turned out to be 448 GF/m^3.

strict deadlines on SONEG regarding investment could increase expansion without giving too much discretionary power to SEEG. However, given the weak administrative capacity and institutions in Guinea, it is unclear whether these rules could have been enforced. In fact, because transaction costs were lower for the private operator, in many cases, it may have been economically and socially reasonable to give greater discretion to SEEG regarding investment. This would also have improved SEEG's incentives regarding system expansion. However, it could also have increased SEEG's ability to capture rents. Increasing expertise at SONEG would have reduced this risk.

The private operator could have been motivated to take greater initiative if its responsibilities were increased. For example, it should have been responsible for day-to-day investment decisions for both the primary and secondary networks. Simultaneously, its autonomy should have been increased to allow it to borrow more easily. Expected results could have been specified in the contract, for example, by fixing precise goals for connections. By taking daily investment decision away from it, SONEG would have been able to focus on long-term planning and on monitoring SEEG. For such a solution to work efficiently, SONEG would have needed greater autonomy and expertise. Taking into consideration the institutional endowment of Guinea, this would have likely involved greater commitment from international donors. Although a concession may not be possible in Guinea, moving towards a system similar to the one in Côte d'Ivoire (see Chapter 7), where most investment decisions are made by a private operator, would probably improve sector performance.

Other immediate steps could also have been taken. In particular, one cause of the problems noted above was that different ministries with divergent views interfered with both SEEG and SONEG, creating confusion and resulting in arbitrary decision-making. At the very least, SEEG and SONEG should have had boards that were independent from short-term government policy and from each other. Further, SONEG, which was a public agency, should have depended on only one ministry or regulating agency.

Steps could also have been taken to reduce information asymmetries. The traditional way of doing this is through using *bidding* procedures to allocate contracts or introducing *yardstick competition*. However, in Guinea, these devices are likely to be only partially effective. Although bidding was used in 1989, only two consortia actually submitted bids. Given the lack of an independent judiciary able to enforce contracts and the dearth of information on the actual situation — about all that was known was that the system was collapsing — this was not surprising. This lack of interest severely limits the government's capacity to pressure the winner with subsequent rounds of bidding. Further, two of the three international companies with a significant presence in Africa — SAUR and Vivendi (Générale des Eaux) — were already partners in SEEG.

Another problem with bidding is that the weak institutional environment in Guinea makes informal renegotiation of contract provisions possible. This severely reduces the information that bidding provides (i.e., the enterprises participating in the bidding have no reason to submit bids based upon estimates of the cost of producing water since they know they can renegotiate the price afterwards). This has been a constant problem in the water sector and particularly in cases (e.g., Guinea and Argentina) in which bidding was conducted over prices.

Yardstick competition is almost impossible, since Conakry is far larger than any other city in Guinea, making comparisons difficult. Unbundling the system geographically (i.e., dividing Conakry into several areas managed by different companies) could allow a regulator to compare some aspects of the performance of different companies. This solution was adopted in Mexico, with unconvincing results so far (see Chapter 5) and, more successfully, in cities like Paris. Although, this deserves serious consideration in Guinea, the small system and the relatively low number of international companies likely to bid on these contracts would make it difficult to implement efficiently.

In conclusion, although the lease was successful in some ways, the actual improvements were not as large as it was hoped they would be. The main problems were weak administrative capacity and the lack of an independent agency capable of restraining arbitrary government action, regulating the private operator and enforcing the contractual arrangement. Although these problems ultimately require long-term solutions, we have outlined several steps that could be taken more immediately. Among these, building expertise at the micro-institutional level (e.g., within SONEG) is likely the key to rapid improvement.

REFERENCES

Arulpragasam, Jeham and David E. Sahn, 1997. *Economic Transition in Guinea: Implications for Growth and Poverty*. New York, NY: New York University Press.

Azarya, Victor and Naomi Chazan, 1987. "Disengagement from the state in Africa: reflections on the experience of Ghana and Guinea". *Comparative Studies in Society and History* 29(1): 106–131.

Brook Cowen, Penelope J., 1996. "The Guinea water lease — five years on: lesson in private sector participation". *Public Policy for the Private Sector* 78: 1–4.

Castilia, 1993. *Renégociation du Tarif Exploitant*. Paris: Castilia.

Clapp, Jennifer A., 1994. "Explaining policy reform implementation in Guinea: the role of both internal and external factors". *Journal of International Development* 6(3): 307–326.

Clapp, Jennifer A., 1996. *Adjustment and Agriculture in Africa: Farmers, the State and the World Bank in Guinea*. New York, NY: St Martin's Press.

Clarke, George R.G., Claude Ménard and Ana Maria Zuluaga, 2000. The welfare effects of private sector participation in urban water supply in Guinea. Processed.

De Merode, Louis (with Charles S. Thomas), 1994. "Implementing civil service pay and employment reform in Africa: the experience of Ghana, Gambia, and Guinea". In: David L. Lindauer and Barbara Nunberg, eds., *Rehabilitating Government: Pay and Employment Reform in Africa*. Washington, DC: World Bank.

Dinar, Ariel and Ashok Subramanian, 1997. *Water Pricing Experiences: An International Perspective*. Technical Paper No. 386, Washington, DC: World Bank.

Durany, Jocelyne and Alain Morel à l'Huissier, 1994. *The Urban Environment in Conakry: Behavior, Attitudes and Practices of Households*. Abidjan: UNDP–World Bank Water and Sanitation Project.

Economist Intelligence Unit. Various years. Country Report: Guinea, Sierra Leone, Liberia. London: Economist Intelligence Unit.

Feachem, R.G., 1980. "Bacterial standards for drinking water quality in developing countries". *The Lancet* August 2: 225–256.

Galal, Ahmed, Leroy Jones, Pankaj Tandon and Ingo Vogelsang, 1994. *Welfare Consequences of Selling Public Enterprises: An Empirical Analysis*. New York, NY: Oxford University Press.

Gélinas, Yves, Harold Randall, Lucie Robidoux and Jean-Pierre Schmidt, 1996. "Well water in two districts of Conakry (Republic of Guinea), and comparisons with the piped city water". *Water Resources* 30(9): 2017–2026.

Jones, Leroy P., Pankaj Tandon and Ingo Vogelsang, 1990. *Selling Public Enterprises: A Cost Benefit Methodology*. Cambridge, MA: MIT Press.

Marston, Lance, Peter A. Thomas and Andrea J. Love, 1986. *A Review of Privatization in the Republic of Guinea/Conakry. Report prepared for Bureau for Private Enterprise, US Agency for International Development.* New York, NY: Center for Privatization.
Mills, Bradford F. and David E. Sahn, 1997. "Labor market segmentation and the implications for public sector retrenchment programs". *Journal of Comparative Economics* 25(3): 385–402.
Republic of Guinea, 1987. *National Recovery Program. Medium-term Development Prospects, 1987–1991.* Conakry, Guinea: Consultative Group Meeting.
SODECI, 1986. *Rapport Annuel.* Abidjan: Société de Distribution de la Côte d'Ivoire.
SODECI, 1986. *Rapport Annuel.* Abidjan: Société de Distribution de la Côte d'Ivoire.
SODECI, 1996. *Rapport Annuel.* Abidjan, Côte d'Ivoire: Société de Distribution de la Côte d'Ivoire.
Triche, Thelma A., 1990. *Private Participation in the Delivery of Guinea's Water Supply Services.* Policy Research Working Paper 477, Washington, DC: World Bank.
United Nations, 1996. *World Urbanization Prospects.* United Nations Population Studies No. 170, New York, NY: United Nations.
World Bank, 1978. *Staff Appraisal Report: Conakry First Water Supply and Sanitation Project.* Washington, DC: World Bank.
World Bank, 1986. *Report and Recommendation of the President. Guinea. Structural Adjustment Program.* Washington, DC: World Bank.
World Bank, 1987. *Project Completion Report: Conakry Water Supply and Sanitation Project.* Washington, DC: World Bank.
World Bank, 1989. *Staff Appraisal Report: Second Water Supply Project.* Washington, DC: World Bank.
World Bank, 1990a. *Program Performance Audit Report. Guinea. First Structural Adjustment Credit.* Washington, DC: World Bank.
World Bank, 1990b. *Republic of Guinea: Country Economic Memorandum. Volume I.* Washington, DC: World Bank.
World Bank, 1990c. *Republic of Guinea: Country Economic Memorandum. Volume II: Sectoral Analysis.* Washington, DC: World Bank.
World Bank, 1990d. *World Development Report 1990: Poverty.* Oxford: Oxford University Press.
World Bank, 1994. *Adjustment in Africa: Reforms, Results and the Road Ahead.* Oxford: Oxford University Press.
World Bank, 1997a. *Staff Appraisal Report: Third Water Supply and Sanitation Project.* Washington, DC: World Bank.
World Bank, 1997b. *World Development Report 1997: The State in a Changing World.* Oxford: Oxford University Press.
World Bank, 1998. *Implementation Completion Report. Second Water Supply Project.* Washington, DC: World Bank.
Young, Crawford, 1982. *Ideology and Development in Africa.* New Haven, CT: Yale University Press.

Thirsting for Efficiency: The Economics and
Politics of Urban Water System Reform
Mary M. Shirley (Ed.)

CHAPTER 9

PROSPECTS FOR PRIVATE WATER PROVISION IN DEVELOPING COUNTRIES: LESSONS FROM 19TH-CENTURY AMERICA

KEITH J. CROCKER AND SCOTT E. MASTEN

INTRODUCTION

Spurred by dramatic changes in generation and transmission technologies, regulatory policymakers and academics have focused attention on the structure and regulation of the telecommunications and electric power industries. These utility sectors have been the source of both significant efficiency enhancements and a range of new and sometimes glamorous products. Whether in response to or as a source of those changes, privatization (where state ownership had existed) and regulatory reform (where it had not) have become commonplace.

The tidal wave of interest in privatization and deregulation that has swept the telecommunications and electric power sectors has largely bypassed the provision of water and sanitation services, however. With a few notable exceptions, the ownership and operation of water systems remain, in both developed and developing economies, the domain of municipal and state enterprises. While the lack of agitation for water privatization in developed economies may indicate satisfaction with public provision, the same inference could not be drawn outside the developed world. According to World Bank estimates, close to two billion people still lack either safe drinking water or adequate sanitation (Briscoe and Garn, 1990). Problems of waste, insufficient quality, misdirected investment, political interference, and corruption plague the water systems of many developing nations.

Successes in mitigating such problems through private sector participation in the telecommunications and power sectors, and in some early experiments with water supply, raise the issue of whether privatization might offer a solution to problems of water supply in developing countries. Several questions need to be answered, however, before such a recommendation could be endorsed. Are the tradeoffs between public and private provision the same for water as for other utilities? If so, why does public ownership and operation dominate water supply in developed economies where other utilities have typically remained private? If not, what attributes of water supply distinguish it from utility services for which private supply has succeeded? Finally, are there conditions or circumstances — economic or institutional — that might make private provision desirable in developing countries even if public provision retains advantages in more developed environments?

In this chapter we look for insights into these questions by reviewing the experience with private water supply of one particular developing country: the 19th-century United States. Like much of the rest of the world, most water in the United States is currently supplied through publicly owned and operated water works. Approximately 80% of U.S. water supply firms were publicly operated in 1970 (Crain and Zardkoohi, 1978; Feigenbaum and Teeples, 1983); in 1987, only 13% of municipalities contracted for their water supplies (Nelson, 1997).[1] The predominance of the public sector in water supply was not always the case, however. Private water firms dominated U.S. water supply for most of the 19th century. Of the sixteen waterworks in operation in the United States in 1800, only one (in Winchester, Virginia) was publicly owned (Baker, 1899, p. 14). As late as 1890, the majority (56%) of the 1878 waterworks then in existence remained private (Baker, 1899, p. 16). Even as public ownership rose, private water companies maintained a foothold in major urban markets into the twentieth century. In 1917, "five important water companies that enjoy public franchises [were] still operating within the limits of greater New York. Indeed, about one fourth of the city's area, and 400,000 of its population, [remained] dependent directly upon the private companies for both domestic service and public fire protection" (Wilcox, 1917, p. 550).

The trend, in New York as elsewhere in the country, was clearly toward municipalization of water service, however. As of 1917, "no water franchise [had] been granted within the ... limits of Greater New York during the [preceding] twenty years" (Wilcox, 1917, p. 558). By the turn of the century, the establishment of new municipally owned waterworks, combined with the conversion of private firms to public control, had tilted the balance to public ownership nationally (Baker, 1899, p. 26).

The shift to public ownership of waterworks was decidedly *not* part of a general trend. American cities throughout the 19th century relied almost exclusively on private capital for initial development of and subsequent investment in all public utility services *except* water and sanitation (Priest, 1993, p. 305; Tarr, 1996, p. 118). At the end of the century, when the majority of waterworks and virtually all sewer systems were municipally owned and operated, ownership of other utilities remained overwhelmingly private. In 1890, only 8% of electric power plants and fewer than 1% of street railways were publicly owned (Bemis, 1899, pp. 505, 565, as reported in Priest, 1993, p. 317). "In street railways ... gas works ... electric light and power plants, private ownership [was] preponderant ... [and in] telephones private ownership [was] universal, or practically so" (Wilcox, 1931, p. 5). By contrast, "In water works public ownership [was] predominant ... In sewerage

[1] In 1958, the figure for private ownership of water utilities was approximately 30% (Anderson, 1988, p. 144).

and drainage systems, public ownership [was] so nearly universal as to remove this utility from the realm of controversy respecting public ownership" (ibid, pp. 4–5).

Why water and sanitation services should have been so much more likely than other utilities to have been publicly owned and operated is far from obvious. Noting the disparity in ownership structures, Delos Wilcox, the most prominent expert on public utilities of the late 19th and early 20th centuries, attributed the acceptance of municipal water works — even by "opponents of general municipal ownership of public utilities" — "partly to the nature of the utility and partly to the fact that the policy of municipal ownership of water works has become well established in this country" (Wilcox, 1910, p. 398). What the features of water were that might explain the acceptance of public ownership of waterworks in the first place, however, Wilcox was unable to say. Eighty years later, a resolution of the puzzle appeared no closer. Priest (1993, p. 314), referring to "the aberrant case of waterworks" concluded, "There is no clear explanation for why, of the variety of public utilities, water supply was the most likely to be publicly owned and managed" (p. 317).

We begin our inquiry into the determinants of waterworks ownership with a brief overview of the regulation-versus-franchise contracting debate and the principles of governance choice that emerged alongside it, followed in the third section by a comparison of this modern conception of the problem of public utility governance with the perspectives and concerns of 19th-century observers. The fourth section describes in more detail the 'puzzle' posed by the municipal ownership of water and sanitation systems, while the fifth section describes a pair of potential 'solutions' to that puzzle, one relating the observed affinity for public ownership of waterworks to differentially large contracting hazards and the other to low costs of public administration for water and sanitation services relative to other utilities. Specifically, we argue that distinctive features of water and sewer services, such as their low marginal service costs, relative ease of resale, and associated public health and nuisance externalities, exacerbate appropriation and adaptation problems for these services. We contrast this with the alternative hypothesis that the relative simplicity of water and sewer systems operations mitigated the incentive and specialization disadvantages commonly attributed to public ownership. We evaluate these hypotheses relative, first, to historical accounts of the attributes of and problems associated with 19th-century water and sewer systems and, second, to variations in the ownership of waterworks over time and among cities. We conclude with a discussion of the policy lessons of our analysis for the privatization of water supply in modern developing economies.

TRANSACTION-COST ECONOMICS AND THE ADMINISTERED CONTRACTS FRAMEWORK

The framework for analyzing public utility governance applied here had its origins in the regulation-versus-franchise bidding debate of the late 1960s and early 1970s. The conventional wisdom at the time was that public utility regulation was a necessary response to natural monopoly cost conditions. According to the standard argument, (i) productive efficiency in an industry with declining long-run average costs requires that a single firm serve the market, but (ii) a sole supplier will restrict output to the monopoly level, generating a deadweight loss. Efficiency can thus only be achieved in such settings if government restricts entry and regulates the price the monopolist can charge to the public.

Harold Demsetz exposed the fallacy of the natural monopoly justification for regulation in his classic article, "Why Regulate Utilities?" (1968). Specifically, Demsetz argued that public utility services could, in principle, be efficiently procured — even where cost conditions dictated a single supplier — simply by awarding a franchise to the firm that offered to serve the market at the lowest price. As long as sufficient ex ante competition existed, the price would be driven to competitive levels, eliminating the motive for regulation.

Demsetz' argument demonstrated that technological conditions dictated *only* that a single supplier serve the market; whether the relation between that supplier and consumers was governed by regulation or a more market-like arrangement was a question to which the traditional theory of natural monopoly and regulation offered no definitive answer. In light of the numerous and well-documented deficiencies of public utility regulation (see, for instance, Posner, 1969), however, the prospect of a 'market solution' to the natural-monopoly problem made franchise contracting an appealing alternative. Indeed, Richard Posner judged franchise contracting so attractive that he raised Demsetz' theoretical critique to a policy prescription, proposing that "municipal authorities ... be permitted — and encouraged — to auction off franchises to the firm promising to provide cable service at the lowest rates to subscribers" (Posner, 1972, p. 126).

To Oliver Williamson and Victor Goldberg, Posner's enthusiasm for franchise contracting appeared premature. In separate but parallel treatments, Williamson (1976) and Goldberg (1976) noted that bidding and contracting problems in the types of environments typically encountered in public utility transactions were likely to be severe. First, although competitive bidding can be an effective way of determining the lowest cost supplier where the price of the good or service being procured is the buyer's only concern, competitive bidding works less well for complex goods and services where the buyer cares about quality, reliability, and other attributes of the procurement. Second, enticing private investment is likely to require long-term contracts; without contractual assurances against ex post appropriation of and conflict over the resulting quasi-rents, potential suppliers will be reluctant to make investments in the durable and non-redeployable production and distribution facilities typically necessary to supply public utility services. Allowing for and effecting adjustments to changing circumstances contractually over such long horizons, however, are likely to entail contract terms and administrative machinery — cost-plus price adjustment, auditing procedures, elaborate and formal dispute resolution processes — that mirror both in character and in costs the administrative apparatus traditionally associated with rate-of-return regulation. As Goldberg put it: "The problems are intrinsic to the service, not to the act of regulation itself" (1976, p. 441). In the end, the choice between franchise contracting and regulation turns on whether court enforcement or regulator administration is the more effective way of bolstering commitment, effecting adaptations, and resolving disputes in the settings that characterize the supply of public utility services.

Though developed in the context of franchise contracting versus regulation, the insights and principles of the administered contracts framework readily extend to the choice between private and public ownership of public utilities. In particular, to the extent that private investments in public utilities remain susceptible to appropriation regardless of whether they are governed by franchise contract or regulation, governments will find it hard to attract private investment in public utilities (see Levy and Spiller, 1994). Even if integration sacrifices the incentive and specialization benefits of private ownership, public ownership may nevertheless be the least cost means of governing utility transactions and investments where the nature of the transaction or of the legal and regulatory systems leave investments vulnerable to governmental appropriation.[2]

19TH CENTURY PERSPECTIVES

The 19th century saw the introduction of many of the utility services we enjoy today — water, sewerage, electricity, telephone, public transportation — plus a few, such as manufactured gas, that have since become obsolete. By the end of the 19th century, the ownership of public utilities had become a major public policy issue, and the development and governance of these systems were of considerable interest to the public.[3] As a result, the historical record contains numerous descriptions of and commentaries on the nature, problems, and organization of public utilities. Two writers in particular, Delos F. Wilcox and Moses N. Baker, provided detailed discussions and accounts of the history and experience of public utility governance in the United States. These and other 19th-century accounts reveal a perception of the tradeoffs inherent in the choice between private and public ownership remarkably congruent with the concerns of the administered contracts framework.

Chief among the advantages attributed to private ownership by turn-of-the-century experts were (i) stronger incentives of private owners to operate utilities efficiently, and (ii) a reduced susceptibility to political influence and bureaucratic encumbrances. According to Baker, "The greatest argument for private ownership, and one that in a measure includes all others, is that under it every endeavor is made to conduct the works in the most prudent and economical manner, in order to insure the greatest possible return on the capital" (Baker, 1899, p. 40). Similarly, Wilcox included at the top of

[2] For a more extensive overview of the attributes of transactions that affect the relative merits of alternative public utility governance structures, see Crocker and Masten (1996). A largely separate literature also exists on privatization or 'contracting out' of public services such as solid waste disposal, social services, and public transportation. Although governance of these services involves many of the issues associated with traditional public utility services, this literature generally approaches the issue of the relative efficiency of public-versus-private provision in more conventional neoclassical production-cost terms. Examples of this approach applied to water include Crain and Zardkoohi (1978) (who find that publicly owned utilities have higher operating costs) and Feigenbaum and Teeples (1983) (who find no cost differences attributable to ownership structures). See also McGuire and Ohsfeldt (1986) and Teeples et al. (1986).

[3] Stanford University's controversial firing of economist Edward A. Ross in 1900 resulted, in part, from Ross' unpopular advocacy of municipal ownership of public utilities.

a list of the advantages of private ownership, "(1) Greater freedom of initiative and adaptability to requirements of progress [and] (3) Incentive of profit to induce economy and efficiency in construction and operation" (1931, p. 6). Private utilities also benefitted from their greater "freedom from civil service red tape in the reward and promotion of efficient employees and freedom from political interference and domination" (Wilcox, 1931, p. 6).[4] At a minimum, Baker maintained, "Private companies ... certainly will not be accused of lowering rates unduly for political effect, as cities sometimes do" (1899, p. 42).

Nearly a century of experience with franchise contracting, however, had educated late 19th-century observers to the potential for conflict between the incentives of private owners and the interests of the communities they served, and to the difficulty of resolving those conflicts contractually, particularly when circumstances warranted adjustments or extensions of the system (Wilcox, 1931, p. 12):

> "Under a good system of supervision the original location and style of utility systems and works might be made to conform reasonably with the general city plan under a private system, but a great difficulty arises when it comes to subsequent readjustments that involve the expenditure of money without much direct benefit to the utilities."

Where the franchise contract established a fixed rate for utility services, public officials chafed that private utilities were unwilling "to adopt the program of expansion to meet requirements based on community policy, independent of immediate or prospective profit" (Wilcox, 1931, p. 6). Contracts that based rates on the utility's costs, on the other hand, led private owners to overinvest, a bias typically associated, since Averch and Johnson (1962), with rate-of-return regulation (Wilcox, 1931, p. 28):

> " ... under private ownership ... the owners of a utility are urged by the most powerful economic motive imaginable to drive persistently, forcefully, adroitly, and effectively to swell the rate base by increasing the capital account. They have no motive to reduce the capitalization ... but every motive to increase it, because increasing it adds to the base upon which their permissible profits are calculated."

To reduce the potential for disagreement, Baker advised utility contractors to exercise care in drafting franchise contracts (Baker, 1899, p. 33):

> "All the conditions of a franchise should be plainly stated, so as to leave as little chance as possible for differences of interpretation. Ambiguity and indefiniteness are seeds from which spring annoying and expensive crops of litigation, neither city nor company being willing to acknowledge itself in the wrong."

At the same time, Baker acknowledged that such precision was likely to be unachievable: "The obligations of the company are so various, and hinge on so many technicalities of law and engineering, that it is often no easy matter to determine them with exactness, much less to enforce them fully and impartially" (Baker, 1899, p. 35). Cities as well as utilities could be expected to take advantage of whatever gaps remained in the contract to evade performance. Perceptions by the public that utilities were

[4] Other advantages on Wilcox' list were territorial flexibility, freedom of association with other utilities, and freedom from debt limits and flexibility in financial arrangements (Wilcox, 1931, p. 6).

earning excessive profits led to "demands for lower rates on every possible occasion" (Baker, 1899, p. 33).[5] Having failed to elicit concessions from the utilities, cities could be expected to "continue as heretofore threats or attempts to annul franchises on the ground of legal flaws" (Baker, 1899, p. 33) or to wield the powers of government to achieve much the same effect (Wilcox, 1931, p. 12):

> "If under its franchise reservations or under its general police power, the city is in a position to act arbitrarily, there is a danger that it sometimes will do so . . . , resulting in uneconomical readjustments which never would be made if the city itself had to bear the cost, as it would under public ownership."

The "dependence of private owners upon public authorities for rights-of-way and franchise grants which terminate or may be revoked" was a deterrent to private investment (Wilcox, 1931, p. 6).

Public ownership stood to avoid some of these problems and could be expected to result in "less friction between consumers and purveyor" (Baker, 1899, p. 46). But municipal control also forfeited the incentives of private owners to operate utilities efficiently and introduced the danger that "political rather than business methods may dominate" (Baker, 1899, p. 37). In principle, according to Baker, there was "no reason why a municipality cannot build and operate water-works as cheaply as a company". The fact that publicly owned utilities often failed to do so was "due to political favoritism, corruption, incompetency, or negligence on the part of public officials, more largely than anything else" (Baker, 1899, p. 47).[6]

THE 'ABERRANT' OWNERSHIP OF WATERWORKS AND SEWER SYSTEMS

A broad look at water utilities in the 19th century reveals two striking trends: their substantial growth in number and the increasing tendency toward municipal ownership. Table 9.1 shows the number of public and private waterworks at five-year intervals between 1800 and 1896. At the beginning of the period, all but one of the sixteen waterworks in existence were privately owned and operated. Following slow but steady

---

[5] According to Wilcox, public demands for rate reductions were often pure rent redistributions: "often the demand for lower rates has arisen from the public recognition of high or excessive dividends as much as from any claim that the rates charged in themselves are unreasonable and excessive" (Wilcox, 1931, p. 28).

[6] Corruption was a particular concern in a period known for political bosses and party machines. Notorious political machines dominated municipal governments in New York City, San Francisco, Denver, Cincinnati, and Philadelphia, among other cities. Whether private or public ownership was more conducive to corruption was an open question, however. Several major cities, including New York, San Francisco, St. Louis, and Grand Rapids, obtained bribery convictions in connection with the awarding of utility franchises (see Wilcox, 1910: chapter V). The ability to check corruption in the operation of utilities (as opposed to awarding of franchises) may also have been greater under public than private ownership: "the difficulty of proving fraud or corruption" could make it harder to dislodge a private company that had secured a franchise than to turn out a "thieving city council or board of public works" (Baker, 1899, p. 48). As Baker observed, "The man who stands ready to secure personal advantages or profit to the detriment of those whom he has been chosen to serve will not be at all particular whether he does this through public or private ownership of municipal monopolies" (Baker, 1899, p. 36).

Table 9.1
Public and private water-works in the United States, 1800–1896

Year	*Public*	*Private*	*Total*	*Percent of total*	
				Public	*Private*
1800	1	15	16	6.3	93.7
1805	2	21	23	8.7	91.3
1810	5	21	26	19.2	80.8
1815	5	21	26	19.2	80.8
1820	5	25	30	16.6	83.4
1825	5	27	32	15.6	84.4
1830	9	35	44	20.5	79.5
1835	15	39	54	27.8	72.2
1840	23	41	64	35.9	64.1
1845	27	43	70	38.6	61.4
1850	33	50	83	39.7	60.3
1855	48	58	106	45.3	54.7
1860	57	79	136	41.9	58.1
1865	68	94	162	42.0	58.0
1870	116	127	243	47.7	52.3
1875	227	195	422	53.8	46.2
1880	293	305	598	49.0	51.0
1885	447	566	1013	44.1	55.9
1890	806	1072	1878	42.9	57.1
1896	1690	1489	3196	53.2	46.8

Source: Baker (1899), p. 16.

growth during the first half of the century, the pace of waterworks construction picked up substantially after the Civil War. Between 1880 and 1896, the number of waterworks increased from 598 to 3196; between 1890 and 1896 waterworks were being constructed at an average of 220 per year. Though the number of private works continued to rise at a brisk pace — nearly quintupling from 305 in 1880 to 1489 sixteen years later — a majority of the new works added between 1880 and 1896 were publicly owned. Reflecting the same trend, a total of 205 works converted from private to public ownership during the century while the reverse occurred in only twenty cases (Baker, 1899, p. 26).

The century also saw the introduction and propagation of urban sewer systems. While no American city had sanitary sewers in 1857, sewer systems had become common in urban areas by the end of the century: two-thirds of urban residents in 1880 and over 80% in 1900 lived in areas served by sanitary sewer systems (see Table 9.2). Of major U.S. cities, only New Orleans and Baltimore managed without sanitary sewer systems into the late 1800s (to the "disgust ... of the average American citizen" (Wilcox, 1910, p. 452)). But whereas early waterworks were often privately owned, sewer systems were, with very few exceptions, municipally owned from the very beginning. The ownership and operation of electric power plants, gas works, street railways, and telephone systems, by contrast, started out and remained overwhelmingly private.

Table 9.2
U.S. population, 1880–1900

Year	*Total*	*Urban*	*With sewers*	*With water treatment*	*With sewage treatment*
1880	50,155,783	14,129,735	9,500,000	30,000	5,000
1890	62,947,714	22,106,265	16,100,000	310,000	100,000
1900	75,994,575	30,159,921	24,500,000	1,860,000	1,000,000

Source: Tarr (1996), p. 194.

Why a comparison of the relative advantages of public and private ownership favored public ownership in the case of water and sewer systems and private ownership in other cases was a mystery to 19th-century observers. Indeed, from very early on, observers questioned the logic of municipal provision of water and sanitation services. In England, Edwin Chadwick, writing more than 100 years before Demsetz (1968), presented "competition for the field" for sewage and water supply as a viable alternative to proposals for public provision (Chadwick, 1859, p. 403):

> "This great field of service might be put up to contractors or companies of competing contractors on this question. 'At what rate will you undertake to abolish the existing cesspools of all sorts ... and ensure the constant removal of all excretia before it can enter into noxious stages of decomposition? ... ' The answer of efficient competitors, for the whole field ... might probably be ... 'We will do it for two pence per week per house ... ' And I will add, that if the whole field of the metropolis were clear of water companies and were put up for competition upon this question, 'At what rate per house will you undertake to provide a supply of pure soft fresh and well aerated spring water for each house?' The answer ... might be ... 'We will do it and keep the distributary apparatus in good, working order for little more than 2d. per week ... .' "

Echoing Chadwick from across the Atlantic and half a century later, Delos Wilcox expressed similar puzzlement with the observed preference for municipal provision of sewer services (Wilcox, 1910, p. 452):

> "It is a little difficult to explain the overwhelming preponderance of municipal ownership and operation of sewer systems. It would seem that a utility which requires a system of underground pipes as extensive as the system needed for the distribution of water, and which is regarded as essential to the health and welfare of all the citizens, would offer an attractive field for private enterprise."

Wilcox did consider one explanation for the divergence in public utility ownership structures: perhaps water and sewer systems were simply easier for public authorities to administer than other utilities. According to Wilcox, the operation of both water and sewer systems were uniquely uncomplicated. "Water is almost alone among public utilities in the fact that it is a natural rather than an artificial product ... If water can be brought from an elevation considerably higher than the highest point in the city which is to be supplied, the power of gravity may be depended upon for the distribution" (Wilcox, 1910, pp. 400–401). Similarly, "Sewers are distinguished from other utilities in the respect ... that once constructed they for the most part operate themselves.

A sewerage system must be constructed on the gravity principle, and excepting in those cities whose topography is such that no natural outlet can be obtained, there is no pumping to be done" (Wilcox, 1910, p. 452).[7] This simplicity of operation, in turn, meant that sewer systems, and to a lesser extent waterworks, required little expertise or oversight. "In the absence of pumping or a purification plant, there is required, in what might be called a normal type of sewerage system, a very small administrative force for operating purposes. This simplicity of administration naturally is conducive to public ownership, and conversely is unattractive to private enterprise" (Wilcox, 1910, p. 453).

In the end, however, Wilcox was not convinced by this explanation. "It is hard to believe," he concluded, "that the almost entire absence of private operation of sewerage systems is the result merely of sanitary considerations or of simplicity of administration" (1910, p. 453).

A MODERN PERSPECTIVE

The patterns of 19th-century utility ownership described in the preceding section pose two puzzles: (i) why, of all public utilities operating at the end of the century, were water and sanitation systems the only ones that came to be predominantly publicly owned; and (ii) what explains the steady (albeit selective) shift toward public ownership of waterworks over the course of the century? The traditional natural-monopoly rationale for public utility regulation provides no guidance. On the surface, at least, the cost structures of water supply and other utilities appear similar. As one modern observer describes it, "The water industry is a classic case of natural monopoly ... Direct competition between firms in the provision of networks of mains and sewers within a given region would entail inefficient duplication of fixed assets. ... The water supply business is capital-intensive, and average asset lives are very long" (Cowan, 1993, pp. 15–16).[8] Nor is it immediately transparent, from a comparative contracting perspective, why franchise contracting would be more difficult for water and sewer services than for other utilities; water and sewer services are not conspicuously more complex or their cost and demand conditions notably more variable or uncertain than those of electricity, gas, or telephone services.

In the following, we take a closer look at the attributes of water and sewer systems for insights into the solution to these puzzles. Our approach remains a transaction-cost one; we look for distinctive features of water and sewer systems that potentially affect the relative efficiency of private and public ownership, base hypotheses on these factors, and then test the hypotheses against observed variations in ownership structures over time and across transactions.

[7] A 1905 Census Bureau survey found that of 154 cities, only 9 reported that all of their sewage was pumped and 9 others that part was pumped (Wilcox, 1910, p. 452).

[8] Cowan adds that "the existence of a network is not in itself a barrier to competition. Firms could compete to supply water ... if they shared access to the pipe networks. The prospect of such competition in service provision is, however, small ... because of the high cost of pumping long distances" (1993, p. 16).

Administrative Simplicity

Although Wilcox rejected administrative simplicity as an explanation for public ownership of water and sanitation, differences in the cost of administering utility operations would constitute a perfectly valid transaction-cost rationale for variations in observed governance arrangements.[9] According to the theory, integrated firms or, in this case, publicly owned utilities, should be able, in principle, to operate as efficiently as private ones simply by replicating what a private company would have done — including employing whatever experts or specialists are needed to design and run the system. In practice, however, public administrators cannot simultaneously (i) retain the authority to intervene selectively in the operations of the system for the benefit of the community, yet (ii) provide operators of the system with the high-powered incentives to run it at its potential.[10] Inevitably, public authorities will find that they must either monitor the effort of employee-operators more intensively or reconcile themselves to less efficient provision of utility services.

In general, the more complex and unfamiliar the activities at issue, the less effective administration by a public authority is likely to be — the greater "the costs of organizing and the losses through mistakes," in Coase's words (1937, p. 395) — and, correspondingly, the more important the judgment of experts is likely to become. To the extent that water and sanitation systems are, in fact, self-operating or in other respects simpler to administer than other public utilities, the incentive and specialization disadvantages of public relative to private ownership could fall enough to tilt the balance in favor of public ownership.[11] Stated as a hypothesis, we would expect public ownership to be more common for utilities whose operations are less complicated, require less technical expertise, and are less demanding of constant attention.

Contracting Hazards

Differences in the relative efficiency of public and private ownership may also arise, of course, because of differences in the hazards of contracting. The challenge, noted above, is to understand why water and sewer services would have been any more difficult or hazardous to contract for than were electricity, telephone, gas, or other privately

[9] The notion that differences in costs of internal administration play a prominent role in integration decisions can be found in "The Nature of the Firm" (Coase, 1937). Since then, Coase has consistently emphasized difference in the costs of organizing activities within firms as the essential determinant of the boundaries of firms (Coase, 1972, 1988). See Masten et al. (1991) for further discussion.

[10] On the limits of internal organization generally, and the impossibility of selective intervention in particular, see Williamson, 1985, pp. 131–156.

[11] Implicit in this argument is an assumption that costs of contracting do not increase — or, at least, do not increase as rapidly as the inefficiencies of public ownership — with the complexity and unfamiliarity of the production process. It is worth noting, in this regard, the distinction between operational and product complexity. Operational complexity is likely to exacerbate the difficulty of monitoring of production processes, whereas product complexity is likely mainly to complicate contracting (and be relatively less important for supervision). On the differential effects of familiarity on organization costs, see Masten et al. (1991, p. 7).

supplied utility services. In the following, we argue that a variety of factors did indeed leave water and sewer transactions more vulnerable than other utility transactions to appropriation hazards. Our development of the hypothesis is divided into two parts: the first addresses pricing and appropriation issues associated with water and sanitation services, and the second, problems of adaptation and contract evasion.

Pricing

The conventional conception of the problem of pricing public utility services consists of identifying a schedule of prices that induces efficient levels of investment, production, and consumption given the information available to the respective parties or, put more simply, that minimizes dead weight loss. In general, the solution to that problem takes the form of a multi-part (or other complex nonlinear) price schedule that relates incremental payments to the cost of service (connection, delivery, use, and so forth) and uses a fixed or inframarginal component to divide resulting surpluses and assure coverage of the utility's capital costs. Like the contracting literature to which it is related, mainstream regulation theory gives little attention to adaptation and enforcement issues, the presumption being that the price schedule established by the 'regulatory contract' covers the full range of relevant contingencies and that both utilities and regulators comply with its terms.

The administered contracts framework, and the contracting perspective that has grown out of it, differs in making the problems of adaptation and enforcement the central concerns. Although pricing enters the analysis, the emphasis is more on price *determination*, that is, on the processes through which prices are arrived at, adjusted, and administered and the costs associated with those processes. Because contracts (or regulatory orders) are incomplete and enforcement is costly and imperfect, transactors may attempt to exploit gaps and ambiguities in an effort to contrive cancellation (Williamson, 1983), evade performance (Goetz and Robert, 1983) or otherwise force a renegotiation of the terms of the transaction.[12] Instead of aligning marginal incentives to minimize deadweight loss, the focus of contract and regulatory design becomes reducing the incentive to engage in wasteful efforts to alter or protect the distribution of quasi-rents established by the initial agreement. Correspondingly, the principal determinants of the efficacy of contracts become (i) the opportunities and tactics the transaction affords transactors to exploit the imperfect enforceability of contract terms, and (ii) the extent to which the terms of the agreement can be fashioned to keep the relationship within the "self-enforcing range" (Klein, 1992).

From this perspective, pricing methods are important primarily because the structure of payment schedules affects the credibility of agreements and the costs of effecting adaptations. Pricing arrangements that tie payments (receipts) to services have at least two advantages over fixed (service-invariant) payments in these respects. First,

[12] Other contracting research representative of this approach includes Goldberg and Erickson (1987), Masten (1988), Crocker and Masten (1991) and Klein (1992, 1996). For a review and comparison of various perspectives on contracting see Masten (2000).

service-based payments foster unilateral adjustments in service provision. Whereas expansions, improvements, or other adjustments in service under fixed-payment schemes require negotiation to bring about, service providers whose receipts vary with service levels will undertake any improvement that increases revenues more than costs.[13] Though similar to standard incentive arguments, the benefit to service-based pricing from this alternative perspective comes about through the reduction in occasions for reopening negotiations.

Second, service-based charges are less attractive than fixed or other inframarginal payments as devices through which to effect redistributions. Whereas proposals to adjust inframarginal payments alone are purely distributive and, thus, likely to be resisted, changes in service-related charges have potential efficiency consequences that affect the expected payoffs to soliciting price concessions in desirable ways: inasmuch as adjustments in per-unit charges distort performance incentives, adjustments to unit charges represent a costly way to redistribute rents relative to changes in inframarginal fees. Conversely, the positive sum nature of proposals to modify misaligned incremental charges is likely to facilitate the achievement of desirable adjustments (cf. Williamson, 1979, p. 251; Crocker and Masten, 1991, pp. 79, 89). Other things the same, pricing schemes that rely relatively heavily on inframarginal payments should thus be more likely to occasion confiscatory pricing demands and thus offer weaker, less credible protections against appropriation.

Despite these commitment and adaptation advantages, service-based pricing is not always possible or practical. Pricing schemes that tie payments to service require additional costs of metering and exclusion. Tying payments to quantity used or delivered, for example, presumes the ability to monitor usage levels at reasonable cost. Services that are costly to meter relative to the value of metering will tend to rely on more discrete payment schemes. Moreover, as we know from the theory of price discrimination, if a service can be stored or resold at low cost, consumers will seek to avoid inframarginal payments by spreading the fixed component of the pricing schedule over time or across consumers. The practicality of access or hook-up fees — which, while inframarginal to usage, are conditional on being connected to the system — will thus depend on the costs of preventing consumers from gaining access to the service through indirect channels such as resale.

Several features of water and sanitation systems complicate the pricing problem for these relative to other utility services. First, the ratio of fixed to variable costs tends to be higher for water and sewer systems than for other utilities. In contrast to utilities such as electrical and telephone service, delivery of which primarily entails installation of aboveground poles, both water and sewer services require extensive underground pipe systems to deliver.[14] Assuring adequate water supplies could also require large

[13] For a more detailed discussion of the relative merits of usage-based versus fixed charges in the context of equipment leasing that emphasizes the automatic adjustment properties of payments linked to usage, see Masten and Snyder (1993).

[14] Contributing to the expense of these systems, water mains and connections "must be laid deep in the ground where they cannot be reached by frost in the coldest weather" (Wilcox, 1910, p. 401).

investments in the construction of aqueducts and reservoirs. At the same time, the often negligible resource cost of water and sewer services combined with the self-operating character of the corresponding systems leave the incremental cost of supplying these services, in many cases, at virtually zero.

Second, water and sewer systems have important public health implications not associated with other utilities. Water is a conduit for diseases, and sewage, if not properly removed and disposed of, becomes a public nuisance and can contaminate ground water and other supply sources, leading to outbreaks and epidemics. Water is also crucial to the effective operation of sewer systems: "The wastes which, unless removed, tend to contaminate the local sources of water supply, cannot be removed except by means of a public system of water works. In other words, in order to remove the cause of the pollution of individual water supplies in cities, it is necessary to have a public supply; a sewage system is impossible without water works" (Wilcox, 1910, p. 399; see also Tarr, 1996).

Given the low marginal costs and public health externalities of water and sanitation services, the appropriate unit price for water and sanitation services will often be so low as to leave metering uneconomic; in many cases, the charges that could be assessed would not cover the cost of installation and reading of meters. Other utilities, by contrast, have either substantial marginal costs of service or low metering costs. Thus, although utilities like gas and electricity have metering costs comparable to that for water, the inefficiencies that would result from failing to assess charges based on usage given the substantial variable costs of generating electricity and manufacturing gas justify metering of these services. And although telephone and telegraph services have low marginal service costs, these services also enjoy exceptionally low metering costs: monitoring of use can be performed from a central office, avoiding the expense of installing and gathering data from on-site meters. Water and sanitation services are thus unique among utilities in having both substantial metering costs and, except in times of drought, (nearly) non-rivalrous consumption.

At the same time that metering costs discourage usage charges, the relative ease with which water and sanitation services can be stored and resold inhibits the ability of water and sanitation utilities to recover capital costs through significant access or connection fees. To the extent, moreover, that access fees *do* deter use, public health and nuisance concerns would also limit the size of fees municipalities would want to charge.[15]

Adaptation and evasion

An implication of the preceding discussion is that water and sanitation utilities will tend to rely more heavily than other utilities on (relatively) inframarginal payments that are unrelated to either use or access. Once a utility can no longer make service contingent on payment, however, the firm becomes dependent on the coercive power of

[15] Note that it is the effect of public health externalities on pricing arrangements, rather than the externalities per se, that motivates integration in this analysis.

the government to tax for its revenue. Dependence on discrete payments from municipal authorities, in turn, increases the utility's exposure to appropriation efforts: lump-sum transfers make tempting targets for redistribution.[16]

A counter argument to the claim that government payments are more susceptible than consumer payments to appropriation is that consumers are more likely to lobby (or riot, depending on the form of government) seeking lower prices if they pay for services directly rather than through the government. Effective lobbying (and rioting) entail the usual collective action problems, however, which would tend to reduce this pressure. Moreover, while the gains to political activity to consumers may be large where the direct cost of the service to them is high, the fact that the 'optimal' usage charges for water and sanitation services are often at or near zero anyway implies little scope for consumers to gain by attempting to force prices still lower. To forestall those efforts, both sides will want to describe in detail the responsibilities of the parties and thereby limit the tactics each has available to evade performance or force a renegotiation. Making contracts harder to evade by increasing the precision of performance obligations is also likely to increase the cost of and frequency with which desirable adjustments will have to be negotiated, however. In short, given the pricing arrangements available to them, parties to water and sewer contracts are likely to find it harder to keep the transaction "within the self-enforcement range."[17]

When a party to a water or sewer contract *does* become dissatisfied with the terms of the agreement, the nature of water and sanitation transactions is likely to provide that party with ample opportunities and a set of tactics to evade or amend the contract. First, compared to utilities such as electrical and telephone service whose delivery primarily entails installation of aboveground poles and wiring, the installation and repair of water and sewage systems has the potential, particularly in urban settings, to cause significant disruptions to traffic and commerce.[18] More important, water is essential for fighting fires. To assure that firefighters have access to water supplies that are adequate in terms of both quantity and pressure, water systems must be capable of providing large volumes of water at high pressure. Historically, the failure of water systems has been responsible for or contributed to a number of major urban conflagrations.

Organizationally, the disruptiveness of installation and repairs of water and sewer systems and the occasional but urgent needs for water to fight fires create an unusually strong need for coordination and cooperation between utilities and municipalities. The potential for disruption and, in the case of fire, destruction from the withholding of cooperation or simply from exercising less-than-full diligence provides a valuable tactic that private suppliers of water and sewer services could exploit in search of price or other concessions from the city. On the other side, city officials could attempt to use

[16] In principle, municipal authorities could both collect revenue (as excise taxes) and finance private utilities as a function of water access and usage. Doing so, however, confronts the same metering and enforcement problems that motivates the need for government transfers in the first place.

[17] To the extent that water and sanitation systems are more durable and technologically stable than utilities, the future efficiency consequences of appropriative actions may also be lower for these systems.

[18] Installation and repair of gas lines are similar to water in this respect.

feigned dissatisfaction (or even interference) with the utility's provision of these services to justify reducing payments or revoking a utility's franchise.[19]

In sum, appropriation and adaptation issues associated with the high fixed to variable costs, resale potential, public health externalities, and disruptiveness of water and sewer systems, and the critical role of water in firefighting, provide a potential basis for distinguishing between these and other public utility systems. Though not as parsimonious as the administrative simplicity explanation, it represents (we believe) a plausible alternative — and potentially testable — explanation for the public ownership of water and sewers.

HISTORICAL ACCOUNTS

In this section, we look again at the turn-of-century public utilities debate to determine the extent to which the factors identified in the preceding hypotheses were of concern to 19th-century policy makers.

Administrative Simplicity

As noted earlier, public utility experts at turn-of-the-century had remarked on the relative simplicity of water and sewer system operations. Whereas other major utility services either were 'manufactured' (like electricity and, for much of the 19th century, gas) or used relatively sophisticated technology (telecommunications), water and sewage systems mainly provided access to existing natural resources, often requiring nothing more than gravity to effect delivery once the infrastructure was in place. Historically,

[19] The uniqueness and distance of suitable water sources may also exacerbate problems of ex ante coordination and hold up relative to other utilities. In contrast to 'artificial' or manufactured services, water, like other natural resources, must be 'produced' where it is found. Thus, even though electricity and gas (which, prior to the commercialization of natural gas in the 1920s and 30s, was extracted or 'manufactured' from coal) required proximate and localized production, the existence of multiple sites on which a facility could be located reduced the potential for hold up by the owner of any particular property. By contrast, water supplies (particularly, surface supplies) are often unique and alternative sources often substantially inferior to the most economical ones. Moreover, "to procure a sufficient supply of water, it is usually necessary for a municipality to go outside of its own limits" (Wilcox, 1910, p. 400), obliging cities to coordinate their action with other governmental agencies. Prominent examples were the acquisition of water by the City of New York from the Catskills and by Los Angeles from the Owens Valley and the Colorado River. Preventing the owner of a resource or right-of-way from holding up access to such a source would often require "the exercise of the power of eminent domain for the condemnation of water rights" (Wilcox, 1910, p. 400). Acquiring water from remote jurisdictions may also be complicated by the continuing vigilance required to protect the integrity of the supply. As Wilcox observed, "If the city procures its water supply from the mountains or from lakes or rivers in a sparsely settled district, it requires special authority to protect its sources of supply from willful or careless pollution" (1910, p. 401). Governments can, and do, acquire property for private parties through eminent domain; for example, the 1872 contract between Birmingham, AL, and a private contractor to construct waterworks provided that the city would "exercise the right of eminent domain on behalf of the contractor in order to enable him to secure a proper source of supply" (Wilcox, 1910, p. 447). The use of the power of eminent domain on behalf of private interests remains a legally and politically controversial area, however.

sewer systems were especially easy to operate, involving, at least until the introduction of waste treatment facilities (which were uncommon for most of the 19th century), little more than allowing waste to flow through a system of pipes to be discharged down-river. As Wilcox described it, "Sewers are distinguished from other utilities in ... that once constructed they for the most part operate themselves" (Wilcox, 1910, p. 452).[20] Waterworks, though somewhat more likely than sewer systems to require some expertise and ongoing attention — especially where supply required pumping or filtration — were, when compared to other utilities, also largely self-operating.

Contracting Hazards

Despite their obvious appreciation for the hazards of franchise contracting generally (see third section above), none of the 19th-century accounts that we have seen proposed a differential contracting hazards rationale for the public ownership of water and sewer systems. Nevertheless, the record contains considerable indirect evidence consistent with that hypothesis.

Pricing and metering

Clearly important to the pricing of water services was the cost of metering (Wilcox, 1931, p. 67):

> "In water systems, the main questions with respect to the distribution of the costs of service among consumers are the following: (1) Shall domestic consumption be on the flat-rate basis, or on the meter basis? ... Meters add very considerably to the capital investment of the utility or of its patrons, except as this may be offset in reduced investment in water sources and water works by reason of a lessened waste or use of water. The meter system also involves operating expenses in meter testing and maintenance and frequent meter reading ... "

Accordingly, few cities required residential consumers to use meters.[21] Yet, despite their similar metering costs, electricity and gas supply were consistently metered: electricity was subject to "very complex" meter rates (Wilcox, 1931, p. 69), while "In gas supply, ... the meter system may be considered universally in use" (ibid, p. 68). Interestingly, the exception to the latter was the pricing of "natural gas in the vicinity of the wells" (ibid), which, unlike manufactured gas, but like water, was a natural resource having low marginal costs.

As would be expected given the public benefits and pricing problems associated with water and sanitation services, municipal works tended to charge lower rates than did

---

[20] This is not to say that the design of sewer systems did not involve important engineering and technological issues. On the contrary, the issue of the most effective and economical sewer design was a matter of significant debate among civil engineers in the late 1800s (see Tarr, 1996). The complexities of sewer systems, however, involved their design rather than their operation.

[21] Landlords vigorously opposed the use of meters because of another manifestation of opportunism: the risk that "a disaffected tenant could and would 'get even' with the landlord by 'letting the water run' " (Wilcox, 1931, p. 67).

private works.[22] While serving as deputy commissioner of water supply for New York City in 1917, Wilcox estimated that a private company operating the city's municipal waterworks would have needed to collect more than twice what the city collected in revenues from fees to earn a fair return (Wilcox, 1917, p. 569). Consistent with this, Anderson (1988), p. 144) concluded that the rates charged by private companies in the late 19th century were more than 40% higher than the rates charged by municipal works, with municipal works subsidizing losses out of tax revenue. The difficulty of recovering capital costs was "all the more important in the case of a utility such as sewerage, where usually under public ownership no rates at all are charged for the service rendered" (Wilcox, 1931, p. 26).[23]

Regardless of whether the low rates municipalities charged reflected legitimate public welfare (efficiency) or more narrow political (distributional) considerations, the evidence suggests that the rates deemed appropriate by municipal authorities could not have sustained private investment without significant direct transfers from the municipal government. Indeed, even at the higher rates charged by private waterworks, communities found it necessary to subsidize private water companies from general revenues. Often, such transfers appear to have been effected through hydrant rates. According to Anderson, rentals on hydrants were set to assure that they "always covered the company's interest obligations incurred on the cost of construction" (Anderson, 1988, p. 143).[24] Consistent with hydrants rates serving this rent transfer function, "disagreements over hydrant rentals were the major source of friction between cities and private companies" (Anderson, 1988, p. 144).

Adaptation and disruption

The need for coordination and cooperation between the cities and water utilities, and the disruptions that lack of cooperation could cause, were major concerns of 19th-century observers. In fact, according to Baker, the fundamental feature of water franchise contracts was the "right to dig up, occupy, and use streets constructed and maintained at public expense, thus interfering more or less with the use of the public highway for street traffic" (Baker, 1899, p. 31). Similarly, Wilcox observed, "Perhaps the most serious way ... in which franchise holding companies tend to disturb public comfort is in the constant tearing up of the streets for the construction or repair of underground

[22] As noted earlier, the greater potential for resale and storage (at least for relatively short periods) of water may also have limited the rates that water utilities could charge. The potential for such practices is illustrated by the practice in London, where, until the late 19th century, water was delivered for two hours every two days and customers stored water in cisterns for use in the interim (Tynan and Cowen, 1999, p. 20).

[23] Wilcox attributed this to a general reluctance on the part of the public to pay for waste disposal: "Additional cause must be found in the unwillingness of citizens generally to purchase a service of this nature of their own free will. Indeed, sewerage differs from other public services in that it is not a utility manufactured or brought to the people, but simply a process of getting rid of their wastes" (1910, p. 453).

[24] The extremely wide range in hydrant rental rates paid across cities — from $15 to $125 per hydrant per year in 1888 (Anderson, 1988, p. 143) — suggests that these rates may have had more to do with covering fixed costs than with marginal costs of hydrant services, which seem unlikely to have varied so widely from one town to the next.

fixtures. It often seems astonishing that business can continue to be done in spite of these long-drawn-out and frequently recurring interferences with the ordinary uses of the city highways" (Wilcox, 1910, p. 122).[25]

Nineteenth-century observers cited two implications for waterworks of the need for water for fire protection. First, water demands for firefighting meant that waterworks had to be much larger than otherwise, raising the fixed costs of water systems. According to Wilcox, the peak-load requirements of fire protection were the primary determinant of the design of water systems and a major factor in their cost (Wilcox, 1910, pp. 400–402):

> "The fire department not only demands a great deal more water for use at particular times than is needed for drinking, but also demands that the water shall be supplied under special conditions of pressure that make the water works much more expensive ... The demand for water in connection with fire protection alone, not so much in the quantity of water used as in the special construction of mains, hydrants and reservoirs and the extraordinary pressure required, constitutes a considerable proportion of the total business of water works."

In the same vein, Baker noted that (1899, p. 39):

> "the cost of the public service depends far less upon the total consumption in a year than upon the comparatively vast quantities which may be demanded on special brief occasions, principally at fires. Thus the use of water for all public purposes may be not over five per cent of the total consumption for the year, but the whole system must be so proportioned as to be capable of delivering an enormous increase of volume during a severe conflagration. Operating expenses may be increased but slightly on this account, but first cost and capital charges may be much more than doubled."

An indication of the importance of water systems for fighting fires is the fact that fire insurance companies as early as 1800 made provision for centralized water systems in their rates (Anderson, 1988, p. 141). At the end of the century, towns with waterworks could expect a reduction in fire insurance premia of between 20 and 50%. (Anderson, 1988, p. 142).

---

[25] Although sewer franchises were rare, the few instances where sewerage services were privately supplied are illustrative of the extent to which cities were concerned with the potential for disruption. Among the provisions of a 1906 contract between the city of Long Branch, NJ (population 8872) and the Long Branch Sewer Company were requirements that the utility return the street to its original condition, and that no excavation should be left open "for any unreasonable length of time". Rather than charging for sewer services directly, payments for sewer services were, like property taxes, based on property assessments. (Wilcox, 1910, pp. 455–456). The franchise contract in Austin, TX (population 22,258) required the sewer company to " 'exercise due care and diligence in unnecessary obstruction to public travel' on the streets, or as to 'any injury or unnecessary interference with any pipes either of gas or water which may be lawfully located beneath the surface thereof when such sewers are laid'. The company was also to take 'every reasonable precaution against accident and danger to persons or property in the execution of the rights and privileges' granted, and was to 'cause all excavations to be properly lighted and guarded at night', and was to restore the streets 'to their former condition as near as may be without unnecessary delay' " (Wilcox, 1910, pp. 457–458). The city, for its part, was required "to pass ordinances with adequate penalties for the protection of the company's sewer connections, catch basins, manholes, and other appurtenances" (Wilcox, 1910, p. 459). "The practice of allowing private sewers to be built in the public streets, unless in conformity with a well-conceived plan for the whole city, is highly objectionable. It results in bad sanitation, poor economy and much possible litigation; for when the city comes to put in a thoroughgoing sewerage system, it is liable to find these private sewers so poorly designed and built as to be almost useless, notwithstanding which the owners insist on having them bought by the city, or at least resist an assessment to meet the cost of suitable new sewers" (Baker, as quoted in Wilcox, 1910, pp. 451–452).

Second, the availability of water for firefighting was clearly a source of tension between cities and private water companies. According to Anderson, "In the latter years of the nineteenth century, many lawsuits, often initiated by insurance companies, were launched against private water companies for recovery of damages for fire losses resulting from insufficiency or failure of the water supply" (Anderson, 1988, p. 144). Private water companies, it was alleged, tended to be "much more conservative in the quality of the fire protection rendered . . . " (Wilcox, 1917, p. 560). In Houston, pressure for a municipal takeover of the city's private waterworks erupted in 1886 after a fire destroyed an important cotton seed mill "while firemen stood by helplessly because the hydrants were dry" (Platt, as quoted in Hausman et al., 1986, p. 15).

In sum, descriptions of the nature and problems of 19th-century public utilities can be fairly read to offer evidence consistent with both the administrative simplicity and contracting hazards solutions to the public utility ownership puzzle.

TRENDS AND PATTERNS IN OWNERSHIP OF WATERWORKS

While the concerns of 19th-century observers appear largely to coincide with the distinctive features and problems associated with waterworks in the preceding analysis, they do not resolve the question of whether and to what extent administrative simplicity and contracting hazards were responsible for differences in the incidence of public ownership. For that, more systematic analysis of trends and variations is needed.

Trends

The increases in both the number and public ownership of waterworks during the 19th century coincides with upward trends in population, industrialization, urbanization and wealth. The population of the United States rose from 5.3 million in 1800 to 76 million in 1900 (U.S. Bureau of the Census, 1961). Whereas roughly 80% of the labor force was employed in agriculture in 1800 (Weiss, 1992, p. 22), that proportion had fallen to 56% or less by 1860 (ibid) and to 40% by 1900 (Lebergott, 1964, p. 510). The percent of the population living in areas classified as urban (population in excess of 2500) rose correspondingly, from 6% in 1800 to 40% 100 years later, while the number of places classified as urban rose from 33 (with an average population below 10,000) in 1800, to 1743 (having an average population just over 17,000) in 1900 (U.S. Bureau of the Census, 1961). By 1904, 150 cities in the U.S. had populations of 30,000 or more; in 1800, there were only three. In 1890, three cities — New York, Philadelphia, and Chicago — had populations in excess of one million.

Not surprisingly, the increases in population, urbanization, and industrialization increased demands on and threatened the quality of original water sources. Given the importance of water to survival and its value in transportation, industrial, and agricultural uses, the availability of reliable water supplies had naturally been an important factor in settlers' choice of location. Lands on or adjacent to water resources tended to be the most valuable and the first settled. In areas of low population density, individual or small-scale cooperative collection or extraction of water by means of wells or available surface sources such as rivers or lakes tended to be adequate and raised few of the problems associated with public utilities: "Except in those portions of the country in which the rainfall is scanty, it is comparatively easy for individuals to supply themselves with water for necessary uses until population becomes congested" (Wilcox, 1910, p. 399). Groundwater, surface water (lakes and rivers), and rainwater collected in cisterns were usually adequate to supply early cities. Where they existed, cooperative water projects tended to be small-scale affairs. "Some of the earliest waterworks plants of the country consisted of a small and short pipeline from a spring on a hillside to a few houses below it. Those supplied with water ... contributed equally to the cost of construction and the small expense necessary to maintenance ... The water supply in such cases is purely a co-operative affair, each paying his proper share of the expense involved, and no attempt being made to realize a profit for the benefit of any one" (Baker, 1899, p. 37).

Population growth, industrial development, and technological change combined over time to undermine those sources, however. Increases in population density diminished the feasibility of individual and small-scale collective water supply. As claims on lands directly adjacent to rivers and lakes were gradually exhausted, newer settlers would find it necessary to locate further from water sources, creating the need to transport water over longer distances and the prospect of disputes over access to water and the rights of way to transport it. At the same time that increases in the size and density of cities were increasing demand for potable water, population growth and urban encroachment threatened local sources with pollution, so that ultimately the exploitation of more distant supplies was required (Wilcox, 1917, pp. 558–559):

> "The continuous encroachment of population on the watersheds ... results in a gradual lessening of the amount of water available from these sources, while at the same time the increased demand for it tends to cause an overdraft upon the wells. ... The possibility of pollution as the growing city flows out to encompass these water-bearing lands tends to weaken the consumer's predilections for his local supply as compared with the supply from the mountains which the city now offers."

To accommodate the increasing need for water, cities, beginning with Philadelphia in 1802, began to develop systems for transporting water from non-local sources (Tarr, 1996, p. 10). The resulting need for large investments — in distant impoundments, aqueducts, and distribution mains to exploit more distant water supplies or, alternatively, in facilities to preserve or purify nearby sources — gave water supply its public utility characteristics.

Ironically, the introduction of large-scale water systems also aggravated some aspects of the pollution problem. Increases in the availability of water strained existing waste disposal methods by increasing the demand for water closets, the adoption of which greatly increased the quantity of sewage to be disposed of. Prior to 1857, sewer systems for the disposal of human waste were nonexistent in the U.S. Those sewer systems that did exist were intended for stormwater collection only, their use for sewage disposal discouraged and sometimes forbidden by city ordinances. For the first three-quarters of the century, city dwellers continued to dispose of their wastes in cesspools and privy vaults or, illegally, in streets, vacant lots, or storm sewers (Tarr, 1996, pp. 8–9). The increasing volume of waste quickly overwhelmed these methods of disposal, however. Overflowing cesspools and improperly disposed of waste created public nuisances; cellars "flooded with stagnant and offensive fluids" at such a rate as to make cleaning "nearly futile". Central water supplies, once contaminated, distributed "fecal pollution over immense areas". Faced with the progressive inadequacy of traditional waste disposal methods, public officials in some states sought to stem the tide of filth and disease by barring installation of water closets in towns without sewers.

As the century progressed, appreciation of the relationship between waste and disease grew, so that, by the late 19th century, "sanitarians in the public health movement, scientists, and engineers increasingly believed that water pollution and epidemic diseases such as cholera, typhoid, and yellow fever were related" (Tarr, 1996, p. 103).[26] Modern epidemiological research attributes significant public health benefits to centralized water and sanitation systems (see Esrey, 1996). Although their understanding of the mechanisms of disease transmission were relatively primitive, public health officials had, by the mid-19th century, generally come to associate urban outbreaks and epidemics of certain diseases with the contamination of water (see Tarr, 1996). Mounting evidence that cities with sewer systems were less prone to epidemics spurred a wave of sewer system construction.

Growth in the size and density of cities also aggravated other urban problems. As cities became more congested, the severity and cost of disruptions from the installation and repair of underground pipes also rose. Like disease, fire hazards also tended to spread at rates directly proportional to the density of population: population growth in cities "resulted in the closer building up of urban areas and in greater fire risks, so that works were imperatively demanded for fire protection" (Baker, 1899, p. 18).

In general terms, these trends in population and congestion are consistent with the contracting hazards rationale for the increase in municipal ownership of waterworks. The factors argued to favor municipal ownership of water systems — the need for large capital investments, public health and nuisance problems, fire hazards, and disruptions to commerce — all became more widespread and severe as the country grew and developed. By contrast, administrative simplicity seems unable to account for these trends. On the contrary, municipal ownership of water utilities appears to have been *increasing* at a time when the operations of waterworks were becoming more complex on average. The last two columns of Table 9.2 report the population of the U.S. living in areas served by sewage and water treatment facilities. In 1880, only about 1 in 10,000 people lived in communities that treated sewage; by 1900, that figure had risen

[26] The need for sewer systems was stimulated by the "growing appreciation of the dangers arising from depositing human wastes in, and drawing water for household use from, holes in the same dooryard" (Baker, 1899, p. 18). Beginning in the 1850s, experts increasingly came to believe in the existence of a relationship between clean water and freedom from epidemic disease (Tarr, 1996, pp. 9, 11). "Public health officials, especially if they believed in the anticontagionist 'filth theory' of disease, viewed overflowing cesspools with water-closet connections as a particularly dangerous threat to a healthful environment." (ibid, p. 10). Sewerage systems appeared to reduce bacterial ailments such as typhus and infant diarrhea in cities (although with deleterious consequences for communities downstream) (ibid, p. 12). Cities with water supply systems had lower death rates during cholera epidemics than places without (Anderson, 1988, p. 141).

to approximately 1 in 76. For water treatment, the corresponding figures were 1 in 1670 for 1880 and 1 in 40 for 1900. Treatment operations were clearly becoming significantly more prevalent over this 20-year period.

Water filtration and treatment, however, increase the complexity of water and sewer operations and, in that respect, make water supply more like other utilities: "The process of filtration, where it is necessary, is equivalent to a process of manufacture. Pure water must be 'produced' where it cannot be procured" (Wilcox, 1910, p. 400). As filtering became more common, the administrative simplicity hypothesis would predict an increase in private ownership. It is impossible to tell with these aggregate data the extent to which works that filtered water were more or less likely to be publicly owned. The general trend during this period, however, was *away from* private ownership. Aggregate trends thus appear to have been more consistent with the exacerbation of contracting problems than with changes in the difficulty of administering water systems.

Ownership Patterns in Large U.S. Cities

In contrast to sewers — and despite similarities in coordination, health, and pricing issues — municipal ownership of waterworks was far from universal. As discussed above, private ownership, which dominated water supply for most of the 19th century, still accounted for 57% of waterworks as late as 1890. The extent of the variation in waterworks ownership is indicated in Table 9.3, which shows the total number and percent of publicly owned waterworks by state in 1896. Of those states with more than one waterworks, the percent that were publicly owned varied from zero (Nevada) to 91% (Nebraska). In ten states, fewer than 25% of waterworks were publicly owned, while in eight states and the District of Columbia municipalities owned in excess of three-quarters of the works.

To gauge the extent to which variations in waterworks ownership were consistent with the contractual hazards and administrative simplicity hypotheses, we assembled a cross-section of data on waterworks ownership in large American cities. (The source for these data is Baker, 1897.) Table 9.4 lists the names and waterworks ownership history of the 37% largest U.S. cities in 1896. The first and third columns list cities served by waterworks that were always private and always public, respectively. The center column identifies cities served by waterworks that were originally private but had converted to public ownership by 1896, along with the year in which the change occurred (in parentheses). Of these cities, only one — New Orleans — changed from public to private, ten years after having been converted from private to public ownership. Fifteen of the cities were served by more than one waterworks and may therefore be listed in more than one column.

Table 9.5 presents summary statistics for 48 of the waterworks in these cities for which there were sufficient data. The most conspicuous contrast is in the relative sizes of the public and private works. By virtually every measure, public works appear to have been roughly four to five times larger than their private counterparts. The exception — county population — is misleading for two reasons. First, a waterworks may not have served the entire county. Second, privately owned waterworks were more likely to operate in cities

Table 9.3

Number and ownership of water-works in the United States in 1896 by state

State	*Number of works*	*Percent public*	*State*	*Number of works*	*Percent public*
Alaska	1	0.0	New Hampshire	53	45.3
Nevada	9	0.0	Washington	51	49.0
Idaho	14	7.1	Kansas	91	49.4
New Mexico	10	10.0	Oklahoma	6	50.0
Maine	57	12.3	Virginia	48	50.0
Montana	16	12.5	New York	342	50.3
California	121	16.7	West Virginia	23	52.1
Arizona	5	20.0	Utah	11	54.5
Arkansas	20	20.0	Wisconsin	92	60.9
Pennsylvania	342	23.9	Wyoming	13	61.5
Connecticut	59	25.1	Vermont	46	63.0
Texas	86	26.3	Colorado	68	63.2
North Carolina	18	27.7	Indiana	111	63.9
Louisiana	10	30.0	Georgia	36	72.2
Alabama	29	31.0	South Dakota	31	74.2
Maryland	29	31.0	Massachusetts	151	74.8
Florida	19	31.6	Ohio	149	76.5
Mississippi	9	33.3	Illinois	201	78.0
South Carolina	12	33.3	North Dakota	10	80.0
Missouri	62	33.8	Michigan	156	80.3
Kentucky	44	34.0	Iowa	170	82.4
Tennessee	28	35.7	Delaware	7	85.7
New Jersey	93	37.6	Minnesota	93	87.1
Rhode Island	15	40.0	Nebraska	91	87.9
Oregon	37	40.5	District of Columbia	1	100.00

Source: Baker (1899), pp. 22, 23, 25.

served by multiple works: 59% of private works versus 35% of utilities that changed from private to public and 18% of the always-public works. Accordingly, the populations of communities served by private works were more likely to be divided among multiple providers. Measured in terms of annual consumption, miles of water mains, and numbers of taps, meters, and hydrants, however, privately owned water systems were much smaller on average than public systems.[27] Thus, even among the country's largest cities, municipal ownership of waterworks appears to have been more prevalent the larger the system.

Other comparisons are also consistent with the contracting hazards hypothesis. Public works were roughly twice as likely to be served by aqueducts, conduits, or tunnels,

[27] Since metering of residential service was very rare, differences in the number of meters most likely reflects industrial development of the city.

Table 9.4

Waterworks ownership in the 37 largest U.S. cities, 1896

Always private	*Private to public (year of change)*	*Always public*
Brooklyn [a]	Albany (1851)	Atlanta
Chicago [a]	Baltimore (1854)	Boston [a]
Denver	Boston (1848) [a]	Brooklyn [a]
Indianapolis	Brooklyn (1895, 1896) [a]	Louisville
Memphis	Buffalo (1868)	Lowell, MA
New Haven	Cambridge, MA (1865)	Milwaukee
New Orleans [a]	Chicago (1851) [a]	Minneapolis
Omaha	Cincinnati (1839)	Richmond, VA
Paterson, NJ	Cleveland (1854)	Philadelphia
Pittsburgh [a]	Detroit (1836)	Pittsburgh [a]
Portland, OR [a]	Kansas City (1895)	Providence
San Francisco	Newark (1860) [b]	Washington
Scranton, PA [a]	New Orleans (1868; returned to private, 1878) [a]	
	New York (1842)	
	Portland, OR (1887) [a]	
	St. Louis (orig. jointly owned) (1835)	
	St. Paul (1882)	
	Syracuse (1892)	
	Worcester (1852)	

[a] Multiple works.
[b] City contracts with private company for the delivery of some water. Under terms of contract, city elected to purchase works from company in 1900.

which increase the capital costs of public works. And although privately owned utilities had much larger reservoir capacity on average, the averages are skewed by a pair of outliers (San Francisco with 7.7 billion gallons and New York City with 4 billion gallons); median reservoir capacity of publicly owned utilities was more than twenty times larger than the private utility median. Finally, private utilities were more than twice as likely to draw some or all of their water needs from wells, which typically involved smaller-scale investments.

Evidence regarding the administrative simplicity hypothesis is mixed. On the one hand, two-thirds of the nine waterworks in our sample that filtered water were privately owned.[28] Thus, although the aggregate trends suggested that public ownership was increasing at the same time that water filtration was becoming more prevalent, at least among the largest cities in the country, private waterworks were more than three times more likely than public works to filter water. On the other hand, even though

[28] A 1899 government survey of 1034 water companies found that 19% of the 375 private water companies in the sample had water filters whereas only 6% of the 659 public works had filters (as reported in Troesken, 1999, p. 935). Troesken's statistical analysis also indicates a positive correlation between private ownership and filtering (ibid, p. 936).

Table 9.5

Mean attributes of 48 waterworks in 37 U.S. cities by ownership, 1896

	Always private (n=17)	*Private to public (n=20)*	*Always public (n=11)*
County population in 1890	271,419	385,379	292,200
Annual consumption (millions of gal.)	9.9	47.3	41.9
Mains (miles)	96.2	369.3	326.4
Taps	8,484.8	43,165.1	41,696.4
Meters	963.5	4,161.5	4,413.1
Hydrants	832.5	3,444.8	2,618.0
Pumping	0.94	0.85	1.00
Pumping capacity (millions of gal.)	19.2	56.8	72.8
Aqueduct, conduits, tunnels	0.12	0.20	0.27
Filter system	0.35	0.10	0.09
Reservoir capacity (millions of gal.)	4,950 (med.: 5)	3,130 (med.: 111)	365 (med.: 117)
Source: surface (proportion)	0.68	0.90	0.91
Source: wells (proportion)	0.41	0.15	0.18
Year current works built	1875	1848	1862
Year of initial operation	1875	1826	1852
Year of change in ownership	–	1864	–
City served by multiple works	0.59	0.35	0.18
Number of works	2.8	2.0	1.7
Number of works (conditional)	4.1	4.0	5.0
Sewer system [a]	0.53	1.00	1.00

[a] Proportion of waterworks reported to serve areas with broad sewer coverage.

the proportions of private and public works that used pumps were not appreciably different, the amount of pumping, as measured by pumping capacity, was larger for public than private works. Thus, while the positive correlation between filtering and private ownership is consistent with the administrative simplicity hypothesis, the higher pumping capacity of public utilities runs counter to this explanation.[29] Finally, whereas all cities with public waterworks also had sewers, only slightly more than half of the communities served by private works had sewer systems, suggesting that cities that already owned and operated sewer systems may have found it less costly to bring water systems under their control.

LESSONS FOR DEVELOPING COUNTRIES

The broad lessons for the governance of public utilities of the transaction-cost/administered-contracts framework are (i) that problems of establishing credible commitments

[29] "If ... water must be pumped from a river or a lake or from wells in the earth, expensive and powerful machinery must be provided, adequate for all emergencies" (Wilcox, 1910, p. 401).

and effecting efficient adaptations are inherent to public utility transactions and can never be entirely eliminated through changes in governance, and (ii) that the prospects for realizing efficiency gains from changes in governance form in any particular case will depend on both the attributes of the transaction and the institutional environment in which it takes place. The key to sound policy prescriptions is to recognize the tradeoffs among governance alternatives and how they relate to the economic and institutional circumstances in which the transaction will take place.

In the cases of electric power and telecommunications, the prevalence and comparative efficiency of private utilities in developed countries such as the United States provided grounds for optimism about the prospects for efficiency gains from privatization of state-owned utilities in developing countries. If nothing else, the utility sectors in these countries represented a model on which to base reforms. By the same token, the fact that the provision of water and sanitation services is *not* generally private in the U.S. should, at a minimum, give policymakers pause: if, despite an institutional environment conducive to private ownership, water and sanitation systems in the United States are overwhelmingly publicly owned and operated, what reason do we have to expect gains in developing countries where the environment for private enterprise is so much less hospitable? Even in sectors such as telecommunications where private enterprise dominates, lack of adequate political and legal supports may make "public ownership of utilities ... the default mode of organization" (Levy and Spiller, 1994, p. 202). That American waterworks had in fact been private for almost a century suggests that factors other than basic institutional endowment played a role in the switch to public ownership. Understanding what led to that change and the variations among systems in ownership structures may help us determine whether and when private ownership makes sense.

Our analysis of early American public utilities identified two possible explanations for the differences and variations in ownership. One is the purported administrative simplicity of water and sanitation systems relative to other utilities. If the operations of water and sewer systems are less technically complex or operationally temperamental than telecommunications or electric power, the loss in efficiency from the attenuation of high-powered incentives that comes with public ownership may be less than it would be for a utility that needs relatively constant and expert attention to function properly. Although our finding that private waterworks were more likely than public ones to operate filtering systems is consistent with the hypothesis, evidence supporting the importance of administrative simplicity as a determinant of waterworks ownership was modest overall. Nevertheless, to the extent that the simplicity of operations affects the effectiveness of public administration, and that modern-day water and sanitation systems require less attention and expertise to operate than other utilities, then, other things the same, fewer efficiency gains should be expected to accrue to private ownership of these systems. Stated conversely, the theory predicts that the potential gains to private sector participation should be largest in those systems in which the diligent application of technical and managerial expertise are most important.

A second explanation relates historical ownership patterns to differential commitment and contracting problems associated with water and sanitation services. Even if the potential for productivity gains from private management of water systems exist, realization of those gains depends on the ability to attract private firms to the business. Our analysis suggests that features of water and sanitation systems exacerbate the evasion and appropriation problems of procuring public utilities services relative to other utilities. These features include (i) high fixed to marginal costs of service provision, health and nuisance externalities, and high metering and exclusion costs that combine to impede usage and access charges and necessitate lump-sum transfers from the state to private suppliers, and (ii) the importance of water to firefighting and the potential disruptiveness to commerce of system installation and repair, which provide both companies and governments tactics for evading performance or forcing renegotiations.

Consistent with the contracting hazards hypothesis, residential water and sanitation services were rarely metered during the 19th century, access charges failed to cover system costs, and conflicts between municipalities and private water companies over municipal transfers were frequent. The problems of franchise contracting for water services increased, moreover, as cities got larger and more congested, increasing, in turn, the necessary capital investments and aggravating the pollution and fire hazards and costs of disruptions to transportation and commerce. Of the 48 waterworks in our sample, publicly owned systems were four to five times larger than their privately owned counterparts.

The contracting and commitment concerns that appear to motivate public ownership of water and sanitation systems in the U.S. have parallels in developing countries. Populations most in need of improved access to water often live in large, congested urban areas where sanitation and pollution are major concerns. The public health benefits of universal supply combined with the low income and high elasticity of demand of urban consumers (for legal connections and use), meanwhile, impedes cost recovery through access and usage charges. Consistent with our analysis, water utilities in developing countries appear to cover less of their costs through direct charges to consumers than do other utilities. According to the 1994 World Bank Development Report, prices cover only about 30% of costs in the water sector in developing countries compared with over 50% in the power sector, 80% in the gas sector, and over 150% in the telecommunications sector (p. 47). Subsidies for water systems in developing countries were estimated at \$18 billion, \$13 billion of which was attributed to 'underpricing' and \$5 billion to illegal connections (World Bank, 1994, p. 121).[30]

Generalizing from the experience of the United States over a century ago to developing economies today is obviously hazardous. Taken at face value, our analysis suggests that the relatively low administrative requirements and large contracting hazards reduce the gains that can be expected from privatization of water and sanitation systems. Other considerations, both systematic and peculiar to the circumstances of a particular country, may alter that conclusion, however. One such consideration is

[30] The existence of significant losses to illegal water connections is itself evidence of the problems identified in our analysis. By contrast, illegal connections to other utilities were not reported and, presumably therefore, not significant.

differences in the nature, integrity and competence of the background legal and political institutions. Although weak institutional supports generally favor public ownership (again, see Levy and Spiller, 1994), two sets of circumstances would seem to strengthen the case for privatization. The first is in countries with unfavorable political institutions (unstable or unified control) but strong, independent legal systems able to enforce the contractual obligations of the government. The second is where the efficient supplier of utility services is a well-known (possibly multinational) firm, for which governmental reneging would have large reputational consequences. In both cases, the enhanced credibility of governmental commitments may be sufficient to tilt the balance from public to private administration.

Finally, we have focused exclusively on full private ownership of public utilities and have not considered 'hybrid' or intermediate forms involving, for example, state ownership and private operation of public utilities. A major reason for this is that such arrangements simply did not exist at any discernable level in the U.S. during the period we studied. More recently, however, some U.S. municipalities have begun to contract with specialist firms for the operation of publicly owned water and sanitation systems. Such arrangements would seem to afford municipalities the benefits of the expertise of specialized firms while avoiding the appropriation hazards of private ownership. Though a full comparative assessment of the merits and limitations of hybrid structures is beyond the scope of this study, service and management contracts suffer many of the same limitations as other contracts and would appear to have the best prospects where maintenance of existing facilities and investment in new infrastructure are of secondary importance.

REFERENCES

Anderson, Letty, 1988. "Fire and Disease: The development of water supply systems in New England, 1870–1900". In: J.A. Tarr and G. Dupuy, eds., *Technology and the Rise of the Networked City in Europe and America*. Philadelphia, PA: Temple University Press.

Averch, H. and L.L. Johnson, 1962. "Behavior of the firm under regulatory constraint". *American Economic Review* 52 (December): 1052–1069.

Baker, Moses N., 1897. *The Manual of American Water-Works, 1897*. New York, NY: The Engineering New Publishing Co.

Baker, Moses N., 1899. "Water-works". In: E.W. Bemis, ed., *Municipal Monopolies*. New York, NY: Thomas Y. Crowell and Company.

Bemis, Edward W., 1899. "Street railways". In: E.W. Bemis, ed., *Municipal Monopolies*. New York, NY: Thomas Y. Crowell and Company.

Briscoe, J. and H. Garn, 1990. *Financing Agenda 21: Freshwater*. Washington, DC: World Bank.

Chadwick, Edwin, 1859. "Results of different principles of legislation and administration in Europe: of competition for the field, as compared competition within the field, of service". *Journal of the Royal Statistical Society* (September): 381–420.

Coase, Ronald, 1937. "The nature of the firm". *Economica* N.S. 4: 386–405. Repr. in: O.E. Williamson and S.E. Masten, eds., *The Economics of Transaction Costs*. Cheltenham: Edward Elgar Publishing, 1999.

Coase, Ronald, 1972. "Industrial organization: a proposal for research". In: Victor R. Fuchs, ed., *Policy Issues and Research Opportunities in Industrial Organization*. New York: National Bureau of Economic Research, pp. 59–73. Repr. in: O.E. Williamson and S.E. Masten, eds., *The Economics of Transaction Costs*. Cheltenham: Edward Elgar Publishing, 1999.

Coase, Ronald, 1988. "The nature of the firm: origin, meaning, influence". *Journal of Law, Economics, and*

Organization 4 (Spring): 3–47. Repr. in: O.E. Williamson and S.G. Winter, eds., *The Nature of the Firm*. New York, NY: Oxford University Press, 1991.

Cowan, S., 1993. "Regulation of several market failures: the water industry in England and Wales". *Oxford Review of Economic Policy* 9 (Winter): 14–22.

Crain, W.M. and A. Zardkoohi, 1978. "A test of the property theory of the firm: water utilities in the United States". *Journal of Law and Economics* 21 (October): 395–408.

Crocker, Keith J. and Scott E. Masten, 1991. "Pretia ex machina? Prices and process in long-term contracts". *Journal of Law and Economics* 34 (April): 69–99.

Crocker, Keith J. and Scott E. Masten, 1996. "Regulation and administered contracts revisited: lessons from transaction-cost economics for public utility regulation". *Journal of Regulatory Economics* 9 (January): 5–39.

Demsetz, Harold, 1968. "Why regulate utilities?" *Journal of Law and Economics* 11 (April): 55–66.

Esrey, Steven A., 1996. "Water, waste, and well-being: a multicountry study". *American Journal of Epidemiology* 143: 608–623.

Feigenbaum, S. and R. Teeples, 1983. "Public versus private water delivery: a hedonic cost approach". *Review of Economics and Statistics* 65: 672–678.

Goetz, Charles J. and Scott Robert, 1983. "The mitigation principle: toward a general theory of contractual obligation". *Virginia Law Review* 69: 967–1024

Goldberg, Victor P., 1976. "Regulation and administered contracts". *Bell Journal of Economics* 7 (Autumn): 426–448.

Goldberg, Victor P. and John R. Erickson, 1987. "Quantity and price adjustment in long-term contracts: a case study in petroleum coke". *Journal of Law and Economics* 31: 369–98. Repr. in: Scott E. Masten, ed., *Case Studies in Contracting and Organization*. Oxford: Oxford University Press, 1996.

Hausman, William J., David M. Kemme and John L. Neufeld, 1986. "The relative economic efficiency of private versus municipal waterworks in the 1890s". *Business and Economic History* 15: 13–27.

Klein, Benjamin, 1992. "Contracts and incentives: the role of contract terms in assuring performance". In: Lars Werin and Hans Wijkander, eds., *Contract Economics*. Cambridge, MA: Basil Blackwell, pp. 149–173.

Klein, Benjamin, 1996. "Why hold-ups occur: the self-enforcing range of contractual relationships". *Economic Inquiry* 34: 444–463.

Lebergott, Stanley, 1964. *Manpower in Economic Growth: The American Record Since 1800*. New York, NY: McGraw-Hill.

Levy, Brian and Pablo T. Spiller, 1994. "The institutional foundations of regulatory commitment: a comparative analysis of telecommunications regulation". *Journal of Law, Economics, and Organization* 10 (Fall): 201–246.

Masten, Scott E., 1988. "Equity, opportunism, and the design of contractual relations". *Journal of Institutional and Theoretical Economics* 144 (February): 180–195.

Masten, Scott E., 2000. "Contractual choice". In: B. Boukaert and G. De Geest, eds., *Encyclopedia of Law and Economics*. Cheltenham: Edward Elgar Publishing.

Masten, Scott E. and Edward A. Snyder, 1993. "United States v. United Shoe Machinery Corporation: on the merits". *Journal of Law and Economics* 36 (April): 33–70.

Masten, Scott E., James W. Meehan, Jr. and Edward A. Snyder, 1991. "The costs of organization". *Journal of Law, Economics, and Organization* 7 (Spring): 1–25.

McGuire, R.A. and R. Ohsfeldt, 1986. "Public versus private water delivery: a critical analysis of a hedonic cost approach". *Public Finance Quarterly* 14 (July): 339–350.

Nelson, Michael, 1997. "Municipal government approaches to service delivery: an analysis from a transactions cost perspective". *Economic Inquiry* 35 (January): 82–96.

Posner, Richard A., 1969. "Natural monopoly and its regulation". *Stanford Law Review* 21 (February): 548–643.

Posner, Richard A., 1972. "The appropriate scope of regulation in the cable television industry". *Bell Journal of Economics and Management Science* 3 (Spring): 98–129.

Priest, George L., 1993. "The origins of utility regulation and the 'theories of regulation' debate". *Journal of Law and Economics* 36 (April): 289–323.

Tarr, Joel A., 1996. *The Search for the Ultimate Sink: Urban Pollution in Historical Perspective*. Akron, OH: The University of Akron Press.

Teeples, R., S. Feigenbaum and D. Glyer, 1986. "Public versus private water delivery: cost comparisons". *Public Finance Quarterly* 14 (July): 351–366.

Troesken, Werner, 1999. "Typhoid rates and the public acquisition of private waterworks, 1880–1920". *Journal of Economic History* 59 (December): 927–948.

Tynan, Nicola and Tyler Cowen, 1999. "The private provision of water: London's historical experience". Fairfax, VA: George Mason University Department of Economics, May, Processed.

U.S. Bureau of the Census, 1961. *Census Population: 1960, Vol. 1. Characteristics of the Population.* Washington, DC: U.S. Government Printing Office.

Weiss, Thomas, 1992. "U.S. labor force estimates and economic growth, 1800–1860". In: R.E. Gallman and J.J. Wallis, eds., *American Economic Growth and Standards of Living before the Civil War.* Chicago, IL: The University of Chicago Press.

Wilcox, Delos F., 1910.*Municipal Franchises: A Description of the Terms and Conditions upon which Private Corporations Enjoy Special Privileges in the Streets of American Cities, Vol. 1.* Rochester, NY: The Gervaise Press.

Wilcox, Delos F., 1917. "The regulation of private water companies in New York City". *The Journal of the New England Water Works* 31: 550–574.

Wilcox, Delos F., 1931. *The Administration of Municipally Owned Utilities.* New York: Municipal Administration Service.

Williamson, Oliver E., 1976. "Franchise bidding for natural monopolies — in general and with respect to CATV". *Bell Journal of Economics* 7 (Spring): 233–262.

Williamson, Oliver E., 1979. "Transaction-cost economics: the governance of contractual relations". *Journal of Law and Economics* 22 (October): 233–262.

Williamson, Oliver E., 1983. "Credible commitments: using hostages to support exchange". *American Economic Review* 73: 519–540.

Williamson, Oliver E., 1985. *The Economic Institutions of Capitalism.* New York: Free Press.

World Bank, 1994. *World Bank Development Report 1994: Infrastructure for Development.* New York: Oxford University Press.

STATISTICAL APPENDIX

Note: see individual case studies for sources described in table footnotes. To ensure comparability across countries, the statistical tables use uniform definitions for all countries throughout. Consequently figures may differ slightly from thc figures in the case studies where authors used definitions more suitable for the individual countries.

Table A.1 CPI (end-of-period) base year (*1990 = 100*)

	1987	1988	1989	1990	1991	1992	1993	1994	1995	1996	1997	1998
Argentina[a]	0.03	0.13	4.14	100	272	339	375	391	404	405	407	411
Chile[a]	59.11	67.80	79.34	100	122	141	158	177	191	205	218	229
México[a]	30.72	65.79	78.96	100	123	142	156	166	225	302	364	422
Peru[a]	0.00	0.04	1.32	100	510	884	1,314	1,626	1,806	2,015	2,187	2,377
Cote d'Ivoire (September)[a]	95.16	101	102	100	102	107	109	143	157	162	171	175
Guinea[b]	50.62	62.46	78.70	100	180	209	219	–	236	241	–	–

Source: [a]IMF-IFC; [b]Ministère du Plan, Guinea.

Table A.2 Average exchange rate (*US dollars/local currency*)

	1987	1988	1989	1990	1991	1992	1993	1994	1995	1996	1997	1998
Argentina[a]	0.00	0.00	0.04	0.49	0.95	0.99	1.00	1.00	1.00	1.00	1.00	1.00
Chile[a]	219	245	267	305	349	363	404	420	397	412	419	460
México	1.38	2.27	2.46	2.81	3.02	3.09	3.12	3.38	6.42	7.60	7.92	9.14
Peru[a]	0.00	0.00	0.00	0.19	0.77	1.25	1.99	2.20	2.25	2.45	2.66	2.93
Cote d'Ivoire[a]	301	298	319	272	282	265	283	555	499	512	584	590
Guinea[b]	428	474	592	660	754	902	955	977	991	1,004	1,095	–

Source: [a]IMF-IFC; [b]Ministère du Plan, Guinea.

Table A.3 Volume of water produced *(millions of cubic meters per year)*

	1987	1988	1989	1990	1991	1992	1993	1994	1995	1996	1997	1998
Buenos Aires, Argentina[a]	1,188	1,208	1,252	1,223	1,252	1,378	1,402	1,351	1,400	1,510	1,433	1,437
Santiago, Chile[b]	459	461	478	462	453	466	469	475	478	483	475	–
México City (D.F.), México[c]	1,159	1,121	1,111	1,089	1,114	1,098	1,113	1,153	1,118	1,100	–	–
Lima, Peru[d]	642	645	654	545	631	527	647	729	667	679	642	705
Coted 'Ivoire[e]	90	95	100	100	100	103	103	106	111	118	–	–
Abidjan[e]	–	–	–	67	67	70	71	73	75	80	–	–
Guinea[f]	–	–	18.94	20.52	21.47	21.70	26.48	33.58	26.77	24.90	–	–
Conakry[f]	–	–	16.43	–	18.58	18.61	20.01	29.60	23.71	22.09	–	–

Source: [a]Abdala (1996) and Aguas Argentinas; [b]EMOS; [c]DGCOH; [d]SEDAPAL, SUNASS; [e]SODECI; [f]World Bank files.

Table A.4 Thousands of water connections *(legal registered connections)*

	1987	1988	1989	1990	1991	1992	1993	1994	1995	1996	1997	1998
Buenos Aires, Argentina[a]	2,059	2,084	2,137	2,187	2,224	2,290	2,286	2,330	2,412	2,555	2,541	2,583
Santiago, Chile[b]	755	784	814	843	870	905	944	985	1,022	1,056	1,092	–
México City (D.F.), México[c]	–	–	–	–	–	1,279	1,279	1,306	1,338	1,477	1,637	1,652
Lima, Peru[d]	579	607	627	654	679	700	732	763	791	823	839	872
Cote d'Ivoire[e]	201	213	225	244	254	268	275	293	319	345	–	–
Abidjan[e]	100	106	112	123	130	138	144	155	167	179	–	–
Guinea[f]	–	12	17.72	17.10	20.31	18.59	19.45	20.06	21.34	23.44	31.00	–
Conakry[f]	–	9.00	13.25	13.18	14.51	13.37	12.88	13.34	15.26	17.34	25	–

Source: [a]Abdala (1996) and Aguas Argentinas; [b]EMOS (Number of Services); [c]CADF (Number of Bills); [d]SEDAPAL, SUNASS (Number of Connections); [e]SODECI (Number of Subscribers); [f]SONEG and Castalia (1993) (Active Connections).

Table A.5 Length of main water systems (*kilometers*)

	1987	1988	1989	1990	1991	1992	1993	1994	1995	1996	1997	1998
Buenos Aires, Argentina[a]	–	–	–	–	–	–	11,914	12,274	14,360	14,716	15,009	15,073
Santiago, Chile[b]	–	7,050	7,085	7,232	7,431	7,593	7,790	8,001	8,185	8,373	8,534	–
México City (D.F.), México[c]	10,487	10,487	10,507	10,536	10,568	10,577	10,597	11,579	11,579	11,586	11,586	–
Lima, Peru[d]	6,049	6,107	6,237	6,537	6,627	6,774	6,903	7,121	7,365	7,791	8,158	8,464
Cote d'Ivoire	–	–	–	–	–	–	–	–	–	–	–	–
Guinea[e]	–	207	–	–	–	–	–	–	–	–	–	–

Source: [a]Aguas Argentinas; [b]EMOS; [c]DGCOH; [d]SEDAPAL, SUNASS; [e]World Bank files.

Table A.6 Surface water as a percentage of total water produced (*percentage*)

	1987	1988	1989	1990	1991	1992	1993	1994	1995	1996	1997	1998
Buenos Aires, Argentina[a]	–	–	–	–	–	–	92	92	90	89	91	92
Santiago, Chile[b]	90	89	89	91	90	90	89	87	89	86	83	–
México City (D.F.), México[c]	33	34	35	34	36	36	38	42	43	44	45	–
Lima, Peru[d]	65	64	65	61	64	61	64	68	62	64	59	–
Cote d'Ivoire	–	–	–	–	–	–	–	–	–	–	–	–
Abidján	–	–	–	–	–	–	–	–	–	–	–	–
Guinea[e]	–	79	–	–	–	–	–	–	–		–	–
Conakry[e]	–	87	–	–	–	–	–	–	–	100	–	–

Source: [a]Aguas Argentinas; [b]EMOS; [c]DGCOH; [d]SEDAPAL; [e]World Bank files.

Table A.7 Unaccounted-for water (UFW) (*percentage of water produced*)

	1987	1988	1989	1990	1991	1992	1993	1994	1995	1996	1997	1998
Buenos Aires, Argentina[a]	–	–	–	–	–	0.44	0.43	0.37	0.34	0.34	0.33	0.34
Santiago, Chile[b]	0.38	0.34	0.31	0.29	0.28	0.32	0.29	0.22	0.20	0.20	0.22	–
México City (D.F.), México	–	–	–	–	–	–	–	–	–	–	–	–
Lima, Peru[c]	0.48	0.46	0.47	0.37	0.42	0.28	0.38	0.42	0.36	0.36	0.35	0.45
Cote d'Ivoire[d]	0.13	0.15	0.15	0.14	0.14	0.13	0.14	0.13	0.16	0.16	–	–
Guinea[e]	–	–	0.51	0.49	0.50	0.47	0.52	0.54	0.49	0.50	–	–

Source: [a]Aguas Argentinas; [b]EMOS; [c]SEDAPAL, SUNASS; [d]SODECI; [e]World Bank files.

Note: In countries with low levels of metering, estimates of UFW and other figures derived from these estimates should be treated with caution.

Table A.8 Estimated billing collection efficiency (*percentage*)

	1987	1988	1989	1990	1991	1992	1993	1994	1995	1996	1997	1998
Buenos Aires, Argentina[a]	–	80.0	–	–	*85.3*	*90.0*	97.8	97.3	95.7	*92.3*	*94.4*	–
Santiago, Chile[b]	92.4	93.0	93.0	93.0	95.5	96.6	97.1	–	–	–	–	–
México City (D.F.), México[c]	–	–	–	–	–	–	–	83.7	72.4	68.4	67	69
Lima, Peru[a]	*67.0*	*66.7*	*69.2*	*63.5*	*77.9*	*84.0*	*87.7*	*94.3*	*98.3*	*96.4*	–	–
Cote d'Ivoire[a]	95.8	95.1	87.6	100.0	100.0	74.6	95.9	76.6	100.0	100.0	–	–
Guinea[a]	12.7	–	89.0	88.4	44.2	39.0	42.1	84.4	84.2	64.0	50.8	–

Source: [a]Estimated using accounts receivable increase/revenue; [b]EMOS; [c]CADF (Amount collected/amount billed); [d]World Bank files.

Table A.9 Number of employees

	1987	1988	1989	1990	1991	1992	1993	1994	1995	1996	1997	1998
Buenos Aires, Argentina[a]	9,251	9,113	9,080	8,360	8,101	7,666	4,678	3,549	3,829	4,268	4,338	4,494
Santiago, Chile[b]	1,696	1,681	1,699	1,717	1,825	1,865	1,905	1,945	1,913	1,859	1,836	–
México City (D.F.), México[c]	14,190	14,178	15,002	15,246	17,837	17,659	17,115	16,710	16,410	16,171	16,193	16,233
Lima, Peru[d]	3,620	3,769	3,572	3,481	3,178	2,465	1,916	1,901	1,686	1,659	1,570	1,555
Cote d'Ivoire[e]	1,665	1,349	1,329	1,326	1,380	1,357	1,327	1,302	1,316	1,294	–	–
Guinea[f]	–	*504*	355	355	339	337	335	493	492	546	547	–

Source: [a]Abdala (1996) and Aguas Argentinas; [b]EMOS; [c]DGCOH, CADF, Delegations (Estimates); [d]SEDAPAL, SUNASS; [e]SODECI; [f]World Bank files (Includes both SEEG and SONEG).

Table A.10 Labor expenses *(millions of 1996 US dollars)*

	1987	1988	1989	1990	1991	1992	1993	1994	1995	1996	1997	1998
Buenos Aires, Argentina[a]	145.64	176.27	136.57	129.78	111.70	127.82	110.94	99.38	105.77	103.04	104.99	109.42
Santiago, Chile[b]	10.28	11.67	15.78	22.78	24.51	31.55	34.84	36.08	41.25	43.24	44.19	–
México City (D.F.), México[c]	–	–	–	–	–	95.57	116.93	126.75	68.65	92.41	92.84	89.69
Lima, Peru[d]	60.84	62.65	56.10	41.65	44.56	37.86	21.38	27.83	31.04	34.78	28.88	27.24
Cote d'Ivoire[e]	22.82	19.83	16.66	19.23	18.09	18.66	17.17	10.55	10.36	10.53	–	–
Guinea[f]	0.73	0.82	1.69	1.77	.23	2.41	2.51	2.96	2.55	2.85	–	–

Source: [a]Abdala (1996) and Aguas Argentinas; [b]EMOS; [c]DGCOH, CADF, Delegations (Estimated using current expenditures in Cuentas Publicas del Distrito Federal (October 1998)); [d]SEDAPAL; [e]SODECI; [f]SONEG (includes salaries and wages for both SONEG and SEEG). Exchange rates are shown in Table A.2. US CPI from the International Monetary Fund International Financial Statistics.

Table A.11 Labor expenses per cubic meter of water produced (*1996 US dollars per cubic meter)*

	1987	1988	1989	1990	1991	1992	1993	1994	1995	1996	1997	1998
Buenos Aires, Argentina	0.12	0.15	0.11	0.11	0.09	0.09	0.08	0.07	0.08	0.07	0.07	0.08
Santiago, Chile	0.02	0.03	0.03	0.05	0.05	0.07	0.07	0.08	0.09	0.09	0.09	–
México City (D.F.), México	–	–	–	–	–	0.09	0.11	0.11	0.06	0.08	0.08	0.08
Lima, Peru	0.09	0.10	0.09	0.08	0.07	0.07	0.03	0.04	0.05	0.05	0.04	0.04
Cote d'Ivoire	0.25	0.21	0.17	0.19	0.18	0.18	0.17	0.10	0.09	0.09	–	–
Guinea	0.04	0.04	0.09	0.09	0.10	0.11	0.09	0.09	0.10	0.11	–	–

Source: See Tables A.10 and A.3.

Table A.12 Thousands of cubic meters sold per staff

	1987	1988	1989	1990	1991	1992	1993	1994	1995	1996	1997	1998
Buenos Aires, Argentina	71.89	74.19	77.18	81.89	86.53	100.55	170.85	239.82	243.11	233.43	221.37	211.07
Santiago, Chile	167.61	180.73	195.17	190.61	179.71	170.05	174.65	191.14	198.73	206.90	200.76	–
México City (D.F.), México	51.45	49.81	46.65	44.98	39.33	39.16	40.99	43.46	42.90	42.84	42.78	–
Lima, Peru	92.21	92.01	96.66	99.14	115.74	154.59	210.49	224.27	253.81	260.58	267.45	249.46
Cote d'Ivoire	46.74	59.84	64.00	64.57	62.34	66.19	66.81	70.66	70.93	77.06	–	–
Guinea	–	17.29	26.27	29.45	31.88	34.44	38.25	31.49	27.78	22.99	–	–

Source: See Tables A.3, A.7, and A.9.

Table A.13 Staff per thousand connections

	1987	1988	1989	1990	1991	1992	1993	1994	1995	1996	1997	1998
Buenos Aires, Argentina	4.49	4.37	4.25	3.82	3.64	3.35	2.0	1.5	1.6	1.7	1.7	1.7
Santiago, Chile	2.25	2.15	2.09	2.04	2.10	2.06	2.02	1.98	1.87	1.76	1.68	–
México City (D.F.), México	11.10	11.09	11.73	11.92	13.95	13.81	13.4	12.8	12.3	10.9	9.9	9.8
Lima, Peru	6.25	6.21	5.70	5.32	4.68	3.52	2.6	2.5	2.1	2.0	1.9	1.8
Cote d'Ivoire	8.27	6.34	5.90	5.44	5.43	5.07	4.8	4.4	4.1	3.8	–	–
Guinea	–	42.00	20.03	20.76	16.69	18.13	17.2	24.6	23.1	23.3	–	–

Source: See Tables A.4 and A.9.

Table A.14 Intermediate inputs per cubic meter produced (*1996 US dollars*)

	1987	1988	1989	1990	1991	1992	1993	1994	1995	1996	1997	1998
Buenos Aires, Argentina[a]	0.17	0.40	0.09	0.10	0.11	0.15	0.11	0.11	0.12	0.10	0.11	0.12
Santiago, Chile[b]	0.04	0.04	0.04	0.04	0.05	0.05	0.05	0.06	0.07	0.07	0.07	–
México City (D.F.), México[c]	–	–	–	–	–	0.20	0.17	0.17	0.12	0.11	0.20	0.18
Lima, Peru[d]	0.03	0.03	0.03	0.05	0.07	0.09	0.10	0.10	0.14	0.12	0.13	0.12
Cote d'Ivoire[e]	0.59	0.51	0.42	0.49	0.48	0.51	0.46	0.24	0.34	0.33	–	–
Guinea[f]	0.06	0.07	0.14	0.14	0.15	0.20	0.17	0.21	0.33	0.34	–	–

Source: [a]Abdala (1996) and Aguas Argentinas; [b]EMOS; [c]DGCOH, CADF (Mat & Suministros, Servicios Gles y Transferencias); [d]SEDAPAL; [e]SODECI; [f]SONEG (operating expenses less salaries and wages, and depreciation and amortization for both SONEG and SEEG). Exchange rates are shown in Table A.2. US CPI from the International Monetary Fund International Financial Statistics.

Table A.15 Investment in millions of 1996 US dollars

	1987	1988	1989	1990	1991	1992	1993	1994	1995	1996	1997	1998
Buenos Aires, Argentina[a]	–	–	–	–	–	65.80	126.35	154.17	279.54	241.29	247.02	150.89
Santiago, Chile[b]	10.53	8.28	11.34	26.71	40.28	59.43	47.85	48.04	55.81	47.49	47.82	–
México City (D.F.), México[c]	–	–	–	–	–	326.45	279.00	279.52	152.01	212.43	271.54	302.80
Lima, Peru[d]	22.71	33.61	27.35	17.68	29.72	29.79	42.85	54.11	39.23	69.85	113.70	134.69
Cote d'Ivoire	–	–	–	–	–	–	–	–	–	–	–	–
Guinea[e]	2.85	7.04	28.73	5.31	21.67	11.37	54.86	0.77	10.86	17.98	–	–

Source: [a]Abdala (1996) and Aguas Argentinas (Investment); [b]EMOS (Investment); [c] DGCOH (Bienes Inmuebles y Obras Publicas), CADF, Delegations (Capital Expenditures) from Cuentas Publicas del Distrito Federal (October 1998); [d]SEDAPAL (Investment); [e]SONEG (Includes investments by SEEG and SONEG). Exchange rates are shown in Table A.2. US CPI from International Monetary Fund International Financial Statistics.

Table A.16 Investment over assets (*percentage*)

	1987	1988	1989	1990	1991	1992	1993	1994	1995	1996	1997	1998
Buenos Aires, Argentina	–	–	–	–	–	0.02	0.05	0.05	0.09	0.07	0.07	0.04
Santiago, Chile	0.02	0.02	0.02	0.05	0.07	0.09	0.05	0.05	0.05	0.04	0.04	–
México City (D.F.), México	–	–	–	–	–	–	–	–	–	–	–	–
Lima, Peru	0.02	0.01	0.01	0.01	0.02	0.04	0.09	0.06	0.04	0.06	–	–
Cote d'Ivoire	–	–	–	–	–	–	–	–	–	–	–	–
Guinea	1.43	0.04	0.91	0.15	0.37	0.12	0.46	0.01	0.07	0.11	–	–

Source: See Tables A.15 and A.27.

Table A.17 Water revenue billed *(millions of local currency)*

	1987	1988	1989	1990	1991	1992	1993	1994	1995	1996	1997	1998
Buenos Aires, Argentina[a]	0.05	0.21	9	85	142	159	141	174	214	239	263	310
Santiago, Chile[b]	5,856	6,716	9,904	14,202	18,040	23,480	28,946	33,894	43,222	47,567	40,767	–
México City (D.F.), México[c]	–	–	–	–	–	–	–	915	1,115	1,667	2,377	2,787
Lima, Peru[d]	0.00	0.01	0.09	10	60	95	171	283	325	368	–	–
Cote d'Ivoire[e]	19,843	17,994	18,613	20,202	20,548	19,292	20,060	22,876	22,060	28,134	–	–
Guinea[f]	876	938	2,900	5,341	6,050	8,515	11,390	16,620	13,508	14,948	–	–

Source: [a]Abdala (1996) and Aguas Argentinas (water revenue + indirect taxes); [b]EMOS (water revenue); [c]CADF (total revenue); [d]1987-1992 MG consultants, 1993-1996 SEDAPAL; [e]SODECI (water revenue + estimated indirect taxes for water); [f]World Bank files and SONEG (water revenue, includes both SEEG and SONEG).

Table A.18 Sewerage revenue *(millions of local currency)*

	1987	1988	1989	1990	1991	1992	1993	1994	1995	1996	1997	1998
Buenos Aires, Argentina[a]	0.03	0.12	5	46	82	111	88	105	119	130	131	151
Santiago, Chile[b]	2,293	3,083	4,409	7,694	8,808	12,143	16,619	22,499	19,492	20,657	–	–
México City (D.F.), México	–	–	–	–	–	–	–	–	–	–	–	–
Lima, Peru	–	–	–	–	–	–	–	–	–	–	–	–
Cote d'Ivoire	–	–	–	–	–	–	–	–	–	–	–	–
Guinea	–	–	–	–	–	–	–	–	–	–	–	–

Source: [a]Abdala (1996) and Aguas Argentinas (sewerage revenue + indirect taxes); [b]EMOS (sewerage revenue + indirect taxes).

Table A.19 Water revenue collected (*millions of local currency*)

	1987	1988	1989	1990	1991	1992	1993	1994	1995	1996	1997	1998
Buenos Aires, Argentina[a]	0.04	0.18	8	73	121	135	135	167	205	228	252	296
Santiago, Chile[a]	5,428	6,226	9,413	13,498	17,145	22,316	27,511	32,213	41,078	45,207	38,745	–
México City (D.F.), México[b]	–	–	–	–	–	471	596	766	808	1,140	1,590	1,930
Lima, Peru[a]	0.00	0.00	0.06	7	41	88	158	261	299	339	–	–
Cote d'Ivoire[a]	18,999	17,121	16,314	20,202	18,949	17,790	18,498	21,096	20,343	25,945	–	–
Guinea[a]	362	388	2,583	4,558	4,969	6,643	8,459	14,597	8,393	9,287	–	–

Source: [a]See Tables A.17 and A.8; [b]CADF revenue collected.

Table A.20 Accounting revenue – all services (*millions of local currency*)

	1987	1988	1989	1990	1991	1992	1993	1994	1995	1996	1997	1998
Buenos Aires, Argentina[a]	0.08	0.34	16	143	247	297	269	301	373	390	434	485
Santiago, Chile[b]	7,908	10,041	13,813	20,720	25,938	34,798	47,115	56,788	62,585	68,498	69,732	–
México City (D.F.), México[c]	na	na	na	na	na	na	na	915	1,115	1,667	2,377	2,787
Lima, Peru[d]	0.00	0.01	0.11	11	67	109	191	312	361	402	434	482
Cote d'Ivoire[e]	28,092	24,983	24,529	27,103	26,926	27,037	26,562	29,539	32,921	38,697	–	–
Guinea[f]	876	938	3,080	4,379	5,852	9,347	12,278	19,441	19,933	21,616	–	–

Source: [a]Abdala (1996) and Aguas Argentinas; [b]EMOS (total revenue); [c]CADF (total revenue); [d]SEDAPAL (total revenue); [e]SODECI (includes taxes); [f]SONEG (includes revenues for both SONEG and SEEG).

Table A.21 Average amount billed per cubic meter of water sold (*US dollars per cubic meter*)

	1987	1988	1989	1990	1991	1992	1993	1994	1995	1996	1997	1998
Buenos Aires, Argentina	0.33	0.35	0.31	0.26	0.21	0.21	0.18	0.21	0.23	0.24	0.27	0.33
Santiago, Chile	0.09	0.09	0.11	0.14	0.16	0.20	0.22	0.22	0.29	0.30	0.26	–
México City (D.F.), México	–	–	–	–	–	–	–	0.37	0.25	0.32	0.43	–
Lima, Peru	0.18	0.13	0.10	0.15	0.21	0.20	0.21	0.30	0.34	0.35	–	–
Cote d'Ivoire	0.85	0.75	0.69	0.87	0.85	0.81	0.80	0.45	0.47	0.55	–	–
Guinea	0.23	0.23	0.53	0.77	0.74	0.81	0.93	1.10	1.00	1.19	–	–

Source: See Tables A.17, A.3, and A.7. Exchange rates are shown in Table A.2.

Table A.22 Average amount collected per cubic meter of water sold (*US dollars per cubic meter*)

	1987	1988	1989	1990	1991	1992	1993	1994	1995	1996	1997	1998
Buenos Aires, Argentina	0.28	0.30	0.26	0.22	0.18	0.18	0.16	0.20	0.22	0.23	0.26	0.31
Santiago, Chile	0.09	0.08	0.11	0.14	0.15	0.19	0.20	0.21	0.27	0.29	0.25	–
México City (D.F.), México	–	–	–	–	–	0.22	0.27	0.31	0.18	0.22	0.29	–
Lima, Peru	0.12	0.09	0.07	0.11	0.15	0.18	0.20	0.28	0.31	0.32	–	–
Cote d'Ivoire	0.81	0.71	0.60	0.87	0.78	0.75	0.74	0.41	0.44	0.51	0.28	
Guinea	0.10	0.09	0.47	0.66	0.61	0.63	0.69	0.96	0.62	0.74	–	–

Source: See Tables A.19, A.3, and A.7. Exchange rates are shown in Table A.2.

Table A.23 Average amount collected per cubic meter of water produced (*US dollars per cubic meter*)

	1987	1988	1989	1990	1991	1992	1993	1994	1995	1996	1997	1998
Buenos Aires, Argentina	0.16	0.17	0.15	0.12	0.10	0.10	0.10	0.12	0.15	0.15	0.18	0.21
Santiago, Chile	0.05	0.06	0.07	0.10	0.11	0.13	0.15	0.16	0.22	0.23	0.19	–
México City (D.F.), México	–	–	–	–	–	0.14	0.17	0.20	0.11	0.14	0.18	0.19
Lima, Peru	0.06	0.05	0.04	0.07	0.08	0.13	0.12	0.16	0.20	0.20	–	–
Cote d'Ivoire	0.70	0.61	0.51	0.75	0.67	0.65	0.64	0.36	0.37	0.43	–	–
Guinea	0.04	0.04	0.23	0.34	0.31	0.34	0.33	0.45	0.32	0.37	–	–

Source: See Tables A.19 and A.3. Exchange rates are shown in Table A.2.

Table A.24 Tariff per cubic meter of household metered water (*1996 US dollars per cubic meter*)

	1987	1988	1989	1990	1991	1992	1993	1994	1995	1996	1997	1998
Buenos Aires, Argentina	–	–	–	–	–	–	–	–	–	–	–	–
Santiago, Chile (with subsidy)[a]	0.05	0.05	0.06	0.06	0.07	0.10	0.15	0.17	0.22	0.23	0.23	0.21
Santiago, Chile (without subsidy)[a]	0.09	0.10	0.11	0.14	0.15	0.16	0.16	0.17	0.22	0.23	0.23	0.21
México City (D.F.), México (0-10 cubic meters)[b]	–	–	–	–	–	–	–	–	–	–	0.12	0.11
(10-20 cubic meters)	–	–	–	0.17	0.15	0.16	0.17	0.17	0.10	0.12	0.12	0.12
(21-30 cubic meters)	–	–	–	0.21	0.18	0.18	0.19	0.19	0.10	0.13	0.12	0.12
(31-50 cubic meters)	–	–	–	0.37	0.42	0.43	0.45	0.44	0.24	0.30	0.30	0.22
(51-60 cubic meters)	–	–	–	0.37	0.42	0.43	0.45	0.44	0.24	0.30	0.30	0.27
Lima, Peru (0-10 cubic meters)[c]	–	–	–	0.08	0.08	0.07	0.08	0.10	0.12	0.13	–	–
(11-20 cubic meters)	–	–	–	0.08	0.16	0.15	0.15	0.19	0.21	0.23	–	–
(21-30 cubic meters)	–	–	–	0.17	0.25	0.23	0.21	0.27	0.30	0.33	–	–
Cote d'Ivoire (0-18 cubic meters)[d]	0.86	0.71	0.63	0.70	0.65	0.67	0.61	0.30	0.33	0.36	0.31	0.30
(19-30 cubic meters)	1.20	0.93	0.83	0.92	0.85	0.88	0.80	0.44	0.47	0.55	0.47	0.46
Guinea (0-20 cubic meters)[e]	–	–	–	–	–	–	–	–	–	0.68	–	–
(21-60 cubic meters)	–	–	–	–	–	–	–	–	–	0.85	–	–
(60-more cubic meters)	–	–	–	–	–	–	–	–	–	0.92	–	–

Source: [a]EMOS (Average CVAP); [b]Gaceta Oficial del Departamento del Distrito Federal 1990-1998; [c]SEDAPAL; [d]SODECI; [e]SONEG. Exchange rates are shown in Table A.2. US CPI from the International Monetary Fund International Financial Statistics.

Table A.25 Fixed charge household metered water (*1996 US dollars per month*)

	1987	1988	1989	1990	1991	1992	1993	1994	1995	1996	1997	1998
Buenos Aires, Argentina	–	–	–	–	–	–	–	–	–	–	2.93	3.04
Santiago, Chile [a]	0.25	0.25	0.25	0.30	0.39	0.57	0.82	1.00	0.74	0.68	0.67	0.63
México City (D.F.), México	–	–	–	–	–	–	–	–	–	–	–	–
Lima, Peru[b]	–	–	–	0.81	0.97	1.01	0.92	1.00	1.05	1.05	–	–
Cote d'Ivoire	–	–	–	–	–	–	–	–	–	–	–	–
Guinea[c]	–	–	–	–	–	–	–	–	–	13.45	–	–

Source: [a]EMOS (average CFAP); [b]Estimated based on 1998 figures; [c]Bimonthly charge - includes the first 20 cubic meters. Exchange rates are shown in Table A.2. US CPI from the International Monetary Fund International Financial Statistics.

Table A.26 Average monthly bill without taxes (for households consuming 30 cubic meters per month) (*1996 US dollars per month*)

	1987	1988	1989	1990	1991	1992	1993	1994	1995	1996	1997	1998
Buenos Aires, Argentina	–	–	–	–	–	–	–	–	–	–	–	–
Santiago, Chile	2.56	2.82	2.91	3.80	4.07	4.78	5.52	6.22	7.44	7.69	7.61	7.01
México City (D.F.), México	–	–	–	7.49	7.99	8.20	8.63	8.39	4.61	5.79	6.29	5.27
Lima, Peru	–	–	–	4.19	5.86	5.54	5.22	6.58	7.36	7.87	–	–
Cote d'Ivoire	29.83	23.91	21.30	23.68	21.92	22.68	20.59	10.72	11.59	13.07	11.19	10.90
Guinea	–	–	–	–	–	–	–	–	–	23.66	–	–

Source: Authors' calculations.

Table A.27 Value of fixed assets without deducting depreciation (*millions of local currency*)

	1987	1988	1989	1990	1991	1992	1993	1994	1995	1996	1997	1998
Buenos Aires, Argentina[a]	1	3	157	1,427	2,362	2,435	2,458	2,656	2,987	3,227	3,478	3,616
Santiago, Chile[b]	87,827	89,412	108,322	143,997	177,565	218,304	348,468	387,919	434,074	477,144	523,693	–
México City (D.F.), México	–	–	–	–	–	–	–	–	–	–	–	–
Lima, Peru[c]	0.02	0.27	5	351	871	930	851	1,985	2,329	2,775	–	–
Cote d'Ivoire	–	–	–	–	–	–	–	–	–	–	–	–
Guinea[d]	621	64,679	14,836	19,585	38,819	75,941	104,937	118,916	146,986	171,396	–	–

Source: [a]Abdala (1996) and Aguas Argentinas; [b]EMOS; [c]SEDAPAL; [d]SONEG (includes both SEEG and SONEG).

Table A.28 Profits less depreciation and amortization, interest and taxes (*millions of local currency*)

	1987	1988	1989	1990	1991	1992	1993	1994	1995	1996	1997	1998
Buenos Aires, Argentina[a]	0	0	11	55	170	9	20	70	131	145	177	184
Santiago, Chile[b]	1,843	5,235	6,218	6,758	10,789	17,196	23,955	30,590	34,139	37,087	36,642	–
México City (D.F.), México[c]	–	–	–	–	–	–	–	(102)	(147)	43	(117)	78
Lima, Peru[d]	0	(0)	(0)	(4)	4	17	52	112	106	162	–	–
Cote d'Ivoire[e]	3,513	2,612	2,672	3,085	2,954	3,096	3,510	5,149	4,217	5,275	–	–
Guinea[f]	10	–	(28)	669	1,058	2,[illegible]82	4,209	7,602	1,971	5,113	–	–

Source: [a]Abdala (1996) and Aguas Argentinas (Investment); [b]EMOS; [c]CADF (Revenue -Cuentas Publicas del Distrito Federal (Labor and Intermediate Inputs)); [d]SEDAPAL; [e]SODECI; [f]SONEG (includes both SEEG and SONEG).

Table A.29 Households connections as percentage of total connections (*percentage*)

	1987	1988	1989	1990	1991	1992	1993	1994	1995	1996	1997	1998
Buenos Aires, Argentina[a]	98	98	98	98	98	98	98	95	94	93	92	91
Santiago, Chile[b]	94	94	94	94	94	94	94	94	94	94	–	–
México City (D.F.), México[c]	67	–	–	–	–	–	–	–	81	82	90	90
Peru[d]	92	90	90	90	91	91	92	92	92	92	92	–
Cote d'Ivoire	–	–	–	–	–	–	–	–	–	–	–	–
Guinea[e]	–	–	97	97	92	91	91	91	92	93	–	–

Source: [a]Abdala (1996) and Aguas Argentinas; [b]EMOS; [c]CADF; [d]SEDAPAL; [e]World Bank files and Castalia (1993).

Table A.30 Households water revenue as percentage of total water revenue (*percentage*)

	1987	1988	1989	1990	1991	1992	1993	1994	1995	1996	1997	1998
Buenos Aires, Argentina[a]	74	74	73	73	73	0	69	66	58	53	56	56
Santiago, Chile	–	–	–	–	–	–	–	–	–	–	–	–
México City (D.F.), México[b]	–	–	–	–	–	–	–	–	29	30	29	32
Lima, Peru[c]	–	–	–	62	61	61	63	62	63	63	–	–
Cote d'Ivoire	–	–	–	–	–	–	–	–	–	–	–	–
Guinea[d]	–	–	44	46	61	60	54	52	63	70	71	

Source: [a]Abdala (1996) and Aguas Argentinas; [b]CADF; [c]SEDAPAL; [d]World Bank files.

Table A.31 Metered connections as percentage of total connections (*percentage*)

	1987	1988	1989	1990	1991	1992	1993	1994	1995	1996	1997	1998
Buenos Aires, Argentina[a]	0	0	0	0	0	1	3	7	8	9	11	14
Santiago, Chile[b]	100	100	100	100	100	100	100	100	100	100	100	100
México City (D.F.), México[c]	–	–	–	–	–	–	–	–	–	54	55	65
Lima, Peru[d]	26	23	16	11	9	7	4	5	6	8	14	28
Cote d'Ivoire	96	96	96	98	98	–	–	–	–	–	–	–
Guinea[e]	5	–	11	30	41	56	–	85	95	98	–	–

Source: [a]Abdala (1996) and Aguas Argentinas; [b]EMOS; [c]CADF (Metering Readings); [d]SEDAPAL; [e]Brook Cowen (1996) and World Bank files.

Table A.32 Welfare gains as a percentage of last year's sales (*percentage*)

Welfare	Year	Total domestic welfare (percentage)	Government (percentage)	Consumers (percentage)	Workers (percentage)	Domestic investment (percentage)	Foreign investment (percentage)
Buenos Aires Argentina	1992	43.99	-5.20	42.99	1.33	4.96	7.14
Santiago, Chile	1988	52.29	44.28	0.75	7.09	0.18	0.00
México City (D.F.), México	1991	–	–	–	–	–	–
Lima, Peru (with privatization)	1994	39.21	0.88	17.65	2.11	18.57	21.53
Lima, Peru (without privatization, but with World Bank loans)	1992	7.06	6.65	9.51	-9.10	0.00	0.00
Cote d'Ivoire	1987	–	–	–	–	–	–
Guinea	1988	111.90	37.43	74.47	0.00	0.00	14.72

Source : Authors' calculations.

Table A.33 Welfare gains (*millions of 1996 US dollars*)

Welfare	Year	Total domestic welfare	Government	Consumers	Workers	Domestic investment	Foreign investment
Buenos Aires, Argentina	1992	1,419.28	(168.65)	1,387.80	43.00	160.10	230.56
Santiago, Chile	1988	284.25	240.67	4.08	38.52	0.98	–
México City (D.F.), México	1991	–	–	–	–	–	–
Lima, Peru (with privatization)	1994	590.26	13.18	265.74	31.77	279.57	324.05
Lima, Peru (without privatization, but with World Bank loans)	1992	42.54	40.06	57.28	(54.81)	–	-
Cote d'Ivoire	1987	–	–	–	–	–	–
Guinea	1988	23.41	7.83	15.58	–	–	3.08

Source: Authors' calculations.

Table A.34 Domestic welfare gain per capita (*1996 US dollars*)

	Population	Domestic welfare gain per capita
Buenos Aires, Argentina	9,424.00	150.6
Santiago, Chile	4,422.80	64.3
México City (D.F.), México	8,531.45	
Lima, Peru (with privatization)	6,954.57	84.9
Lima, Peru (without privatization, but with World Bank loans)	6,954.57	6.1
Cote d'Ivoire	6,315.58	–
Guinea	2,022.36	11.6

Source: Authors' calculations.

Table A.35 Consumer gain per capita (*1996 US dollars*)

	Population 1996 (000)	Domestic welfare gain per capita
Buenos Aires, Argentina	9,424.00	147.26
Santiago, Chile	4,422.80	0.92
México City (D.F.), México	8,531.45	–
Lima, Peru (with privatization)	6,954.57	38.21
Lima, Peru (without privatization, with World Bank loans)	6,954.57	8.24
Côte d'Ivoire	6,315.58	
Guinea	2,022.36	7.70

Source: Authors' calculations.

AUTHOR INDEX

SUBJECT INDEX